MEDITERRANEAN QUATERNARY RIVER ENVIRONMENTS

REFEREED PROCEEDINGS OF AN INTERNATIONAL CONFERENCE
UNIVERSITY OF CAMBRIDGE / UNITED KINGDOM / 28 – 29 SEPTEMBER 1992

Mediterranean Quaternary River Environments

Edited by

JOHN LEWIN
Institute of Earth Studies, UCW Aberystwyth, Aberystwyth, Dyfed, UK

MARK G. MACKLIN
School of Geography, University of Leeds, Leeds, UK

JAMIE C. WOODWARD
Department of Environmental and Geographical Sciences, Manchester Metropolitan University, Manchester, UK

A.A.BALKEMA / ROTTERDAM / BROOKFIELD / 1995

Cover photograph: A left bank section in the Quaternary gravels of the Alfios River basin in the western Peloponnese, Greece. Recent channel incision, resulting from alluvial gravel extraction, and rapid lateral erosion have revealed a large number of tree trunks. These represent a former riparian woodland which has been buried by coarse alluvial materials. Photograph taken by Jamie Woodward in September 1994.

100064619X

Published by
A.A. Balkema, P.O. Box 1675, 3000 BR Rotterdam, Netherlands (Fax: +31.10.4135947)
A.A. Balkema Publishers, Old Post Road, Brookfield, VT 05036, USA (Fax: 802.276.3837)

ISBN 90 5410 191 1

Contents

PART 3: GEOCHRONOLOGY, CORRELATION AND CONTROLS OF QUATERNARY RIVER EROSION AND SEDIMENTATION

Preface

The river systems of the Mediterranean region bear witness to an exceptional and varied history of environmental change during the Quaternary Period. The triple effects of tectonics, climate and human activity have produced complex sequences of river sediments and landforms. It is perhaps not surprising, therefore, that the debate concerning the relative importance of the underlying controls of river erosion and sedimentation has been especially contentious and protracted in the Mediterranean region. Elucidating these alluvial sequences often requires meticulous and detailed field research, and commonly involves several disciplines including geology, geography and archaeology.

As this volume goes to press at the end of 1994 it is the 25th anniversary of Claudio Vita-Finzi's pioneering text *The Mediterranean Valleys: Geological Changes in Historical Times* which was published in 1969. Our own work in the Mediterranean began in the mid 1980s in the Voidomatis basin as part of the Klithi Archaeological Project – an investigation of Upper Palaeolithic settlement and environmental change in Epirus, northwest Greece, directed by Geoff Bailey of the Department of Archaeology in Cambridge. The need for the *MQRE* conference and the present volume became apparent in the early 1990s as the large number of research projects initiated in the previous two decades had produced an extensive and rather dispersed literature (including many unpublished reports) on Mediterranean river histories. Even a cursory inspection of this material reveals a wide range of approaches and more than a little controversy.

This book aims to fill a considerable gap in the literature by providing an up-to-date review of current knowledge of Quaternary river behaviour from contrasting tectonic, climatic and cultural settings within the Mediterranean region. The volume also illustrates the use of different methodologies for studying past river environments – across a variety of time and space scales – and highlights many of the problems associated with establishing age control and identifying causality in alluvial sequences.

This book represents the first overview of Mediterranean river histories and alluvial palaeoenvironments for over twenty five years and is the first to combine a full range of expertise from many disciplines and field data from most countries of the Mediterranean. Results from over fifty river basins are presented which will provide a benchmark for Quaternary river studies in the Mediterranean region. This volume was generated by a conference held at Peterhouse in the University of Cambridge in September 1992 which was kindly hosted by the Sub-department of Quaternary Research. The nineteen research reports involve thirty six scientists representing the interpretations of the leading researchers who work in the region. Only eight of these papers have an all-British authorship, reflecting the truly international flavour of the Cambridge meeting. In our editorial role we have linked these detailed case studies with an extended introductory review and by three overview chapters. A concluding chapter provides some suggestions for future research in Mediterranean Quaternary river environments.

All the papers have been subjected to peer review and were refereed by at least two specialists. Naturally, many of the papers were reviewed by conference participants and contributors to this volume. We are very grateful to the following people (especially those who reviewed more than one paper!) for cheerfully undertaking this task and for providing many useful comments: Tjeerd van Andel, Mike Anketell, Graeme Barker, Chris Caseldine, David Chester, Richard Collier, Peter Friend, Philip Gibbard, David Gilbertson, Adrian Harvey, Chris Hunt, Peter James, Mike Kirkby, Henry Lamb, Anne Mather, Phil Owens, David Passmore, Tim Quine, Keith Richards, Neil Roberts, Allan Straw and Jim Terry. From the outset we have resisted the temptation to simply 'stack and bind' the conference proceedings and hope we have provided some helpful context for the original symposium papers.

John Lewin, *Aberystwyth*
Mark G. Macklin, *Leeds*
Jamie C. Woodward, *Manchester*

October 1994

CHAPTER 1

Quaternary fluvial systems in the Mediterranean basin

MARK G. MACKLIN
School of Geography, University of Leeds, Leeds, UK

JOHN LEWIN
Institute of Earth Studies, UCW Aberystwyth, Aberystwyth, Dyfed, UK

JAMIE C. WOODWARD
Department of Environmental and Geographical Sciences, Manchester Metropolitan University, Manchester, UK

1 INTRODUCTION

This introductory chapter has three principal aims. First, in broad physiographic terms, to define and delimit the Mediterranean basin. Second, to examine the nature and controls of Quaternary environmental change (natural and anthropogenic) in the region and consider how this has affected fluvial processes and drainage basin development. Third, to provide both a valley-reach and catchment-scale synopsis of alluvial settings in the Mediterranean basin, highlighting some of the distinctive elements of past and contemporary river environments.

The Mediterranean basin, demarcated by the watersheds of rivers that border and drain to the Mediterranean Sea (Fig. 1) (excluding the Nile whose catchment extends well to the south of the region), lies between 30-47°N and between 5°W–37°E. It covers an area of similar proportion to the conterminous USA or Australia and, in terms of geology, geomorphology, climate, vegetation and culture, is equally diverse. The Köppen definition of a Mediterranean climate is one where winter rainfall is more than three times summer rainfall. This definition, however, includes areas such as Iran and Iraq that have winter precipitation of a cyclonic origin but which have no Mediterranean littoral. The distribution of the olive, where it is grown without the aid of irrigation, together with the northern limit of the palm (Fig. 2) have also frequently been used as the yardstick for the typical Mediterranean climate (Braudel 1972). Using these criteria it is clear that the area with a 'truly' Mediterranean climate is quite restricted although, conveniently for the purpose of this review, its boundary largely coincides with the watershed of river systems that flow into the Mediterranean Sea (Fig. 1).

The present day configuration of the Mediterranean basin results from the interplay between three major series of relief-forming factors. These are crustal mobility (directed in both horizontal and vertical directions), periodic climate and sea-level change and, in more recent times, human action. These allogenic, or extrinsic controls of river behaviour and development are now reviewed.

2 TECTONICS AND TECTONIC HISTORY OF THE MEDITERRANEAN BASIN

The Mediterranean basin is not only a zone of transitional climate and vegetation, but also forms the boundary zone between the Eurasian, African and Arabian plates (Fig. 3). The interaction of these plates has produced the Alpine fold belt that extends from Gibraltar to the Middle East. It has, however, an extremely complex and variable structure and is composed of a number of smaller secondary or microplates that have, in some cases, very different tectonic and stratigraphic histories to the adjoining Eurasian and African cratons (Dewey et al. 1973).

During the Cenozoic the Mediterranean region has been affected by generally north-south compression that culminated in the Late Miocene. This was followed by a tensional phase during which large areas of the crust foundered and sank (Dewey et al. 1973). In the Aegean, for example, roughly north-south extension initiated in the Middle Miocene has resulted in a tectonic province characterized by approximately east-west trending horsts and grabens. The complex and dynamic situation along the boundary of the Mediterranean basin is the result of two types of horizontal relative motion affecting the Eurasian and African plates (Smith and Woodcock 1982). If it is assumed that the African plate is fixed, the Eurasian plate has moved to the east, as a consequence of different spread velocities since the opening of the Atlantic, and also to the south. Plate movement, however, is complicated by the presence of partially independent microplates or small blocks. The first type of motion produces strike-slip right lateral movement along the Azores, Gibraltar and Anatolia faults (Fig. 3). The second produces collision of the two plates, and formation of orogenic belts at the Calabrian and Hellenic subduction zones and also along the compressional belts of the Coast of Cadiz, the Tellian Atlas and the Dinaric Alps (Fig. 3). In all of these areas, oceanic crust of the African plate is being underthrust beneath the margin of the European plate. Along the Alpine chain, Pyrenees and Betic ranges, both sides of the contact are formed of continental crust.

The result from this tectonic framework as far as landforms are concerned is to produce three rather distinct environments around the Mediterranean. First, the Precambrian African plate underlying North Africa. Generally this is a low elevation desert environment. In the

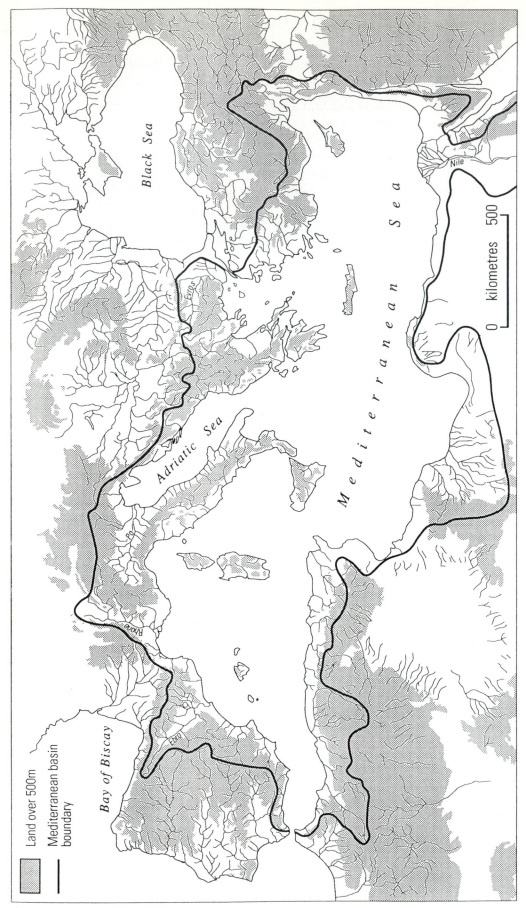

Figure 1. Relief and river drainage network of the Mediterranean basin.

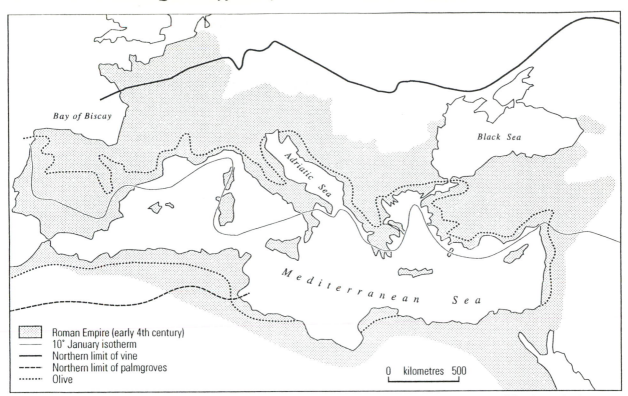

Figure 2. Delimitation of the Mediterranean basin based on the distribution of the olive, the northern limits of the vine and palmgroves, and the 10°C January isotherm (after Braudel 1972).

eastern Mediterranean it is diversified by rifting in Sinai and the Jordan Valley, whilst the higher mountains of Lebanon reach 3086 m (west) and 2659 m (east) of the Bequa'a Valley. Second, the folded and partly metamorphosed Variscan Massifs in the Iberian Peninsula, Corsica and Sardinia. In eastern Spain these are covered by flat-lying or gently folded Mesozoic and Cenozoic sediments. The Spanish Meseta (tableland) can also be elevated (2592 m near Madrid). Third, the Alpine fold belts in the Maghreb, Pyrenees, the Apennines and Sicily, and the Alps proper extending eastward to Greece and Turkey. A major characteristic of these Mediterranean lands is their high relief. Before Alpine orogenesis, these areas formed part of an extensive shelf sea in which limestones were deposited. Their uplift has produced limestone mountains and karstic features, as in former Yugoslavia from which many of the technical terms for limestone landforms are derived. These hard-rock massifs commonly are juxtaposed with syn-orogenic and post-orogenic zones of marine (molasse) and subaerial (flysch) sediments, whilst active Late Cenozoic uplift has elevated such highly erodible materials to give high rates of erosion, as in the 'young' landscapes of the Italian Apennines and Calabria where there is highly active slope instability (Ergenzinger 1992).

Tectonics and structural controls have exerted a strong influence over fluvial systems in the Mediterranean basin in terms both of large-scale drainage basin morphology (size and shape) and river development. In tectonically active parts of the Mediterranean (Fig. 4), faulting, fold-

ing and tilting have resulted in significant recent drainage network disruption including drainage reversal, stream capture, dissection or ponding (e.g. Bailey et al. 1993; Harvey and Wells 1987). The Ebro and Po rivers, two of the largest river systems in the Mediterranean basin, show a strong relation to structural controls (Figures 1 and 3) and can be described as intra-orogen in type (*sensu* Summerfield 1991; p. 419, Table 16.5) draining along strike between mountain belts. More typically, in areas subject to active normal faulting or compressional folding (e.g. northeastern Mediterranean and Betics), Mediterranean rivers are of a trans-orogen type (Summerfield 1991) that drain across the strike of a mountain belt and have a characteristic basin and mountain range morphology. In these areas active structures produce a series of local folds, or uplifted blocks, separated by troughs with river profiles that alternate between stretches of downcutting of uplifted structures and aggradation in the subsiding regions between them. Climatically-induced cycles of incision and aggradation are superimposed on this but tectonic processes provide the underlying long-term determinants on the pattern of river erosion and sedimentation in the landscape (e.g. Lewin et al. 1991).

3 THE MEDITERRANEAN CLIMATE AND QUATERNARY CLIMATE RECORD

3.1 *Present day Mediterranean climate*

The present day wet-winter, dry-summer climate of the

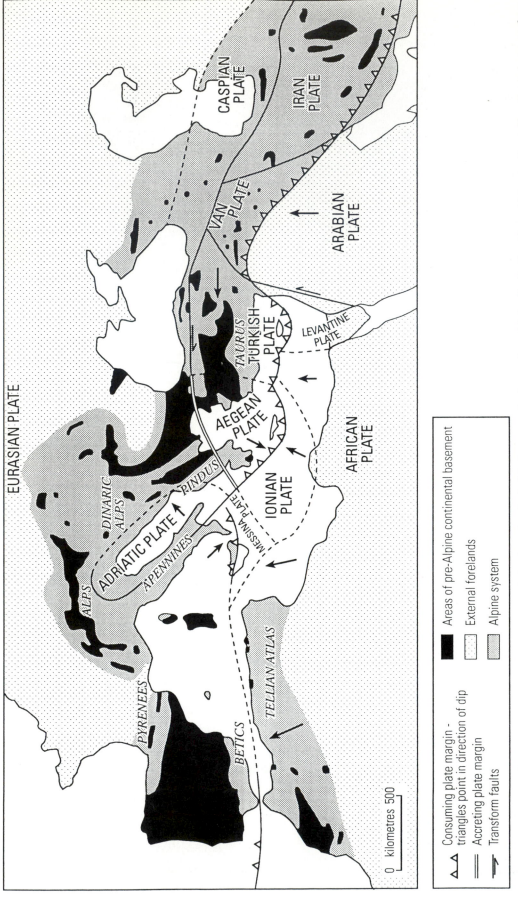

Figure 3. The present lithospheric plate configuration and types of plate boundary in the Mediterranean basin together with the distribution of pre-Alpine continental basement. Arrows refer to plate motions relative to the Eurasian plate (after Dewey et al. 1973 and Windley 1984).

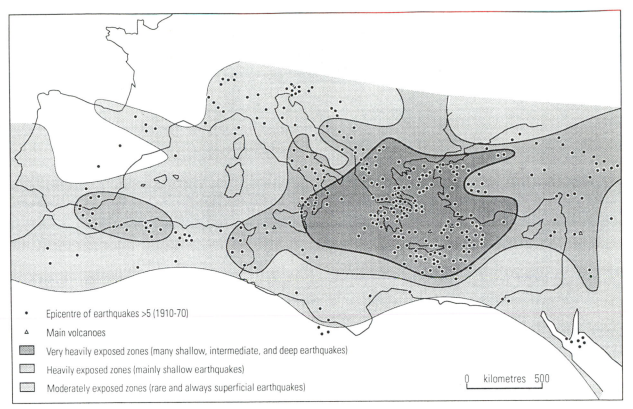

Figure 4. Present day seismicity and volcanism in the Mediterranean basin (after Grenon and Batisse 1989).

Mediterranean results from the seasonal expansion and contraction of the circumpolar vortex and the consequent displacement and withdrawal of the upper westerlies that guide the tracks of the rain-bearing depressions. The Mediterranean basin lies within the boundary between subtropical and mid-latitude atmospheric patterns and in common with southern California, western and southern Australia, the central Chilean coast and the Cape Town region of South Africa, which are similarly interposed between temperate maritime type and arid subtropical desert climates, it is particularly sensitive to a shift in this boundary. What sets the Mediterranean lands apart from these other west coast subtropical regions of the world is the influence of the Mediterranean Sea, which results in the extension of this climate type for more than 3000 km into Eurasia and North Africa. The intrusion of this large body of water into this continental area has an important effect on the climate of the bordering land. During the summer the effect is mainly one of moderating temperatures while in the winter the juxtaposition of cold and dry continental air with a relatively warm sea causes strong evaporation and atmospheric instability and this leads to local cyclone genesis (Gat and Magaritz 1980). There are, however, marked regional contrasts in annual rainfall (Fig. 5) and rainfall regime (Fig. 6) within the basin. These are partly determined by large-scale synoptic disturbances (e.g. cyclones advected from the North Atlantic or anticyclones that form over adjacent continental land masses), partly by interaction between orographic and land-sea controls, and partly by more local effects

(Wigley and Farmer 1982). Rainfall amount in the Mediterranean, and duration of the rainy season, decreases from west to east and also from north to south across the basin (Fig. 5). The summer drought increases in the same direction both in length and intensity (Fig. 6). This primarily reflects the location of the southern and eastern Mediterranean littoral close to the southern margin of the depression tracks during the winter months. Rainfall amounts are also augmented by altitude most notably along the northern part of the basin on the western flank of mountain areas such as the Apennines, Dinaric Alps and Pindus Mountains (Fig. 5). Conversely, areas to the lee of these relief features tend to be correspondingly drier.

There exists, as a consequence of large-scale atmospheric-oceanic circulations in the region, a marked contrast between the pressure conditions over the eastern and western (Italian Peninsula and westwards) parts of the Mediterranean in winter and summer (Wigley and Farmer 1982). In summer both areas are comparatively stable though the western basin is mainly under the influence of a strong high pressure ridge which pushes east from the Azores subtropical high over the Mediterranean. The eastern basin falls under a low pressure area which extends from the Persian Gulf northwest towards Greece and which is associated with the Indian summer monsoon. In winter, pressure over the eastern basin is generally much higher than the west, with the former area affected by the Siberian high and associated polar continental air mass. In contrast, the western basin in winter is

Figure 5. Average annual precipitation in the Mediterranean basin (after UNESCO-WMO 1970).

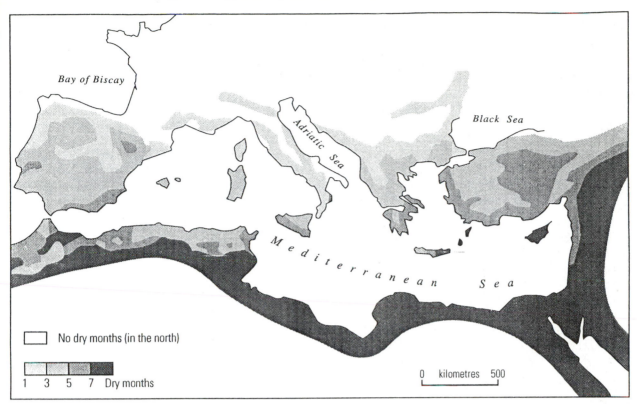

Figure 6. Length of the dry season in the Mediterranean basin (after Grenon and Batisse 1989).

normally an area of almost permanent low pressure – the result of relatively high sea temperatures and incursions of moist polar maritime air from the northeast Atlantic. Winter rain is mainly associated with cyclonic disturbances that originate in the Mediterranean basin (very few surface cyclones can be traced back to Atlantic origins) though their development is strongly influenced by orographic effects (e.g. the Mistral and Bora winds and associated weather). The vast majority of rainfall in the Mediterranean, especially in the southern and eastern parts of the basin, occurs when the upper westerlies are in their low index or blocked mode, and the polar front jet stream exhibits a strong meridional character allowing southerly transport of cold air which favours cyclogenesis (Perry 1981). On rare occasions monsoon air masses may bring summer rain to the eastern Mediterranean (Gat and Lagaritz 1980).

3.2 Mediterranean Quaternary Palaeoclimates

In general terms, climate in the Mediterranean during historic and prehistoric times has varied between relatively dry and relatively humid phases, distinguished by differences in amplitude and duration. As outlined above, the main factors which control weather patterns in the region are sea surface temperatures, in both the North Atlantic and Mediterranean, and the strength and latitudinal position of the mid-tropospheric circumpolar westerlies which principally determine the geographic distribution and amount of rainfall. Analysis of instrumental

rainfall records, in conjunction with studies of recent changes in atmospheric circulation, are particularly useful in this respect as they provide a framework for interpreting past variations of climate in the Mediterranean and their likely synoptic controls. Winstanley (1973), for example, has demonstrated a link between increased winter-spring rainfall over the Mediterranean in the middle decades of this century and a low zonal index associated with a general cooling in the northern hemisphere. He attributes this to an 'expansion' of the circumpolar vortex and to changes in the strength and wave pattern of the upper westerlies. With a relatively weak zonal circulation there are relatively short, large amplitude waves in the circumpolar westerlies and the strongest westerly flow tends to be shifted southwards. Weather systems in the mid-latitudes are slower moving and periodic anticyclones develop which 'block' the eastward moving depressions and deflect them to the north and the south resulting in heavy rainfall over the Mediterranean. Regional variations of seasonal rainfall are determined by the preferred longitudinal position of the troughs and ridges. Recent enhanced blocking frequency over northwest Europe resulting in an increase in the percentage of the total Mediterranean precipitation falling in the warm seasons (North and Jones 1977), has also been shown to correspond with negative sea surface temperatures in the western Atlantic (Perry 1981).

Recent climate changes in the Mediterranean basin, especially the emerging link between rainfall and variation in the position and intensity of semi-permanent features of both atmospheric and oceanic circulation, provide a useful basis for interpreting Quaternary climate fluctuations in the region. These climatic changes have been reconstructed primarily from pollen and plant-macrofossil records, and lake level changes. Although, in practice, the relatively small number of sites investigated, and the lack of geochronological control beyond the effective dating range of radiocarbon, has restricted detailed palaeoclimatic reconstruction to the Last Glacial to Interglacial transition through to the present day (e.g. Pérez-Obiol and Julià 1994). Even over this period, pollen- and lake level-based climate reconstructions do not always match, especially during cold stages of the Quaternary when vegetation and climate analogues are particularly problematic (Prentice et al. 1992). Furthermore, poor dating precision and the relative insensitivity of many vegetation formations to modest, short-term, climate events usually results in long-term, major changes in climate (trends over several millennia) being more reliably recorded than lower order climate variability.

Lake level data for the Mediterranean as whole (Fig. 7) show high lake levels around the time of the Last Glacial Maximum (LGM) (c. 18,000 years BP), indicating that the climate was wetter than today (Harrison and Digerfeldt 1993). This is, however, at variance with pollen

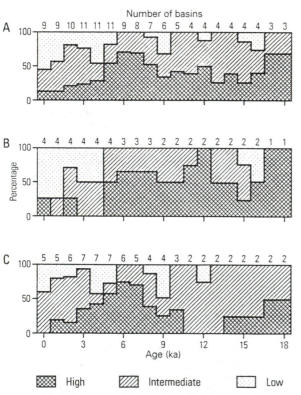

Figure 7. Lake-level variations during the last 18,000 years in (a) the Mediterranean region as a whole, (b) the western Mediterranean (the Iberian Peninsula), and (c) the eastern Mediterranean (the Balkan Peninsula). The histograms indicate the relative number of lakes with high, intermediate and low levels in each region at 1000-year intervals (after Harrison and Digerfeldt 1993).

evidence for widespread *Artemisia* steppe in the region at this time, considered to reflect increased aridity (cf. Bottema 1978; Bailey et al. 1983). Prentice et al. (1992) have addressed this apparent contradiction and suggested that increased seasonality of precipitation was a key factor, with the LGM climate of the Mediterranean region characterised by cold winters, intense winter precipitation and summer drought. Markedly reduced winter temperatures resulted from strong westerly advection from the cold North Atlantic and the development of a fixed anticyclone over the northern European ice sheet. The drying and cooling effect of these may have been counteracted in winter by increased storm frequency under a southward-shifted jet stream. This may account for greater runoff and high lake levels, while at the same time a growing season soil moisture deficit and low winter temperatures would have maintained an open vegetation.

Lake levels in the Mediterranean region began falling soon after 18,000 years BP (Fig. 7) though there are some systematic differences between lake level changes in the Balkans and in the Iberian Peninsula (Harrison and Digerfeldt 1993), and also between the western (Maghreb) and eastern (Libya – Egypt) North African coast where arid and humid phases during the Holocene are opposed (Rognon 1987). In the western Mediterranean, the peak of Late Glacial aridity occurred at 16,000-15,000 years BP and the lake levels were already high again by 12,000 years BP. In the east, maximum Late Glacial aridity occurred between 10,000-11,000 years BP. These dry conditions seem to have resulted from the periodic northern movement and strengthening of the subtropical high pressure cells and with the general poleward shift of the climatic zones during deglaciation. The northerly displacement of the jet stream axis as the ice sheets receded, and the low sea surface temperatures, would have removed the additional source of winter precipitation. In the western Mediterranean conditions remained significantly wetter than at present throughout the early to middle Holocene. An abrupt transition to more arid conditions occurred after 5000 years BP. In the east there was a slower return to moister conditions with high lake levels at 6000 years BP and a more gradual transition to drier conditions between 5000 years BP and the present (Harrison and Digerfeldt 1993). Along the southern Mediterranean littoral a wet phase in the Maghreb after 6000-5000 years BP corresponded to a period of extreme aridity along the eastern North African coast. Conversely, an arid phase in the Maghreb around 12,000-6000 years BP was mirrored by rainfall extension in the Egyptian-Libyan desert (Rognon 1987). These differences in regional Holocene climate histories are probably explained by gradual changes in the number and location of meanders in the jet stream which directly control the storm tracks and precipitation in the southern Mediterranean area (Rognon 1987). In addition, enhanced early to middle Holocene precipitation in the eastern Mediterranean is believed to have been caused by a northern shift in the Inter Tropical Convergence Zone (ITCZ), and its associated rains, related to an 'orbitally' strengthened monsoonal system (Kutzbach and Street-Perrott 1985). Indeed, increased summer rainfall may have been a key factor in the relatively high lake levels over most of the Mediterranean basin during the first half of the Holocene (Roberts 1989). Pollen and lake level evidence both suggest increasing summer dryness after 6000 years BP with the replacement of subhumid deciduous forest by drought adapted sclerophyll evergreen trees and shrubs. The lake level data support the hypothesis of Huntley and Birks (1983) which argues that this was primarily controlled by climatic change.

To summarise, Quaternary climate fluctuations in the Mediterranean basin can be related to shifts in the boundaries, and influence of, mid-latitude and subtropical weather systems, which resulted in significant changes in the seasonality and geographic distribution of precipitation in the region. Humid and arid phases of climate have been a characteristic of both the Late Pleistocene and Holocene, although the synoptic conditions that controlled precipitation were significantly different during these two periods. Thus high lake levels during the LGM appear to have been in response to an increased winter storm frequency under a southward-shifted jet stream, displaced by the orographic barrier presented by the north European ice sheet. High lake levels during the early to middle Holocene, by contrast, can be explained by warmer sea surface temperatures, in both the Mediterranean and North Atlantic, increasing atmospheric moisture availability and precipitation, and also more frequent tropical air mass incursions into the region. Conversely, lower precipitation during the Last Glacial to Interglacial transition, and over the last 5000 years, can both be related to a northerly shift of the main upper westerly flow reducing the length of the rainy season. Greater aridity during the period of deglaciation reflects significantly lower sea surface temperatures than at present, reducing evaporation and atmospheric instability that leads to cyclone genesis over the sea. Climate during the middle to late Holocene, however, appears to have been spatially and temporally more variable due to periodic expansion and contraction of the circumpolar vortex and variation in its strength and wave pattern in the manner outlined by Winstanley (1973). Generally lower lake levels during this period indicate a progressive contraction of the circumpolar vortex and movement of the belt of westerlies northwards away from the region, although since around 3000 years BP it is difficult to differentiate between aridification and anthropogenic desertification.

4 RIVER RESPONSE TO QUATERNARY CLIMATE CHANGE

4.1 *Climate change: sensitivity of geomorphic and vegetational systems*

The climate of the Mediterranean basin over the Quaternary Period has been controlled by slow (10^3–10^6 years) insolation changes driven by orbital cycles and also by abrupt (10^1–10^2 years) shifts in atmospheric-oceanic circulation. These latter, decadal to century-scale oscilla-tions in the thermohaline circulation of the North Atlantic have been attributed to outflows of glacial meltwater (Broecker and Denton 1989) and icebergs (Bond et al. 1992) during the transition from the last ice age to the present interglacial, and in the Holocene to variations in solar radiation, influencing temperature, precipitation and salinity in that part of the North Atlantic where deep water formation is important (Stuiver and Brazuinas 1993). There is also considerable empirical evidence to suggest a link between short term climate change (decades or less) and volcanic activity (e.g. Baillie and Munro 1988). Geomorphic and vegetational systems do, however, differ greatly in their sensitivity to climate change particularly with respect to magnitude and duration. For example, vegetation generally responds to mean conditions with the distribution of vegetation types at the sub-continental scale reflecting average climate. Thus, major vegetation changes between interglacial and glacial stages have been primarily controlled by, and tracked, long-term insolation changes resulting from orbital forcing. River systems, on the other hand, can respond immediately to variations in the character, frequency and magnitude of extreme events, such as floods, that need not necessarily modify vegetation within a catchment. Quaternary climate change has had a profound effect on sea level, vegetation development and catchment hydrology, all of which directly influence river processes and environments. These major controls of short and long-term fluvial dynamics are now briefly examined.

4.2 *Sea-level change*

It is well known that the alternating growth and shrinkage of Pleistocene ice sheets resulted in large (of the order of 120-150 m) global (eustatic) sea-level change. The coast of the Mediterranean during the LGM was significantly different from that of the present (Fig. 8). With sea level lowered to around 120 m, large plains existed off the coast of Tunisia, fringed most of Italy, southern France and much of Greece (van Andel 1989). During periods of falling sea level rivers adjusted through drainage network extension across the newly exposed shelf surface and by limited channel degradation near the former coastline. At times of sea-level rise, network contraction and valley back-filling took place. Away from the coast, however, the steep valley floor gradients of most Mediterranean rivers ensured that base-level changes, associated with fluctuations in sea level, had very little impact on channel and floodplain development inland. The main effects were to shift the position of deltas found at the mouths of a number of major rivers including the Ebro, Nile, Po and Rhône, and to change rates and patterns of sediment delivery to the shelf-slope depositional systems. In the Gulf of Argos, Greece, for example, fluctuating Quaternary sea levels resulted in the formation of thick wedges of coastal sediments comprising marine sediments intercalated with floodplain and alluvial fan deposits (cf. van Andel et al. 1990). Relative sea level changes related to alternating climatic conditions are complicated by changes resulting from tectonic activity. For example, in

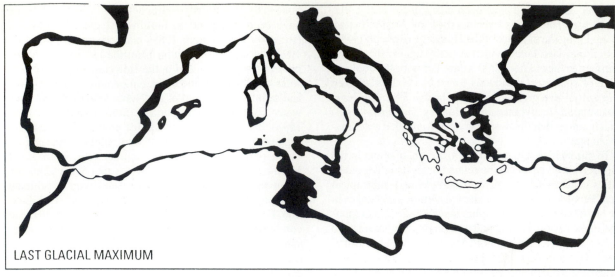

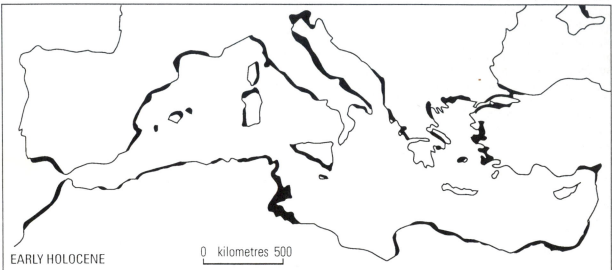

Figure 8. Late Quaternary coastal palaeogeography of the Mediterranean (land exposed at low sea levels shown in black) – Mercator projection (after van Andel 1989).

Greece conglomerates which are of Late Pliocene or early Quaternary age can be found at over 1000 m above present sea level. In western Crete, coastal archaeological features dated to around 2000 years BP have been up-lifted as much as 9 m (Mourtzas 1990).

4.3 *Vegetation change*

Vegetation in the Mediterranean basin, as elsewhere in the world, is largely controlled by climate in the absence of human interference. Pollen studies (e.g. Pons and Reille 1988; Wijmstra 1969) indicate alternations between forest and open vegetation communities over the Quaternary in response to shifts from 'warm' to 'cold' climate modes. Generally speaking, glacials (stadials) in the Mediterranean were characterised by *Artemisia*-dom-inated steppe, interstadials by forest steppe to open pine-oak forest and interglacials by mesic or evergreen oak

forests (e.g. Tzedakis 1993). The critical climate variable governing these vegetation changes is believed to be differences in precipitation with changes in temperature being of secondary importance. In the northern part of the basin, and in mountain areas, temperature changes and a resistance to cold (as well as drought) are likely to have been equally important. The Mediterranean basin, most notably in the mountains of the Iberian, Italian and Balkan Peninsulas, also provided refugia for the tem-perate tree flora of Europe during glacial periods (Bennett et al. 1991).

The Mediterranean forest, typified by evergreen trees such as cork-oak (*Quercus suber*), holm-oak (*Quercus ilex*), Aleppo pine (*Pinus halepensis*), stone pine (*Pinus pinea*) and olive (*Olea europaea*) forms a narrow peri-pheral coastal belt transitional between the mid-latitude deciduous (summer green) forest and the desert biome (Fig. 9). Towards temperate environments (in latitude,

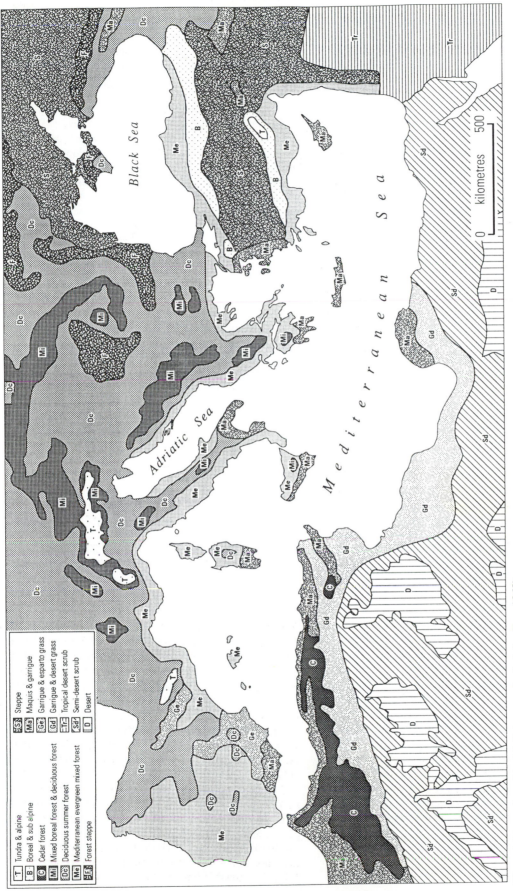

Figure 9. Vegetation of the Mediterranean basin (after Eyre 1968).

altitude and inland location) deciduous trees mix with the holm oak. Towards low latitudes, holm oak is replaced by trees more resistant to drought (thuja) and to cold at altitude (conifers, including cedar) (Quezel 1977). Of the major plant formations found in the Mediterranean basin today, only sclerophyll evergreen forest, scrub (maquis) and dry heath (garrigue) are distinct to that ecotype. All are adapted to survive through the long summer drought and are fire resistant. The Mediterranean maquis and garrigue are not generally considered to be natural formations, except in particular drought-prone areas where maquis could have been the primary vegetation (Tomaselli 1977). More usually they are stages in the degeneration of the Mediterranean sclerophyll evergreen forest, a degeneration almost always due to human intervention (see below).

Given that erosion is largely controlled by vegetation cover, significant changes in biomass, litter amounts and stem density in response to climatic (or human) perturbation may be expected to lead to accelerated hillslope runoff and erosion and increased sediment delivery to rivers. The impact vegetation change has on the fluvial system, however, depends on three factors (Thornes 1989). First, there is the location of vegetation change within a catchment as this affects the relevant delivery ratio through the nature of possible intervening stores. Second, the density of vegetation is an indication of the effectiveness of vegetation cover as a protection against raindrop impact and runoff. Above 70% cover the effect of changes in cover is far less significant than below that level, beyond which soil loss increases dramatically. Third, a steady state can neither be reached, nor maintained, if the relaxation time of the system is longer than the mean recurrence time of the disturbance to it. This third factor is probably the most contentious and certainly the least well understood. Until recently it was generally assumed that climate change during the late Quaternary was relatively gradual, occurring over millennia, and that vegetation would probably have had enough time to adjust to new climatic conditions. However, recent results from the European Greenland Ice Core Project (GRIP) and the US Greenland Ice Sheet Program II (GISP 2), together with analysis of sea floor sediments in the North Atlantic, have shown that climate change was neither smooth nor gradual, but occurred very rapidly with a change from interglacial to glacial conditions taking place, in some instances, in less than a decade (e.g. Bond et al. 1993). An abrupt or step-functional change of this nature would undoubtedly have exceeded both vegetation and soil systems' capacity for adjustment, triggering a period of landscape instability with high erosion rates, valley floor aggradation and river metamorphosis (cf. Knox 1972). Unfortunately, even though this scenario appears intuitively very probable, there are at present no well-dated, high resolution Late Pleistocene fluvial chronologies in the Mediterranean to test this hypothesis.

4.4 *Hydrological change*

There have been considerable changes in catchment hydrology and river regime over the Quaternary Period in the Mediterranean basin. Present day river regimes in the region bear the stamp of the Mediterranean climate with a strong late autumn, winter or early spring peak and minimum flow in the summer unless rivers are fed by groundwater in limestone areas. This runoff pattern largely results from rainfall alone (Fig. 10), though prominent early spring peak flows in the mountain areas of the Iberian Peninsula, Balkans and Anatolia relate to storage of some winter precipitation as snow which is then released in spring to give a thaw peak (Beckinsale 1969). During the summer months thunderstorms (frequently orographically enhanced) can cause localised flooding especially in small mountain basins with high relief. On rare occasions, northward intrusion of tropical air masses in the eastern Mediterranean can also generate large summer floods (Gat and Magaritz 1980). Some of the most notable recent floods, however, such as the famous Tunisian flood of 1969 (Stuckmann 1969), the flood disaster in southeast Spain in 1973 (Bork and Bork 1981) and the 1977 floods in Piedmont and Liguaria (Perry 1981), occurred during the transitional seasons, especially in September and October when the air-sea surface temperature difference is at a maximum and potential atmospheric instability is high.

General circulation modelling (GCM) simulation experiments (Kutzbach and Guetter 1986) indicate that temperatures in the Mediterranean region during the LGM were 5-10°C lower that at present in winter and 1-3°C lower in summer. They also show a year-round strengthening, and equatorial shift, of the jet stream and increased winter precipitation along the jet stream axis at or over the latitude of the Mediterranean (COHMAP 1988). This is likely to have had two principal effects on catchment hydrology. First, it probably increased seasonal snow cover in mountain areas, particularly in the northern Mediterranean littoral and provided favourable conditions for ice accumulation and glacier development (Fig.11). Second, in internally draining, tectonically-controlled basins, higher runoff created sizeable lakes or significantly increased water levels where lakes were already present (Bottema 1978; Pons and Reille 1988). In rain-fed basins principally draining the southern and eastern margins of the Mediterranean basin, and unaffected by snow or ice, seasonal flow fluctuations would have been especially pronounced with a considerably extended dry season in the summer half of the year. Glacierized catchments and those with significant seasonal snow cover would have had a strong spring/early summer nival flow peak. In contrast to present hydrological regimes, winter rather than summer would have been the time of minimum flow in these systems. On the west facing flanks of coastal mountains, autumn storms bringing heavy rain on shallow snow could have induced very large floods. Overall, river regime in the Mediterranean basin during cold stages of the Quaternary was probably not only more seasonal than at present but also exhibited greater spatial and temporal variability.

In the eastern Mediterranean Sea the formation of extensive layers of black mud rich in organic matter,

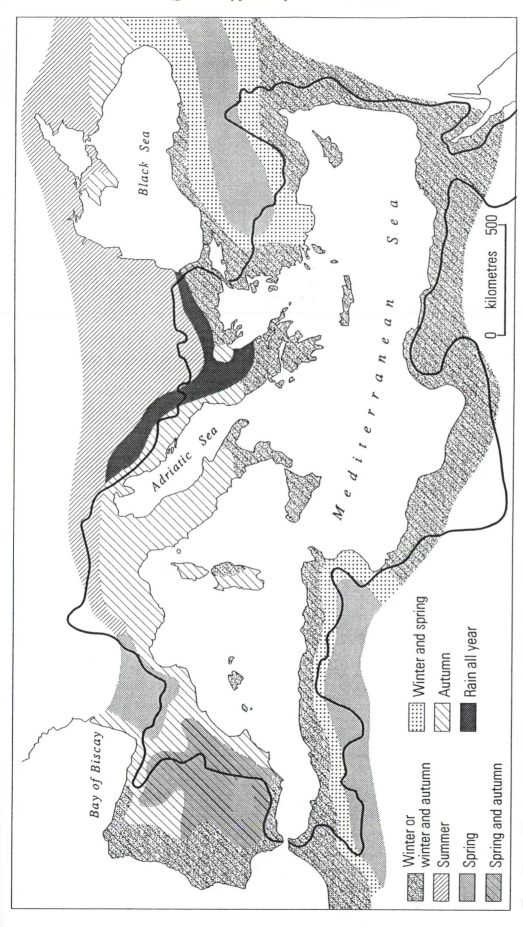

Figure 10. Rainfall regimes in the Mediterranean basin (after Huttary 1950). This map also provides a useful approximation of the range of river regimes found across the Mediterranean region. The Mediterranean watershed is also shown (see Fig. 1).

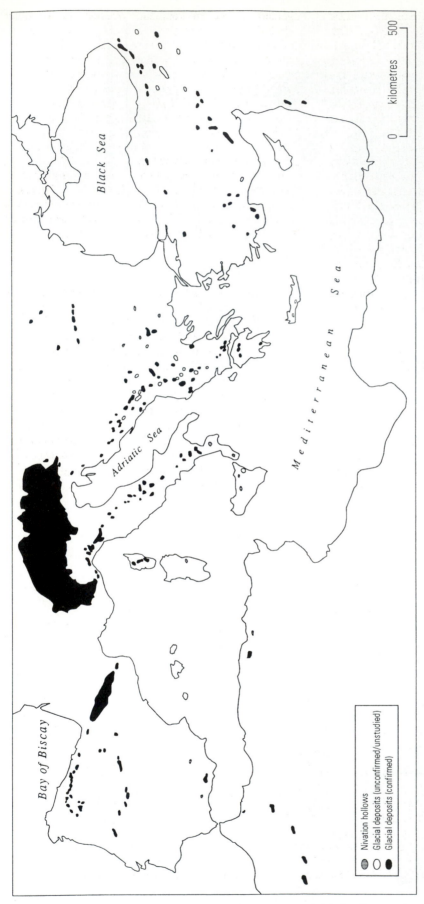

Figure 11. Pleistocene glacial features and sediments in the Mediterranean basin and southern-central Europe (after Messerli 1967).

called 'sapropels' has been shown by Rossignol-Strick and others (1982; 1985) to be related to periods of high discharge from the River Nile, caused primarily by increased monsoonal precipitation and runoff from the northern Ethiopian highlands. Such floods deliver large volumes of freshwater which, being less dense that sea water, form a low salinity surface layer that stratifies the ocean and prevents thermohaline convection. This results in oxygen depletion in bottom waters allowing organic-rich sapropel muds to be deposited on the sea bed. In this way, Mediterranean sapropels constitute a unique proxy record for the longer-term variations in the African monsoon (Rossignol-Strick 1985). Eleven discrete basin-wide sapropels have been identified in the eastern Mediterranean Sea and were formed at various periods during the last 465,000 years. The two most recent sapropels, deposited in the intervals 11,800-10,400 and 9000-8000 years BP, have been related to major flooding along the Nile as a consequence of the intensification and northward displacement of the African monsoon during the Late Glacial to Holocene transition (Rossignol-Strick et al. 1982).

5 HUMAN IMPACT

5.1 *Historical and geographical context*

One of the singular characteristics that sets the Mediterranean basin apart from other parts of the world which also have a Mediterranean climate, such as central Chile, western and southern Australia, South Africa and California, is the long history of human interference and man-

agement of the environment. Whereas all the other 'Mediterranean' areas experienced major human modification only during and after the era of European colonization (post 1500 AD), environmental change in the Mediterranean basin induced by human action can be traced back to the Neolithic agricultural revolution in the Near East around 8000 BC. Indeed by about 5000-4000 BC agro-ecosystems had been established along the entire Mediterranean coast, in all areas suitable for farming, and since then they have been continuously subject to human influence (Ammerman and Cavalli-Storza 1971).

Human environmental impact can be considered as inadvertent when arising, for example, from deforestation or land use change, or direct when interference is planned, as in the case of river regulation for irrigation or construction of embankments for flood protection.

5.2 *Inadvertent change*

Beginning in the southeastern Mediterranean, extensive deforestation in advance of agriculture created an open landscape as early as 3000 BC, which proved highly susceptible to erosion. The first major basin-wide assault on the forest occurred with the establishment of the Roman Empire and by the time of the end of the Roman Period more than half of the Mediterranean forests had been devastated (Tomaselli 1977). Since then the forest balance has hardly changed, at least in the most populated regions where forests have remained concentrated on ground unfit for cultivation. In the Middle Ages there was renewed clearance notably in Arab-held Spain and Sicily and large parts of the eastern Mediterranean governed by the Byzantine or Ottoman Empires. By the middle of the

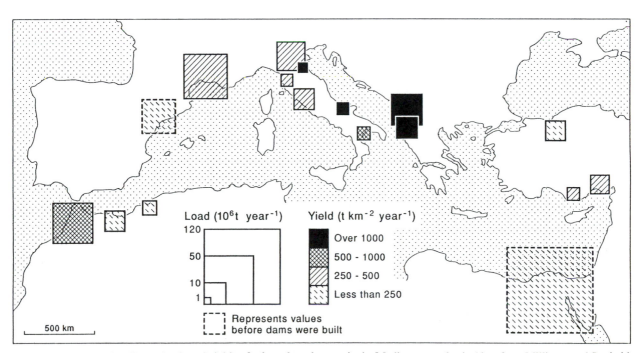

Figure 12. Suspended sediment loads and yields of selected catchments in the Mediterranean basin (data from Milliman and Syvitski 1992 and other sources).

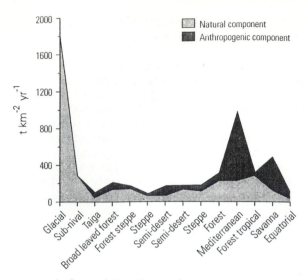

Figure 13. Suspended sediment yield in the mountains of various climatic-vegetation zones showing the relative importance of natural and anthropogenic components (after Dedkov and Moszherim 1992).

nineteenth century at least three quarters of the Mediterranean forest had disappeared (Tomaselli 1977).

Not surprisingly, therefore, large-scale vegetation destruction and land misuse, which accelerated rapidly in classical times, are generally held to be the principal causes of environmental degradation in the Mediterranean region (cf. van Andel et al. 1986). Even in the absence of humans, high relief and steep slopes, large areas of friable soils and unconsolidated sedimentary rock, and high-intensity rainfall events are all factors that make Mediterranean river basins naturally prone to erosion. Suspended sediment load data is not available for many Mediterranean rivers especially those draining southshore catchments between Algeria and the Nile. Elsewhere, however, a few nations have long-established sediment monitoring networks which provide some useful insights into regional contrasts in fluvial suspended sediment transfer rates within the Mediterranean environment (Fig. 12). Jovanonviv and Vukcevic (1957) have mapped sediment yield data in the former Yugoslavia and report values ranging from <100 to over 600 t km^{-2} yr^{-1} with the latter figure associated with severe sheet and gully erosion in the humid montane zone to the east of Albania. Elsewhere the highest suspended sediment yields are often associated with those steepland terrains that experience high intensity precipitation events (Fig. 12).

By removing, or modifying, the natural vegetation cover, which usually provides effective protection against erosion, human activity has led to a reduction in evapotranspiration and increased surface water drainage during storms causing accelerated rates of soil erosion, piping, gullying, gravitational movement and, in some cases, deflation. This is well illustrated in Figure 13 which shows that present suspended sediment yields in the Mediterranean have the highest anthropogenic com-

ponent of any climatic-vegetation zone, underlining the inherent fragility of the Mediterranean ecosystem and its vulnerability to human disturbance. The widespread anthropogenic acceleration of erosion rates is partly reflected by the presence of gully networks and badlands terrain in many parts of the Mediterranean region. Table 1 lists some published examples of gully and badlands development from across the region which span a wide range of precipitation regimes (c. 90 to 2000 mm). In most cases the erodible substrates are unconsolidated silt-rich materials of Tertiary age (see Bryan and Yair 1982; Campbell 1989). Changes in catchment vegetation cover (e.g. excessive burning, overgrazing, increases in the size of cultivated areas) are one of the most commonly cited causes of increased runoff and channel network extension, though short-term climate fluctuations or possibly the occurrence of threshold-exceeding storm events may also be important (Alexander et al. 1994). Despite the erodible nature of many Mediterranean soils, the marked regional variations in suspended sediment yield discussed above indicate that many areas are undergoing relatively little fluvial erosion at the present time. In some situations, however, these figures may be accounted for by low sediment delivery ratios and high rates of colluvial and floodplain storage. Alternatively, in some areas, a longer-term sediment exhaustion effect is evident where the pace of soil removal has outstripped soil formation during the historical period. Sparsely vegetated bedrock slopes and thick sequences of stratified slope deposits containing large volumes of fine sediment are a characteristic feature of the Mediterranean landscape. Recent work by Gilman and Thornes (1985) for example, suggests that, despite the widespread presence of gully systems, only limited amounts of soil erosion have taken place in parts of southeast Spain since the Bronze Age.

In general flooding and flood peaks, river sediment loads, rates of valley floor and coastal alluviation also usually increase following disturbance, or destruction, of vegetation by human activity (cf. Butzer 1982). At the same time, erosion and compaction of soil causes reduced soil depth, soil water storage and infiltration capacity, often resulting in a decrease in groundwater recharge and lower dry season flows. The appearance of Mediterranean landscapes bears witness to the contrasting impact of accelerated erosion on different terrains. On limestones, the removal of shallow soils may produce bare rock surfaces. Extensive grazing by goats may extend the effects of human occupancy widely into mountain terrains (cf. McNeill 1992). On readily-erodible bedrock, including flysch deposits, vegetation degeneration may lead to extensive gullying and slope instability. Fine sediments generated by these processes may become a major component of alluvial materials accumulating in the lower parts of river catchments (Macloed and Vita-Finzi 1982; Woodward et al. 1992).

5.3 *Planned change*

Since antiquity the three primary elements of river regulation in the Mediterranean basin have been irrigation,

Table 1. Some examples of badland and 'semi-badland' development on a range of lithologies and under various rainfall regimes in the Mediterranean region (largely after Campbell 1989 with additions).

Reference	Mean annual rainfall (mm)	Location	Materials
Alexander (1982)	450	Agri basin, Basilicata, Italy	Plio-Pleistocene marine clays, silt clays, interbedded sands, soft shales and mudstones
De Ploey (1974)	350	Kasserine area, central Tunisia	Clays, loams and sandy lithosols developed on Cretaceous marls
Harris and Vita-Finzi (1968)	1500	Kokkinopilos, Louros valley, Epirus, NW Greece	Red silts and clays of uncertain age and provenance
Harvey (1982)	170-350	Almeria-Alicante region, Spain	Cenozoic (mostly Tertiary) and Triassic marls, silts, shales and sandstones
Imeson et al. (1982)	300	Rif Mountains, Morocco	Villafranchian, Pliocene and sub-recent marine sediment, alluvial and colluvial deposits
Woodward et al. (1992)	2000	Pindus Mountains, Epirus, NW Greece	Miocene flysch sediments: alternations of sandstones and fissile siltstones
Yair et al. (1982)	90	Northern Negev, Israel	Palaeocene marls and soft shales

land reclamation and flood control. Deliberate, or direct, modification of the river environment and river basin hydrology in the Mediterranean region probably began in Egypt where there is evidence that floodplain agriculture was supplemented by elaborate gravity-fed irrigation systems as early as 3,000 BC (Hamdan 1961). Irrigation was, and still is, required by nearly all agriculture in the semi-arid and arid areas of the Mediterranean basin. In land adjoining mountain areas such as the Sierra Nevadas of Spain, the Taurus Mountains of Anatolia and the peaks of Lebanon, snowmelt has also been used for irrigation. Floodwater farming (often called rainwater harvesting) has also been extensively practised in certain parts of the region (cf. Evenari et al. 1982). Irrigation has often been associated with the reclamation of mountain slopes by terracing in mountainous regions such as Greece where agricultural land is limited in extent (cf. Pope and van Andel 1984). In contrast, flood-prone and naturally waterlogged valley bottoms, lake basins, coastal lowland and deltaic flats required land drainage and flood abatement measures. In some of the larger river and coastal wetlands, such as the Ebro Delta and Pontine Marshes, this proved to be impracticable until the present century. In smaller river systems, structural approaches to flood control have long been employed and were common practice among the Homeric Greeks (Braudel 1972). In the face of repeated disastrous floods, frequently exacerbated by catchment deforestation and poor agricultural practices, two basic forms of regulation were used. The first utilized flood storage dams in smaller tributaries of the catchment to reduce flood peaks and to slow down runoff to the main stream. Stored water could also be used for irrigation purposes. The second type of flood control was designed to protect floodplain areas directly through the construction of artificial levées and embankments or to accelerate runoff by channel straightening.

Although irrigation, land reclamation and flood control have all increased agricultural productivity in the Mediterranean basin, and provided additional land for settlement and farming, designed modification of channels and drainage networks not in sympathy with natural river dynamics have had a number of adverse environmental impacts. Ultimately, irrigation systems in drylands suffer from two undesirable effects; salinization and waterlogging of the soil. As a result large areas of once fertile soil deteriorate into waste land. Problems can also arise from flood irrigation works. For example, construction of artificial levées along rivers with large sediment loads frequently leads to aggradation resulting in channels becoming perched well above the level of their former floodplain. This not only makes drainage of riparian fields almost impossible but also significantly increases the risk of disastrous floods if embankments are breached. Even agricultural terraces, though of great value in soil and water conservation in steepland environments, by decoupling slope-channel sediment transport significantly reduce river loads which in the long term promotes channel incision and lowering of valley bottom water tables (cf. Wise et al. 1982; van Andel et al. 1986). To summarise, the Mediterranean basin is an ecologically fragile environment which has been deeply marked, and often irrevocably damaged by millennia of human activity. Echoing the views of Grenon and Batisse (1989), perhaps nowhere else has nature done so much for humankind and human's have in turn so transformed nature.

6 FLUVIAL PROCESSES AND ALLUVIAL SETTINGS FOR MEDITERRANEAN QUATERNARY RIVER ENVIRONMENTS

6.1 *Background*

By world standards rivers draining to the Mediterranean Sea, with the notable exception of the exogenous River Nile, are relatively minor. This primarily reflects tectonic plate configuration in the region with areas of uplift tending to be located close to the coast thereby constricting inland drainage. The Ebro (84,230 km^2), Po (70,090 km^2) and Rhône (95,590 km^2), by far the largest rivers in the region which have their headwaters entirely within the Mediterranean basin, are dwarfed in terms of discharge, sediment load and drainage area by the major rivers of the World such as the Amazon (7,050,000 km^2), Mississippi (3,248,000 km^2) and Yangtze (1,175,000 km^2). Nevertheless, active tectonics (including volcanism), periodic climate change and human activity over the Quaternary Period have, to varying degrees, induced significant variation in rates and patterns of sediment and water yield, and created a highly diverse suite of active and relict Mediterranean riverine landscapes. Although a comprehensive model of river response to environmental change has yet to be defined, several broad generalizations can be made about fluvial processes, sedimentation styles and resultant alluvial architectures in the region.

6.2 *Steepland river systems*

Taking the 500 m contour as the mountain-lowland boundary (Fig. 1), it is clear that most of the Mediterranean basin (especially its northern and western margins) is drained by steepland river systems. These usually have steep, boulder- or cobble-bed channels which are very often deeply incised into older alluvial or colluvial fills and, in some instances, bedrock. These entrenched channels are termed Wadis in the Near East and North Africa, Fuimares in Italy and Ramblas in Spain. They carry high sediment loads derived from channel scour, bank erosion or direct inputs (e.g. mass movements) from valley slopes. In the wetter and cooler mountain environment on the northern Mediterranean littoral (e.g. Iberian Peninsula, Italy, Balkans) high magnitude floods, resulting in rapid channel changes, are often associated with snowmelt and summer thunderstorms. A significant number of these catchments (Fig. 11) were glaciated in the Würm and in earlier Pleistocene cold stages resulting in the formation of extensive outwash trains in some areas (e.g. Lewin et al. 1991). Formerly glaciated drainage basins have had to adjust to markedly different flood and sediment delivery regimes in the Holocene. Many have followed a 'paraglacial' (*sensu* Church and Ryder 1972) course of development with decreased sediment supply rates resulting in progressive channel incision and the formation of a staircase of valley floor terraces. However, in steepland catchments severely modified by forest clearance and agriculture, rates of valley floor incision have slowed, or have even been reversed, during the late Holocene as a result of accelerated slope erosion. It is also very likely that mountain drainage basins in the Mediterranean have been affected by second-order, relatively modest, climatic fluctuations during the Holocene, most probably shown by changes in flood regime and mass movement rates. Unfortunately the provision of dating control for Holocene alluvial fills in steepland river systems within the region, at the moment, is generally inadequate to be able to test a causal link between river activity and recent climate change.

Upland and mountain streams in the presently semiarid (mean annual rainfall 250-500 mm) and arid (mean annual rainfall 50-250 mm) environments of North Africa and the Near East have ephemeral regimes characterized by rare, high magnitude, short-lived flood events. Following intense convective storms, hydrographs can rise and fall almost instantaneously with 'walls of water' often reported racing down channels (Schick 1988). These floods commonly transport very large volumes of clastic bedload (Laronne and Reid 1993) and yield high amounts of suspended sediment. Evaporation and transmission losses are often high, resulting in a marked decline in flow rates downstream and rapid depletion of sediment loads. Semi-arid and arid upland river environments in the Mediterranean basin have been especially sensitive to Quaternary climatic variations, primarily in response to precipitation-related changes in vegetation cover and sediment yield (e.g. Macklin et al. 1994). In general terms, a shift to a wetter climate in these areas increased vegetation density, reduced sediment delivery to valley floors and resulted in enlargement and incision of trunk channels. Conversely, slope degradation and valley floor aggradation would appear to be more characteristic of periods with lower rainfall and decreased vegetation cover. Emptying of hillslope sediment reservoirs, and greater flashiness of runoff resulting from increased areas of exposed bedrock are, however, likely to have provided negative feedback mechanisms limiting valley floor aggradation (cf. Bull 1991). Sediment exhaustion effects may therefore have induced degradation under stable bioclimatic conditions. One particularly distinctive type of steepland Mediterranean river system are those which drain active volcanoes (Fig. 4). Quaternary-age volcanoes in the Mediterranean basin are restricted to the Hellenic and Calabrian areas (Fig. 4), though the only river systems affected by explosive volcanic activity that have been studied in detail are the Alcantara and Simeto Rivers which drain Mount Etna, Sicily (Chester and Duncan 1982). Here, several episodes of volcanism-induced sedimentation and erosion are documented dating back to Middle Pleistocene times. Aggradation occurred during short periods when eruptions produced sediment-choked streams, and these were separated by longer periods of non-deposition and incision as streams adjusted to a diminished sediment load.

6.3 *Alluvial fans*

Where steepland river systems issue from faulted or tectonic mountain fronts, or at valley-tributary junctions, alluvial fans commonly develop. Alluvial fans form important sedimentary environments and are one of the most widespread fluvial depositional landforms in the Mediterranean basin. The majority of the larger alluvial fans in the region, particularly those developed at the mountain-coastal plain junction (e.g. Crete, Nemec and Postma 1993; southern and southeast Spain, Harvey 1990), have been shown to be relict Pleistocene forms which presently display low activity rates. Currently active fans tend to be restricted to steep mountain environments and intensely dissected badlands, sometimes in regions of tectonic activity (Sorriso-Valvo and Sylvester 1993). It is generally believed that alluvial fans in the Mediterranean region accumulated during Pleistocene cold stages with relatively high rates of weathering in their mountain catchments and a greater effectiveness of storm runoff than today (Rohdenburg and Sabelberg 1980). Climatic amelioration and stabilisation of slopes by vegetation saw a reduction in sediment supply, fan trenching and, in some cases, limited progradation of the distal fan area (Harvey 1992).

6.4 *Basin and range environments*

Downstream of the mountain-piedmont junction the recurrent relief pattern in the middle and lower reaches of many Mediterranean drainage basins (particularly those in current or former compressional or extensional tectonic terrains) is one of mountain and basin topography (cf. Bloom 1991). Narrow, deep bedrock canyons cut in uplifted blocks, alternate with lower gradient alluvial plains accommodated in downthrown basins within which active gravel-bed, braided river systems are commonly developed. Many of the larger Late Pleistocene 'pluvial' lakes (e.g. Padul, Spain; Ioannina, Greece) were also formed in intermontane basins where, as the result of long-term subsidence, thick interbedded fluvial, paludal and lacustrine sedimentary sequences are preserved. Faulting and folding associated with active tectonics has generally restricted the development of extended alluvially-formed channel reaches in the Mediterranean basin. Indeed, extensive alluvial channel systems only occur on coastal plains and in the larger drainage basins such as the Ebro and Po which are developed in transorogen tectonic settings. Even in these areas of relatively low seismic activity, slow ongoing deformation can result in the disruption of drainage networks. This is well shown in the middle Ebro Valley near Zaragoza where localised uplift has resulted in progressive channel entrenchment and partitioning of an extensive alluvial plain into a series of smaller alluvial basins separated by bedrock-controlled reaches (Ramirez Merino et al. 1992).

6.5 *Coastal alluvial plains and deltas*

At the coast, in the absence of serious tidal scour in the Mediterranean Sea, impressive deltas have been formed at the mouths of many river systems. The four major Mediterranean deltas (in order of size) are the Nile (delta area 22,000 km^2), the Po (770 km^2), the Rhône (720 km^2) and the Ebro (350 km^2). Sedimentary sequences that comprise these emerged deltas, and those accumulated on adjacent continental shelves, constitute the thickest and most extensive Pliocene-Quaternary deposits in the Mediterranean basin (Got et al. 1985). Deep penetration seismic profiling, and drilling by the oil industry, have established Pliocene-Quaternary series thicknesses of more than 5000 m and 3500 m on the Po and Nile delta shelves, respectively, and depocentres of between 1000-2300 m in the Ebro and Rhône deltas. In the indented coastlines of Greece and Turkey, historical alluviation has wrought considerable changes to the coastline and the development of coastal alluvial plains has transformed the settings of many historical and earlier archaeological sites (cf. Vita-Finzi 1978). For example, the ancient site of Troy, in Anatolia, western Turkey, is now well inland (Kraft et al. 1980), whilst the battle site of Thermopylae north of Athens (at the time, 480 BC, a narrow coastal pass around 100-200 m wide at the foot of Mount Kallidromon) is now 4 km from the sea in the Gulf of Malia (Kraft and Rapp 1988).

The Ebro (Maldonado 1975), Nile (Warne and Stanley 1993), Po (Got et al. 1985) and Rhône (Duboul-Razavet 1954) deltas have experienced rapid growth in historic times. In the case of the Ebro and the Po, progradation of the delta front in the Medieval and post-Medieval periods approached 5-7 km per century (Got et al. 1985; Maldonado 1972). Accelerated deposition in these and many other coastal delta plains resulted in the silting up of a considerable number of ancient harbours including Ephesus (Eisma 1978; Erinç 1978) and Ostia Antica, the former port of Rome. This has been attributed to deforestation and poor agricultural practices in catchment areas (Brückner 1986), and the construction of artificial levées inland increasing sediment delivery to the coast. Interestingly, recent work on the Rosetta and Damietta promontories of the Nile Delta has indicated that the switch from a predominantly prograding to an eroding state can take place relatively rapidly on such major delta systems. Such a change took place on the Nile Delta as recently as 1900 following precipitation changes in East Africa and a marked decrease in Nile flood magnitude (Frihy and Khafagy 1991).

7 INTRODUCTORY REMARKS AND STRUCTURE OF THE VOLUME

Nineteen papers on river basins from more than half of the countries (Algeria, France, Greece, Italy, Libya, Spain, Tunisia and Turkey) that border the Mediterranean Sea are included in this volume (Fig. 14). In Tables 2(a) to 2 (d) their climate, geomorphology and tectonic regime are summarised together with the time periods and topics of investigation. Two studies, by James and Chester in southern Portugal and Roberts in north-central Turkey, fall outside the watershed of the Mediterranean basin *sensu stricto*. However, southern Portugal is encompassed by the classic definition of the Mediterranean based on the distribution of the olive (Fig. 2). From Table 2 it is clear that coverage of the northern Mediterranean basin in this book is more comprehensive that the southern and eastern Mediterranean littoral. Significant omissions in this respect are river systems in the Middle East (Israel, Lebanon, Syria), western Maghreb (Morocco), Cyrenaica (Libya) and Egypt, as well as the former Yugoslavia and Albania. Nevertheless, a wide spectrum of river environments and histories are presented which, in terms of climate, tectonic activity and human impact, are generally representative of the Mediterranean basin as a whole. Climatically they extend from the hyper-arid Tripolitanian valleys of western Libya to the humid, formerly glaciated river basins of the northern Pindus Mountains, Greece. With reference to tectonics, the vo-

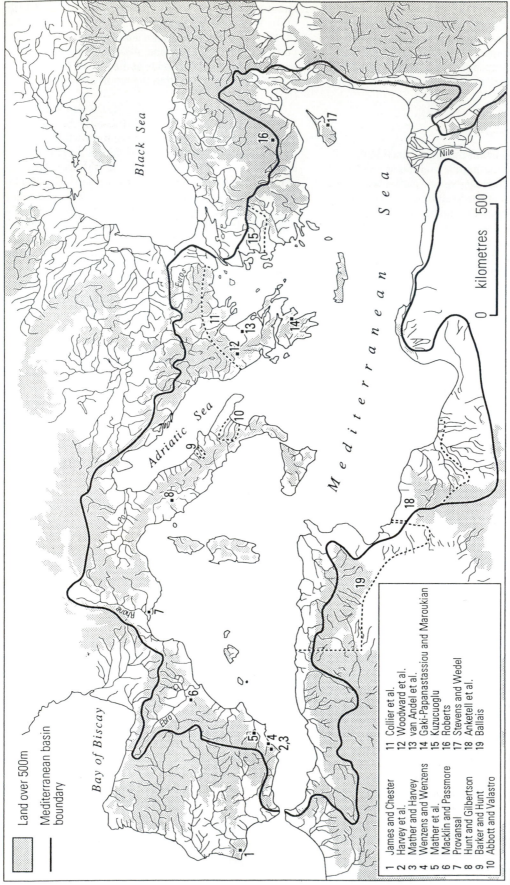

Figure 14. Map of the Mediterranean basin showing locations of study areas in the volume.

Legend:
- Land over 500m
- Mediterranean basin boundary

1 James and Chester
2 Harvey et al.
3 Mather and Harvey
4 Wenzens and Wenzens
5 Mather et al.
6 Macklin and Passmore
7 Provansal
8 Hunt and Gilbertson
9 Barker and Hunt
10 Abbott and Valastro
11 Collier et al.
12 Woodward et al.
13 van Andel et al.
14 Gaki-Papanastassiou and Maroukian
15 Kuzucuoglu
16 Roberts
17 Stevens and Wedel
18 Anketell et al.
19 Ballais

Table 2(a). List of authors, climatic and geomorphic details of study basins, time periods and topics of investigation in the Iberian Peninsula.

Author(s)	Country	River basin(s)	Climate	Present tectonic activity	Fluvial setting	Time period of investigation	Topic of investigation
James & Chester (Chapter 22)	Portugal	Arade, Enxerim, Lombos, Odelouca, Quarteira	Semi-arid to sub-humid	Moderate	Basin and range, coastal alluvial plain	Quaternary	River-sediment catchment source linkages, alluvial soil chronosequences
Harvey et al. (Chapter 23)	Mainland Spain	Aguas, Feos, Sorbas	Semi-arid	High	Basin and range	Middle-late Quaternary	Relative dating and correlation of river terraces using soil chronosequences
Mather & Harvey (Chapter 6)	Mainland Spain	Aguas, Feos	Semi-arid	High	Basin and range	Pliocene-Quaternary	Tectonics and long term drainage network evolution
Wenzens & Wenzens (Chapter 5)	Mainland Spain	Almanzora	Semi-arid	High	Basin and range	Pliocene-Quaternary	Influence of tectonics on river drainage pattern development
Mather et al. (Chapter 7)	Mainland Spain	Mula	Semi-arid	High	Basin and range	Quaternary	Impact of climate, tectonics and base-level change on dissectional and aggradational behaviour of river systems
Macklin & Passmore (Chapter 10)	Mainland Spain	Guadalope	Semi-arid	Low	Basin and range	Quaternary	Alluvial archaeology, geochronology, climate-river interactions

Table 2(b). List of authors, climatic and geomorphic details of study basins, time periods and topics of investigation in France and the Italian Peninsula.

Author(s)	Country	River basin(s)	Climate	Present tectonic activity	Fluvial setting	Time period of investigation	Topic of investigation
Provansal (Chapter 14)	France	Arc	Sub-humid	Low	Coastal alluvial plain, delta	Holocene	Relationship between climate, sea-level and land-use change and delta plain development
Hunt & Gilbertson (Chapter 15)	Mainland Italy	Feccia	Sub-humid	Moderate	Basin and range	Late Holocene	Catchment landuse change and valley floor alluviation
Barker & Hunt (Chapter 13)	Mainland Italy	Biferno	Sub-humid	High	Basin and range	Quaternary	Influence of tectonics, climate and human activity on river erosion and sedimentation
Abbott & Valastro (Chapter 18)	Mainland Italy	Basento, Bradano, Cacchiavia, Cavone	Sub-humid	High to very high	Steepland, basin and range	Late Quaternary	Relationship between regional tectonics, cultural and climatic sequences, and valley floor aggradation and incision

Table 2(c). List of authors, climatic and geomorphic details of study basins, time periods and topics of investigation in Greece and Cyprus.

Author(s)	Country	River basin(s)	Climate	Present tectonic activity	Fluvial setting	Time period of investigation	Topic of investigation
Collier et al. (Chapter 3)	Mainland Greece, Evvia, Greece	Arakthos, Luros, Peneios, Sperchios	Semi-arid to humid	High to very high	Steepland, alluvial fan, basin and range, fan-deltas	Pliocene-Quaternary	Tectonics and river drainage development, sediment fluxes
Woodward et al. (Chapter 11)	Mainland Greece	Voidomatis	Humid	High to very high	Basin and range, steepland	Middle-late Quaternary	Mountain glaciation and fluvial sedimentation linkages, Upper Palaeolithic settlement
van Andel et al. (Chapter 12)	Mainland Greece	Peneios	Semi-arid to sub-humid	Very high	Basin and range	Early-middle Holocene	Neolithic floodwater farming
Gaki-Papanastassiou & Maroukian (Chapter 8)	Mainland Greece	Berbadiotis, Dafnorema, Havos, Koulouvinas, Ksirias, Megalo Rema	Sub-humid	Very high	Steepland, alluvial fan	Quaternary	Tectonic, climatic and anthropogenic influences on river erosion and sedimentation
Stevens & Wedel (Chapter 20)	Cyprus	Pouzi, Tremithios, Xeropouzi	Semi-arid	High	Alluvial fan, basin and range, coastal alluvial plain	Quaternary	Facies sequences and allogenic/autogenic controls of fluvial sedimentation

Table 2(d). List of authors, climatic and geomorphic details of study basins, time periods and topics of investigation in Turkey and North Africa.

Author(s)	Country	River basin(s)	Climate	Present tectonic activity	Fluvial setting	Time period of investigation	Topic of investigation
Kuzucuoglu (Chapter 4)	Turkey	Bakirlik, Gediz, Gökmen, Havran, Karakoc, Kozak	Sub-humid	Very high	Basin and range, coastal alluvial plain	Miocene-Quaternary	Influence of tectonics on drainage network development
Roberts (Chapter 19)	Turkey	Ibrala	Semi-arid	High	Alluvial fan, basin and range	Late Quaternary	Alluvial fan development and climate change
Anketell et al. (Chapter 21)	Libya	Gerrim, Ghan, Ghobeen, Mansur, Merdum, Suf Al Jim, Tininai	Semi-arid to extremely arid	Low to moderate	Steepland, alluvial fans, basin and range, coastal, alluvial plains	Quaternary	Influence of vulcanicity, tectonics, climate and human activity on river development, fluvial stratigraphic sequences and their correlation
Ballais (Chapter 17)	Tunisia Algeria	Bir el Amir, Cheria-Mezeraa, El Akarit, El Hattab, Es Sgniffa, Jenain, Kebir-Milliane, Krima, Leben, Medjerda, Seradou	Humid to arid	Moderate to high	Steepland, basin and range, coastal alluvial plain	Holocene	Climatic and anthropogenic controls of river alluviation and erosion

lume includes papers on seismically quiescent river basins, such as those that drain the northern flank of the Iberian cordillera, as well as river systems presently experiencing high rates of uplift or subsidence (e.g. Greece or southeast Spain). The intensity and duration of human impact has also varied widely in the river basins examined, ranging at one extreme from thinly-populated sub-Saharan Libya, which has been little affected by human activity, to the more populous Italian Peninsula whose rivers have been regulated since antiquity and catchments totally transformed by many thousands of years of intensive farming. Alluvial settings that are characteristic of the Mediterranean basin (see Section 6) including alluvial fans, basin and range environments and coastal alluvial plains are well represented. Steepland river systems (developed within upland/mountain environments), coastal deltas and large alluvial rivers (e.g. Ebro, Po and Rhône) are not considered in detail. Notwithstanding these minor limitations, largely new results from over 50 river basins are presented which in itself represents something of a benchmark for Quaternary river studies in the Mediterranean region.

Three major themes provide a common thread to the papers in the volume. Firstly, there is the influence of tectonics and base-level changes on drainage network evolution. Secondly, human-river environment interactions have assessments of the archaeological record, with particular reference to the impact of land use change on soil erosion and valley floor alluviation. Thirdly, dating and correlating fluvial stratigraphies are used to establish the causes of river aggradation and incision. Reflecting these themes, the volume has been divided into three parts. Papers in Part 1, 'the impact of Quaternary tectonic activity on river behaviour', focus on the seismically active area of the Betics, southeast Spain and the Aegean basin. In Part 2, 'archaeology and human-river environment interactions', two further sub-themes emerge. First, investigations (primarily concerned with the Palaeolithic) which have sought either to reconstruct fluvial envi-

ronments contemporary with human settlement in river valleys, or have used fluvial stratigraphies to provide a temporal framework for the archaeological record. Second, studies (primarily concerned with the Neolithic and later periods) that have tried to quantify the response of river systems to human disturbance. Papers in part 3, 'geochronology, correlation and controls of Quaternary river erosion and sedimentation', explore approaches for dating fluvial sequences and identifying the major intrinsic and extrinsic controls of river development. Finally, in the concluding section, some of the research needs in Quaternary river studies in the Mediterranean basin are highlighted and suggestions are made for possible future lines of inquiry in the region.

8 ACKNOWLEDGEMENTS

The authors would like to thank Lois Wright and Alison Manson of the Graphics Unit in the School of Geography, University of Leeds, for drawing the diagrams in this chapter.

REFERENCES

Alexander, D. 1982. Difference between 'calanchi' and 'biancane' badlands in Italy. In R.B. Bryan & A. Yair (eds), *Badland geomorphology and piping.* GeoBooks, Norwich: 71-88.
Alexander, R.W., A.M. Harvey, A. Calvo, P.A. James & A. Cerda 1994. Natural Stabilisation Mechanisms on Badland Slopes: Tabernas, Almeria, Spain. In A.C. Millington & K. Pye, *Environmental Change in Drylands: Biogeographical and Geomorphological Perspectives.* Chichester, John Wiley & Sons, 85-111.
Ammerman, A.J. & Cavalli-Storza 1971. Measuring the rate of spread of early farming in Europe. *Man* 6: 674-688.
Bailey, G.N., P.L. Carter, C. Gamble & H.P. Higgs 1983. Asprochaliko and Kastritsa: Further investigations of

Palaeolithic settlement and economy in Epirus (North-west Greece). *Proceedings of the Prehistoric Society* 49: 15-42.

Bailey, G.N., G.C.P. King & D.A. Sturdy 1993. Active tectonics and land-use strategies: a Palaeolithic example from Northwest Greece. *Antiquity* 67: 292-312.

Baille, M.G.L. & M.A.R. Munro 1988. Irish tree rings, Santorini and volcanic dust veils. *Nature* 332: 344-346.

Beckinsale, R.P. 1969. River Regimes. In R.J. Chorley (ed) *Water, Earth and Man*. London, Methuen, 455-471.

Bennett, K.D., P.C. Tzedakis & K.J. Willis 1991. Quaternary refugia of north European trees. *Journal of Biogeography* 18: 103-115.

Bloom, A.L. 1991. *Gemorphology: A Systematic Analysis of Late Cenozoic Landforms*. New Jersey, Prentice Hall.

Bond, G. and 13 others 1992. Evidence for massive discharge of icebergs into the North Atlantic ocean during the last glacial period. *Nature* 360: 245-249.

Bond, G. and 6 others 1993. Correlations between climate records from the North Atlantic Sediments and Greenland Ice. *Nature* 365: 143-147.

Bork, H.R. & H. Bork 1981. Oberflächenabflub und Infiltration. Erzebruine von 100 Starkregensimulationen im Einzugsgebiet der Rambla del Campo Santo (SE – Spanien). Landschaftsgenese und Landschaftsökologie 8. Cremlingen – Destedt.

Bottema, S. 1978. The Late Glacial in the eastern Mediterranean and the Near East. In W.C. Brice (ed), *The Environmental history of the Near and Middle East since the last Ice Age*. London, Academic Press, 15-28.

Braudel, F. 1972. *The Mediterranean and the Mediterranean world in the age of Philip II*. London, Collins.

Broecker, W.S. & G. Denton 1989. The role of ocean-atmosphere reorganization in glacial cycles. *Geochimica et Costmochimica Acta*. 53: 2465-2501.

Brückner, H. 1986. Man's impact on the evolution of the physical environment in the Mediterranean Region in historical times. *Geojournal* 13.1: 7-17.

Bryan, R.B. & A. Yair 1982. *Badland geomorphology and piping*. GeoBooks, Norwich.

Bull, W.B. 1991. *Geomorphic Response to Climate Change*. Oxford, Oxford University Press.

Butzer, K. 1982. *Archaeology as Human Ecology*. Cambridge, Cambridge University Press.

Campbell, I.A. 1989. Badlands and badland gullies. In D.S.G. Thomas (ed) *Arid zone geomorphology*. London, Belhaven Press: 158-183.

Chester, D.K & A.M. Duncan 1982. The interaction of volcanic activity in Quaternary times upon the evolution of the Alcantara and Simeto Rivers, Mount Etna, Sicily. *Catena* 9: 319-342.

Church, M. & J.M. Ryder 1972. Paraglacial sedimentation: a consideration of fluvial processes conditioned by glaciation. *Geological Society of America Bulletin* 83: 3059-3072.

COHMAP Members, 1988. Climatic changes of the last 18 000 years: observations and model simulations. *Science* 241: 1043-1052.

Dedkov, A.P. & V.I. Moszherin 1992. Erosion and sediment yield in mountain regions of the world. In D.E. Walling, T.R. Davies & B. Hasholt (eds) *Erosion, Debris Flows and Environment in Mountain Regions*, Proceedings of the Chengdu Symposium, July 1992. IAHS Publication No. 209: 29-36.

De Ploey, J. 1974. Mechanical properties of hillslopes and their relation to gullying in central semi-arid Tunisia. *Zeitschrift für Geomorphologie, Supplementband* 21: 177-190.

Dewey, J.F., W.C. Pitman III, W.B.F. Ryan & J. Bonnin 1973.

Plate tectonics and the evolution of the Alpine system. *Geological Society of America Bulletin* 84: 3137-3180.

Duboul-Razavet, C. 1954. Contribution à l'étude géologique et sedimentologique du Delta du Rhône. *Méin. Soc. Géol. Fr.* 76.

Eisma, D. 1978. Steam deposition and erosion by the eastern shore of the Aegean. In W.C. Brice (ed) *The environmental history of the Near and Middle East since the Last Ice Age*. London, Academic Press, 67-81.

Ergenzinger, P. 1992. A conceptual geomorphological model for the development of a Mediterranean river basin under neotectonic stress (Buonamico basin, Calabria, Italy). In D.E. Walling, T.R. Davies & B. Hasholt (eds) *Erosion, Debris Flows and Environment in Mountain Regions*, Proceedings of the Chengdu Symposium, July 1992. IASH Publication No. 209: 51-60.

Erinç, S. 1978. Changes in the physical environment in Turkey since the end of the last Glacial. In W.C. Brice (ed). *The environmental history of the Near and Middle East since the Last Ice Age*. London, Academic Press, 87-119.

Evenari, M., L. Shanan & N. Tadmor 1982. *The Negev: The Challenge of a Desert*. Cambridge, Mass, Harvard University Press.

Eyre, S.R. 1968. *Vegetation and Soils: A world picture*. Chicago, Aldine Publishing Company.

Frihy, O.E. & A.A. Khafagy 1991. Climate and induced changes in relation to shoreline migration trends at the Nile Delta promontories, Egypt. *Catena* 18: 197-211.

Gat, J.R. & M. Magaritz 1980. Climatic variations in the eastern Mediterranean Sea area. *Naturuissenschaften* 67: 80-87.

Gilman, A. & J.B. Thornes 1985. *Land Use and Prehistory in south-east Spain*. London, Allen and Unwin.

Got, H., J.C. Aloïsi & A. Monaco 1985. Sedimentary processes in Mediterranean deltas and shelves. In D.J. Stanley & F.C. Wezel (eds), *Geological Evolution of the Mediterranean Basin*, 355-376.

Grenon, M. & M. Batisse 1989. Specific Characteristics and Permanent Features. In M. Grenon & M. Batisse (eds). *Futures for the Mediterranean Basin: The Blue Plan*. Oxford, Oxford University Press, 2-13.

Hamdan, G. 1961. Evolution of irrigation agriculture in Egypt. In: *History of land use in arid regions. Arid Zone Research* 17: New York, UNESCO, 119-142.

Harris, D. & C. Vita-Finzi 1968. Kokkinopilos – A Greek badland. *Geographical Journal* 134: 537-546.

Harrison, S.P. & G. Digerfeldt 1993. European Lakes as palaeohydrological and palaeoclimatic indicators. *Quaternary Science Reviews* 12: 233-248.

Harvey, A .M. 1982. The role of piping in the development of badland and gully systems in southeast Spain. In R.B. Bryan & A. Yair (eds) *Badland geomorphology and piping*. Geobooks, Norwich, 317-336.

Harvey, A.M. 1990. Factors influencing Quaternary alluvial fan development in southeast Spain. In A.H. Rachocki & M. Church (eds) *Alluvial Fans: A Field Approach*, Chichester, John Wiley and Sons, 247-269.

Harvey, A.M. 1992. Controls on sedimentary style on alluvial fans. In P. Billi, R.D. Hey, C.R. Thorne & P. Tacconi (eds) *Dynamics of Gravel-bed Rivers*, Chichester, John Wiley and Sons, 519-535.

Harvey, A.M. & S.G. Wells 1987. Response of Quaternary fluvial systems to differential epirogenic uplift: Aguas and Feos river systems, southeast Spain. *Geology* 15: 689-693.

Huntley, B. & H.J.B. Birks 1983. *An Atlas of Past and Present Pollen Maps for Europe 0-13,000 years ago*. Cambridge, Cambridge University Press.

Huttary, J. 1950. Die Verteilung der Niediershläge auf die Jahreszeiten im Mittelmeergbeit. *Meteorolgische Rundschau* 3: 111-119.

Imeson, A.C. F.J.P.M. Kwaad & J.M. Verstraten 1982. The relationship of soil physical and chemical properties to the development of badlands in Morocco. In R.B. Bryan & A. Yair (eds) *Badland geomorphology and piping.* GeoBooks, Norwich: 47-70.

Jovanovic, S. & M. Vukcevic 1957. Suspended sediment regimes on some watercourses in Yugoslavia and analysis of erosion processes. *International Association of Hydrological Sciences,* Publication No. 43, 337-359.

Knox, J.C. 1972. Valley alluviation in southwestern Wisconsin. *Annals Association of American Geographers* 62: 401-410.

Kraft, J.C., I. Kayan & O. Erol 1980. Geomorphic reconstructions of the environs of ancient Troy. *Science* 209:775-782.

Kraft J.C. & G.R. Rapp 1988. Geological reconstructions of ancient coastal landforms in Greece with predictions of future coastal changes. In P.G. Marinos & G.C. Koukis (eds) *The Engineering Geology of Ancient Works, Monuments and Historical Sites.* Balkema, Rotterdam Volume 3: 1545-1556.

Kutzbach, J.E. & F.A. Street-Perrott 1985. Milankovitch forcing of fluctuations in the level of tropical lakes from 18 to 0 kyr BP. *Nature* 317: 130-134.

Kutzbach, J.E. & P.J. Guetter 1986. The influence of changing orbital parameters and surface boundary conditions on climate simulations for the past 18,000 years. *Journal of Atmospheric Sciences,* 43: 1726-1759.

Laronne, J.B. & I. Reid 1993. Very high rates of bedload sediment transport by ephemeral desert rivers. *Nature* 366: 148-150.

Lewin, J., M.G. Macklin & J.C. Woodward 1991. Late Quaternary fluvial sedimentation in the Voidomatis basin, Epirus, northwest Greece. *Quaternary Research* 35: 103-115.

Macklin, M.G., D.G. Passmore, A.C. Stevenson, B.A. Davis & J.A. Benavente 1994. Responses of Rivers and Lakes to Holocene Environmental Change in the Alcañiz Region, Teruel, northeast Spain. In A.C. Millington & K. Pye (eds) *Environmental Change in Drylands: Biogeograpaphical and Geomorphological Perspectives.* Chichester, John Wiley and Sons, 113-130.

Macleod, D.A. & C. Vita-Finzi 1982. Environment and Provenance in the development of recent alluvial deposits in Epirus, northwest Greece. *Earth Surface Processes and Landforms* 7: 29-43.

Maldonado, A. 1972. El delta del Ebro. Estudio Sedimentologico y estratigrafico. *Boll. Estrat. Barcelona* 1.

Maldonado, A. 1975. Sedimentation, stratigraphy and development of the Ebro Delta, Spain. In M.L. Broussard (ed) *Delta Models for Exploration,* Houston Geological Society, 311-338.

McNeill, J.R. 1992. *The Mountains of the Mediterranean World: An Environmental History.* Cambridge, Cambridge University Press.

Messerli, B. 1967. Die Eiszeitleche und die gegenwartige vergletscherung im Mittelmeeraum. *Geographica Helvetica* 22: 105-228.

Milliman, J.D. & J.P.M. Syvitski 1992. Geomorphic-tectonic control of sediment discharge to the ocean: the importance of small mountainous rivers. *Journal of Geology* 100: 525-544.

Mourtzas, N.D. 1990. Holocene vertical movements and changes of sea-level in the Hellenic Arc. In P.G. Marinos & G.C. Koulkis (eds) *The Engineering Geology of Ancient Works, Monuments and Historical Sites.* Balkema, Rotter-dam, Volume 4: 2247-2262.

Nemec, W. & G. Postma 1993. Quaternary alluvial fans in southwestern Crete: sedimentation processes and geomorphic evolution. *Special Publication International Association of Sedimentologists* 17: 235-276.

North, & P.D. Jones 1977. Does the Mediterranean region still qualify for classification type Cs? *Climate Monitor* 6: 50-53.

Pérez-Obiol, R. & R. Julià 1994. Climate change in the Iberian Peninsula recorded in a 30,000 yr pollen record from Lake Banyoles. *Quaternary Research* 41: 91-98.

Perry, A. 1981. Mediterranean climate – a synoptic reappraisal. *Progress in Physical Geography* 5: 107-113.

Pons, A. & M. Reille 1988. The Holocene and Upper Pleistocene pollen record from Padul (Granada, Spain). A new study. *Palaeogeography, Palaeoclimatology and Palaeoecology* 66: 243-263.

Pope, K.O. & T.H. van Andel 1984. Late Quaternary alluviation and soil formation in southern Argolid: its history, causes and archaeological implications. *Journal of Archaeological Science* 11: 281-306.

Prentice, I.C., J. Guiot & S.P. Harrison 1992. Mediterranean vegetation, lake levels and palaeoclimate at the last glacial maximum. *Nature* 360: 658-660.

Quezel, P. 1977. Forests of the Mediterranean Basin. In *Mediterranean forests and maquis: ecology, conservation and management.* MAB Technical Notes 2, Paris, UNESCO, 9-32.

Ramirez Merino, J.I., A. Olive Davo & M.H. Pascual Muñoz 1992. Evidence geomorfologicas de la existencia de actividad neotectonica durante el Pleistoceno en us Sector de la zona central de la cuenca del Ebro. *Estudios de Geomorfologia en España*, 643-651

Roberts, N. 1989. *The Holocene: an Environmental History.* Oxford, Basil Blackwell.

Rognon, P. 1987. Aridification and abrupt climatic events on the Saharan northern and southern margins, 20 000 years BP to present. In W.H. Berger & L.D. Labeyrie (eds) *Abrupt Climatic Change: Evidence and Implications.* Dordrecht, D., Reidel Publishing Company, 209-220.

Rohdenburg, H. & U. Sabelberg 1980. Northwestern Sahara margins: Terrestrial stratigraphy of the upper Quaternary and some palaeoclimatic implications. In E.M. Sr. Van Zinderen Bakker & J.A. Coetzee (eds) *Palaeoecology of Africa and the Surrounding Islands,* 267-276.

Rossignol-Strick, M. 1985. Mediterranean Quaternary Sapropels, and immediate response of the African Monsoon to variation of insolation. *Palaeogeography, Palaeoclimatology, Palaeoecology,* 49: 237-263.

Rossignol-Strick, M., W. Nesteroff, P. Olive & C. Vergnand-Grazzini 1982. After the deluge: Mediterranean stagnation and sapropel formation. *Nature* 295: 105-110.

Schick, A.P. 1988. Hydrologic Aspects of floods in Extreme Arid Environments. In V.R. Baker, R.C. Kochel & P.C.Patton (eds) *Flood Geomorphology,* Chichester, John Wiley and Sons, 189-203.

Smith, A.G. & N.H. Woodcock 1982. Tectonic synthesis of the Alpine-Mediterranean region: A review. In: *Alpine Mediterranean Geodynamics, American Geophysical Union,* Geodynamics Series, 7, 15-38.

Sorriso-Valvo, M. & A.G. Sylvester 1993. The relationship between geology and landforms along a coastal mountain front, northern Calabria, Italy. *Earth Surface Processes and Landforms* 18: 257-273.

Stuckmann, G. 1969. Les Inondations de Septembre-Octobre 1969 en Tunisie. Partie II. *Effels morphologies,* Paris, UNESCO.

Stuiver, M. & T.F. Braziunas 1993. Sun, ocean, climate and atmospheric [14]C: an evaluation of causal and spectral relationships. *The Holocene* 3 (4): 289-305.

Summerfield, M.A. 1991. *Global Geomorphology.* Longman Scientific and Technical.

Thornes, J.B. 1989. the Palaeo-ecology of Erosion. In J.M. Wagstaff (ed) *Landscape and Culture: Geographical and Archaeological Perspectives.* Oxford, Blackwell, 37-55.

Tomaselli, R. 1977. Degradation of the Mediterranean maquis. *In Mediterranean forests and maquis: conservation and management.* MAB Technical Notes 2, Paris, UNESCO, 33-72.

Tzedakis, P.C. 1993. Long-term tree populations in northwest Greece through multiple Quaternary climatic cycles. *Nature* 364: 437-440.

UNESCO-WMO 1970. *Climatic Atlas of Europe.*

van Andel T.H., C.N. Runnels & K.O. Pope 1986. Five thousand years of land use and abuse in the southern Argolid, Greece. *Hesperia,* 55: 103-128.

van Andel, T.H. 1989. Late Quaternary sea-level changes and archaeology. *Antiquity* 63: 733-745.

van Andel, T.H., E. Zangger & C. Perissoratis 1990. Quaternary Transgressive/Regressive Cycles in the Gulf of Argos, Greece. *Quaternary Research* 34: 317-329.

Warne, A.G. & D.J. Stanley 1993. Archaeology to refine Holocene subsidence rates along the Nile delta margin, Egypt. *Geology* 21: 715-718.

Wigley, T.M.L. & G. Farmer 1982. Climate of the eastern Mediterranean and the Near East. In J.L. Bintliff & W. van Zeist (eds) *Palaeoclimates, Palaeoenvironments and Human Communities in the Eastern Mediterranean Region in Late Prehistory* B.A.R. International Series 133, Oxford, British Archaeological Reports, 3-37.

Wijmstra, T.A. 1969. Palynology of the first 30 metres of a 120 m deep section in northern Greece. *Acta Botanica Neerlandica* 18: 511-527.

Windley, B.F. 1984. *The Evolving Continents.* Chichester, John Wiley and Sons.

Winstanley, D. 1973. Rainfall Patterns and General Atmospheric Circulation. *Nature* 245: 190-194.

Wise, S.M., J.B. Thornes & A. Gilman 1982. How old are the badlands? A case study from southeast Spain. In R.B. Bryan & A. Yair (eds), *Badland geomorphology and piping.* Geo-Books, Norwich: 259-277.

Woodward, J.C., J. Lewin & M.G. Macklin 1992. Alluvial sediment sources in a glaciated catchment: the Voidomatis basin, northwest Greece. *Earth Surface Processes and Landforms* 17: 205-216.

Yair A, P. Goldberg & B. Brimer 1982. Long term denudation rates in the Zin-Havarim badlands, northern Negev, Israel. In R.B. Bryan & A. Yair (eds), *Badland geomorphology and piping.* GeoBooks, Norwich: 279-91.

PART 1

The impact of Quaternary tectonic activity on river behaviour

CHAPTER 2

The impact of Quaternary tectonic activity on river behaviour

JOHN LEWIN

Institute of Earth Studies, UCW Aberystwyth, Aberystwyth, Dyfed, UK

The history of geomorphology has been intimately bound up with a developing understanding of the nature and timing of tectonic activity. Early geological mapping led to many instances of perceived correspondence between drainage patterns and geological structure and lithology, and these became incorporated into formal theory collated and extended by W. M. Davis in his 'cycle of erosion' (Chorley et al. 1964). For convenience, this incorporated rather simple tectonic assumptions, with initial uplift followed by progressive erosion and landform evolution under assumed conditions of stability. Geomorphological studies in the first half of the twentieth century, commonly undertaken in relatively stable tectonic environments, were especially marked by searches for 'initial surfaces' and sequences of landform development over long timescales.

In some environments this broad approach was less easy to sustain, particularly where a separation of tectonics and (subsequent) erosion was impossible. This was long apparent to geologists working in dynamic belts of Tertiary crustal activity, as in the Alps where erosion and sedimentation proceeded simultaneously. In areas like Japan, New Zealand or coastal California, the recency of the landsurface became appreciated, not least because of the clear linkages possible between contemporary earthquake disturbances and cumulative developments in the longer term. Even in the relatively stable mid-latitudes where earlier theory had flourished, the field evidence for glaciotectonic activity attracted a great deal of fruitful research.

The 1960s and 1970s were characterised by divergent tendencies in geomorphology and main-stream geology. The former became – and to a large extent has remained – concerned with process characterisation based on direct contemporary observation, both in the field and in laboratory simulations. Longer-term research concerns have focused on areas where dateable evidence is forthcoming, especially for late Quaternary events where radiocarbon dating has proved the key to depositional sequences in glacial, lacustrine, fluvial and marine materials. As discussed elsewhere in this volume, major concerns have been to relate the sedimentary record to global climatic change (as evident, for example, in deep ocean sediments, or in ice cores), or to link rates of sedimentation to human activities on an historical timescale.

Meanwhile, geological studies have been revolutio-

nised in a quite different way with the development of plate tectonic concepts involving large-scale global perspectives in space and time. Though by no means simple to apply locally in many instances, this new theory and its supporting developments provided geomorphologists with a new regionally-actualistic framework in which landform developments could be set. In some areas, patterns of river development would be rapid and contiguous with structural history; in others tectonic interactions would be minimal over extended geological timescales, or at least the rate of landform development was effectively much slower as far as major landforms were concerned.

This, then, brings us to the present and to what an appropriate approach to the relationship between tectonic activity and river behaviour should be. This has to be informed about the extent and regional nature of crustal mobility, and by the emerging timescales and rates over which allied changes in tectonics and erosion occur. Relevant studies may range from small-scale models which may provide insights into channel pattern changes (Ouchi 1983), through regional tectonic modelling (e.g. King et al. 1993), to continent scale appraisals (as in Summerfield 1991).

The Mediterranean region provides an especially significant area in which a growing awareness of relationships between tectonics and river development may be explored. Studies by Vita-Finzi and others (see Vita-Finzi 1986) have shown how earth movements have impacted on the landscape, and how they may be recorded in the archaeological and historical record. The Mediterranean – unlike the scenes of many earlier geomorphological studies which had such impact on theory – has been a zone of persistent Cenozoic tectonic activity, and crustal deformation in places has involved hundreds of metres of Quaternary elevation. It is possible to elucidate changes to river environments in these varied real-world situations so as considerably to enhance our global understanding both of river development, and the ways in which long-term geological perspectives may be linked to the hazardous experiences of human beings.

In Chapter 3, Collier, Leeder and Jackson take a new look at the drainage patterns of mainland Greece and the ways in which they have developed in this tectonically active area. They suggest four major drainage domains, and show how drainage development in faulted terrains may take place. They also calculate Quaternary erosion rates and sediment fluxes for particular basins, and are able to suggest how varying rates can be explained by alternative and altering controls – such as tectonics, relief, rock type, climate and human activity.

Northwestern Anatolia, Turkey, is examined by Kuzucuoglu in Chapter 4. Here both volcanic and tectonic activity have proceeded alongside drainage development so that the latter has been greatly transformed as uplift, rifting and basin subsidence have occurred. Volcanism is particularly helpful in providing extra information on landform development that is not frequently used or available in a European context.

Three contributions which follow in Chapters 5 to 7 are set in different river basins in southeast Spain – those of the Huércal/Overa, the Sorbas and the Mula. The Betic Cordillera form a most interesting area in which interactions between tectonics, sedimentation and river development can be explored using modern methods, helped in part by detailed analysis of relatively good sedimentary exposures. These papers show very clearly how basin-to-basin equivalence cannot be assumed given differentiation of tectonic history; comparisons between these chapters allow appreciation of a number of themes common to this volume, including the relative influence of a whole range of controls on landform development rather than simply that of tectonics.

Chapter 8, the final one in this section of the volume, is again concerned with Greece though this time with the eastern Peloponnese and the development of landscapes on limestone and flysch or other relatively easily eroded sediment, in part in relation to archaeological evidence. Landform development under conditions of tectonic activity and strong lithological contrasts is a theme of considerable significance in the Mediterranean basin.

Together, the following six chapters illustrate the strong necessity of incorporating tectonic understanding into any analysis of Mediterranean alluvial environments. But the Afro-European boundary area is a structurally complex one, which has produced contrasting geological interpretations and not a few controversies. Just as it is not always easy to relate short term channel dynamics observations to longer term alluvial history, so also the linkages between seismic data and structural evolution are not necessarily straightforward. The timing and rate of tectonic activity are also often not directly known, and circular arguments are possible when these derive from alluvial sequences which themselves may provoke interpretation in tectonic terms.

Yet an essential outcome of recent research, some examples of which are contained in the present volume, is to look at Mediterranean river environments in a fresh light and following the lead set by advances in structural geology and neotectonics. It is then possible to appreciate the regionally differentiated dynamicism of Mediterranean landforms, and to move towards a better understanding than was possible in traditional geomorphological models. Many of the fluvial features of the Mediterranean landscape – the rock gorges, the massive scarp fronts and alluvial fans, and the high-relief but 'soft rock' eroding terrains – will then be better understood.

REFERENCES

Chorley, R.J., A.J. Dunn & R.P. Beckinsale 1964. *The History of the Study of Landforms 1: Geomorphology before Davis.* London, Methuen.

King, G., D. Sturdy & J. Whitney 1993. The Landscape Geometry and Active Tectonics of Northwest Greece. *Geological Society of American Bulletin* 105: 137-161.

Ouchi, S. 1985. Response of Alluvial Rivers to Slow Active Tectonic Movement. *Geological Society of America Bulletin* 96: 504-515.

Summerfield, M.A. 1991. *Global Geomorphology.* Harlow, England, Longman.

Vita-Finzi, C. 1986. *Recent Earth Movements.* London, Academic Press.

CHAPTER 3

Quaternary drainage development, sediment fluxes and extensional tectonics in Greece

R. E. Ll. COLLIER and M. R. LEEDER
Department of Earth Sciences, University of Leeds, Leeds, UK

J. A. JACKSON
Bullard Laboratories, Madingley Road, University of Cambridge Cambridge, UK

ABSTRACT: Mainland Greece offers the opportunity to study the relationship between topography generated by the vertical displacements and tilting associated with active faulting and drainage basin evolution. Four distinct drainage domains may be distinguished in Greece, each having a separate geologic and tectonic history, and a different drainage character in terms of scale and predominant direction of flow. In southern and east-central Greece, where active extension has generated a number of tilted fault blocks, drainage basin length scales and areas are related to the spacing and geometry of the major normal faults. Drainage pathways respond to lithological variations in the substrate and to the sequence of activity on different fault segments through time.

Preliminary estimates of late Quaternary denudation rates are made for three catchment areas in central and southern Greece. Sediment fluxes can be calculated from either the volumes of material eroded from defined drainage basins or from the volume of the sedimentary sequence deposited beyond a catchment during some known time interval. Lithological influence on erosional flux rates can exceed the influence of drainage basin relief. Stream behaviour on alluvial fans and fan deltas is sensitive to changes in climate, through discharge rates and sediment yields. Fans deriving sediment from the footwall of the Skinos normal fault crossed a critical stream power threshold from aggradation to incision some time after about 6000 years BP, when climatic conditions became more arid. In the Maliakos Gulf, the sedimentary record of flux rates indicates a significant increase in mechanical denudation rates between 4500-2470 years BP and the period 2470 years BP to present. This can be ascribed to anthropogenic effects of deforestation and increased goat and sheep grazing.

KEYWORDS: Quaternary drainage development, drainage domains, extensional tectonics, active faulting, denudation rates, sediment yield.

1 INTRODUCTION

The northern margins of the Mediterranean are geologically young and tectonically active. For these reasons the area is a good one in which to examine the growth of drainage catchments and their interactions with the changing topography. We discuss the present day drainage of Greece and aspects of its Quaternary evolution in terms of major tectonic controls upon drainage basin development and river behaviour. Active extensional tectonics have produced new topography rapidly in the Quaternary compared to rates of both erosion and deposition. The size of the areas made available for drainage basin development is controlled by fault spacing, fault overlap and by the tilting produced by extension. Tectonic processes thus largely control the magnitude, position and development of drainage basins and, with climate and local geology, control the flux of sediment fed into the fluviatile system. Variations in the resistance of rock type to erosion can also be shown to influence drainage patterns.

Most of the extension in central Greece has occurred since the Early Pliocene, around 5 Ma (Kissel and Laj 1988; Mercier et al. 1989), although in places it may have begun in the Middle Miocene (Mercier et al. 1979). Active high-angle normal faults cut a basement dominated by resistant massive Mesozoic limestones (with less resistant flysch and ophiolitic horizons) that were deformed by WSW-directed thrusting during the local phase of the Alpine deformation. Normal faulting controls the directions of regional tilt in syn-rift Neogene basins, some of which are now inactive and uplifted in the footwalls of younger faults (Roberts and Jackson 1991; Leeder et al. 1991).

2 DRAINAGE DOMAINS OF GREECE

The major river catchments of Greece are outlined in Figure 1 and have been grouped into four distinct drainage domains. Each domain forms a region in which drainage basins are broadly consistent in character, in terms of scale and any predominant direction of flow. As will be seen, each domain approximates to a region having a particular geological and tectonic history. An attempt will therefore be made to relate the sequence of structural events that governs the typical drainage basin geomorphology in each case.

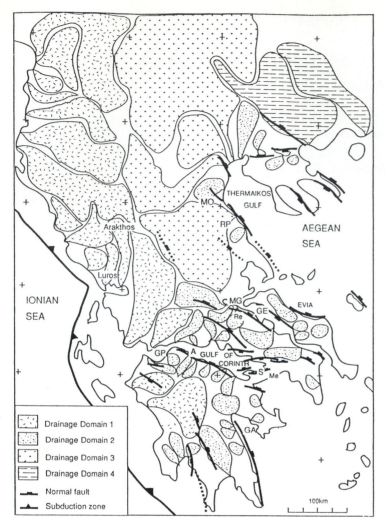

Figure 1. Drainage domains and major Neogene normal fault systems in Greece. The major catchment areas outlined are grouped according to regions showing consistent scale, and general flow direction, of drainage basins. The axis of the Hellenic trench to the southwest is marked by the heavy, barbed line. MO, Mount Olympos; RP, River Peneios; MG, Maliakos Gulf; Re, Renginion; GE, Gulf of Evia; GP, Gulf of Patras, A, Aigion; GA, Gulf of Argos; S, Skinos footwall; Me, Megara.

2.1 *Drainage domain 1*

Drainage domain 1 consists of a large, westward flowing system draining the thrust belt of the Hellenides mountain chain. The principal drainage divide that runs NNW-SSE through Greece and the Peloponnese, and which marks the eastward limit of domain 1, follows the highest topography and roughly corresponds with the axis of the crustal root inferred by Makris (1977) from Bouguer gravity anomaly data. This root formed by crustal thickening during the Late Mesozoic/Early Tertiary episode of shortening and thrust faulting. We therefore deduce that this drainage divide is a response to the Late Mesozoic/Early Tertiary orogenesis rather than the presently active tectonic regime. This is supported by the simple observation that the drainage divide does not follow a present day tectonic boundary of any significance.

Locally, both lithological variations in erodibility and the position of thrust-parallel, Tertiary flysch basins have influenced the present drainage network. Additionally, the N-S Arakthos and Luros rivers (Fig. 1) follow faults that are active as strike slip or normal faults (such as the

earthquake near Ioannina of 67.5.1 in Anderson and Jackson 1987). South of the Gulf of Amvrakia NNW-SSE faulting controls the drainage between Amphilokia and Messolonghi (Fig. 2).

Domain 1 is cut by the east-west Gulfs of Patras and Corinth, which are active extensional grabens. These developed during the regional extension of Pliocene-Quaternary age and post-date the major Hellenide collision event. Some modification of the western drainage domain has also occurred in westernmost mainland Greece and on the Ionian Islands (Fig. 2) where thrusting, reverse fault reactivation of normal fault planes, and subordinate strike-slip faulting have altered slopes and drainage patterns. Crustal shortening in this area in the Late Neogene and Quaternary has been related to outer-arc compression due to the subduction of the Mediterranean plate under western Greece (Mercier et al. 1976; Mercier et al. 1979; Underhill 1989). The east-west length scales of catchments in this western area (Fig. 2) are controlled by the spacing of thrust and reverse faults active during the Plio-Quaternary.

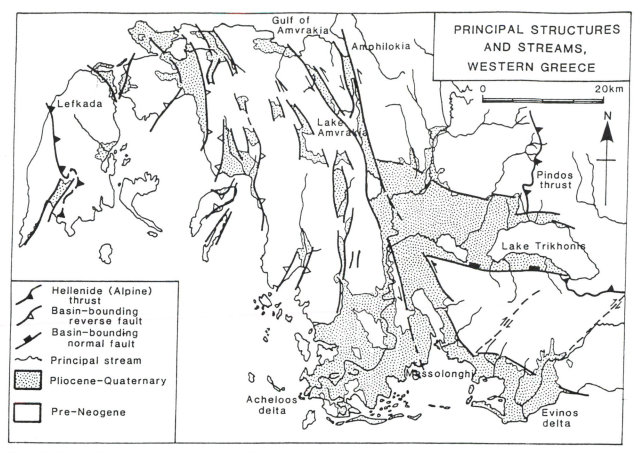

Figure 2. Map highlighting the dependence of principal stream position and the location of Neogene sedimentary basins upon fault geometries in western Greece. Structural information after Underhill (1989).

2.2 Drainage domain 2

A southern and east-central tilt-block domain is characterized by numerous small catchments of normal fault footwall and hangingwall slopes and axial (fault-parallel) trends (Fig. 1). This domain is the most tectonically active in mainland Greece. Relief and erosion rates are variable due to rapid extensional displacements, footwall uplift, block rotations, variable lithology and the consequent erosional interactions (discussed in later sections).

The small catchments of domain 2 are developed in response to slopes generated by normal faulting. We concentrate here on the effects of larger normal faults, as minor intrabasinal normal faults rarely exhibit tilts or vertical displacements of sufficient magnitude or scale to significantly modify regional drainage patterns. 'Large' normal faults in central Greece are those that move in earthquakes of surface wave magnitude M_s 5.0-6.0 or greater, and which seismological observations suggest are approximately planar to mid-crustal depths of 8-12 km (Jackson 1987; Jackson and White 1989). Fault segments are typically 10-20 km in length, with throws of between one and a few metres in any single seismic event (e.g. Soufleris et al. 1982; Jackson et al. 1982; Lyon-Caen et al. 1988). A number of offset fault segments may make up a boundary fault system to a basin, such as along the

southern coast of the Gulf of Corinth or the west coast of the Thermaikos Gulf. For more detailed information on the geometric characteristics of active normal faults and normal faulting mechanisms, the reader is referred to Jackson and White (1989) and Leeder and Jackson (1993).

In domain 2, lengths of transverse drainage basins are generally short and related to fault spacing (typically 20-30 km across strike). Fault spacing may itself be related to the thickness of the brittle seismogenic layer, or the effective elastic strength of the lithosphere (Jackson and White 1989; Roberts and Jackson 1991).

2.3 Drainage domain 3

A north-central domain comprises large catchments draining eastwards from areas dominated by Pelagonian Zone basement and cover rocks and Pindos Zone flysch units of the Hellenide massif (Fig. 1). Miocene to Quaternary sediments of the Mesohellenic, Trikala and Larisa basins and ophiolitic intrusives are also being eroded, as are Prealpine metasediments which occur in the north of the domain (Bornovas and Rondogianni-Tsiambaou 1983). An important characteristic of this area is that it is seismically almost inactive. Most of the active faulting is to the east and south of the area. The easterly flow direction in domain 3 can be understood by reference to the structural evolution of the Hellenide contractional belt. Collision and shortening began in mid-Cretaceous time and continued, at least episodically, until Late Eocene times. Thrusting migrated westwards through time, as is recorded by the decreasing westward age of flysch basins, and the age of the youngest strata deformed in successive thrust sheets of the external Hellenides (Jacobshagen et al. 1978). Ramping and rotations on successive thrusts led to the easterly tilts reflected in the present drainage pattern. The continuous NNW-SSE drainage divide between domains 1 and 3 suggests that there was no significant antecedent drainage from west to east or *vice versa* prior to thrust-stack development, as none has been inherited.

The north to south channel trends in the western part of domain 3 reflect the generation of a major normal fault-bounded basin, the Mesohellenic molasse basin (or 'trough'), in the Oligocene-Miocene (Papanikolaou et al. 1988). Old and now inactive and degraded normal fault escarpments still govern drainage patterns in this area. The basin axis trends north-south.

What is clear is that the downstream courses of domain 3 rivers have been strongly influenced by Neogene-Quaternary normal fault geometries and subsiding basins on the eastern edge of the area (Fig. 1). From the west, the River Peneios reaches the Thermaikos Gulf through a major topographic offset in its border fault system. Interestingly, the normal faults active during the Quaternary extensional deformation in this region parallel earlier thrust structures and may exploit structural weaknesses inherited from the Hellenide structural grain of similar strike. The modification of downstream river positions, though, is consistent with tilting controlled by the developing footwall highs of the active normal faults.

2.4 *Drainage domain 4*

The north-eastern domain 4 (Fig. 1) is characterized by large catchments that drain parallel to the structural trend of the Hellenides and which reach the Aegean along the axes of extensional or transtensional basins (after Pavlides and Kilias 1987). It appears that in this area original strike-parallel drainage basins that were developed in response to thrusting have been maintained in their original orientation by the coincidence of structural strike and imposed slopes in the Neogene-Quaternary extensional phase.

3 DRAINAGE BASIN DEVELOPMENT

3.1 *Footwall and hangingwall drainage*

The asymmetry of a tilted fault-block terrain is manifest in short, steep footwall slopes on the eroded fault faces, and long, gentle hangingwall slopes. The short slopes of the eroded fault faces are drained by numerous small catchments, whilst fewer, larger basins drain the gentler

back-tilted slopes. The initially steep slope of any fault face or hillslope will become progressively reduced by erosion (Carson and Kirkby 1972; Wallace 1978; Bucknam and Anderson 1979; Hanks et al. 1984). At the same time, gullies cut into the top of the fault scarp will propagate drainage networks back up the relaxed upper slopes and into areas of pre-existing slopes with surface drainage (Wallace 1979; Stark 1991). Erosion will thus modify the tectonically generated slopes of a developing tilt-block, causing migration of the central drainage divide.

The contrast between footwall- and hangingwall-sourced drainage is well shown on the island of Evia, off the NE coast of mainland Greece (Fig. 3). The island resembles a twisted horst block, with the central drainage divide shifting markedly along its length. In the NW part of the island the steep, short and small footwall drainage basins of coastal fault A flow SW, whilst the large drainage basin X eroded into Neogene sediments dominates the NE-facing back-tilted dip slopes. The 20 km wide zone in the central part of the island where fault polarity reverses also sees a complete reversal of drainage basin type, with the footwall drainage to fault B now towards the N or NE, and the larger dip slope drainages Y and Z towards the SW.

So far we have assumed that slope gradient and length are the main controls on drainage basin morphology and size, but the nature of the rock type being drained, specifically its resistance to erosion, is also of great importance. Where variations in rock type occur along the length of a fault we can directly assess this variable. The generalisation that catchment and fan size are primarily related to structural position breaks down when variations in rock type along a fault segment cause differential drainage basin growth and thus a change in sediment flux.

Figure 4 shows a series of catchments draining the footwall uplands to the active coastal faults south of Skinos in the eastern part of the Gulf of Corinth. The eastern catchments drain easily eroded serpentinites and high sediment fluxes have caused the growth of the volumetrically large fan deltas and alluvial fans A-G (it should be noted that water depths increase and therefore fan delta volumes increase for given surface area, towards fans A, B and C). The large fans F and G (in Fig. 4) are preferentially located at a prominent concave bend in the northernmost fault. This bend may have influenced catchment growth by locally funnelling runoff down the contour curvature. The westernmost catchment, H, also disgorges at a fault bend but drains mostly resistant limestones and the alluvial fan H is correspondingly small. This is true even though the catchment is one of the largest of the whole footwall and despite the fact that gradients and relief are significantly steeper.

Streams draining the Skinos fault footwall are currently incising into the fan deltas and alluvial fans A-G. The alluvial fans and fan deltas have surface profiles which can be extrapolated as asymptoting to present, Holocene marine highstand, base level (Leeder et al. 1991). The switch from fan deposition and construction

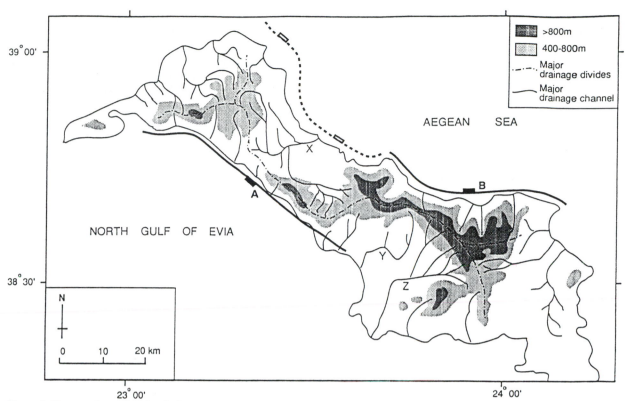

Figure 3. Topography and major drainage of Evia (after Leeder and Jackson 1993). The dashed line adjacent to the northern coast of Evia denotes a possible subdued westwards continuation of the coastal fault system that has a larger displacement in the east. See text for discussion of features A, B, X, Y and Z.

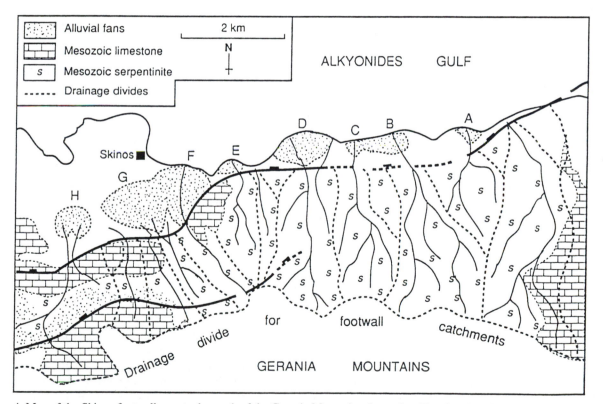

Figure 4. Map of the Skinos footwall area to the north of the Gerania Mountains (located on Figs 1 and 5). Simplified lithological divisions, alluvial fans, fan deltas and footwall-draining catchments are shown (after Leeder and Jackson 1993). See text for discussion of fans A-H, of which A-E are affected by marine erosion so that their present subaerial extent cannot be directly compared to fans F-H.

to stream erosion implies that either there has been a change in relative base level or that a geomorphic threshold has been crossed: the threshold of critical stream power (Bull 1979; Bull 1991). As has just been stated, the fan surfaces are consistent with deposition graded to present sea level. Any tectonic shift in relative base level would be expected to lead to fan aggradation, not incision, as these fans are in subsiding, hangingwall locations. Therefore it appears that since present sea level was attained c. 6000 years BP, climatic change has led to lower discharge rates and sediment yields. The consequence is that the threshold between stream deposition and incision has been crossed. Wave action has reworked and truncated the front of fan deltas A–E, removing sediment longshore and basinwards, and eroding cliffs that are now up to 15–20 m high. As runoff has fallen, and sediment supply rates from hillslopes decreased, coastal erosion has been able to rework each fan-front and cause further headward stream incision. This incision has exceeded any sediment accumulation 'captured' in response to tectonic hangingwall subsidence.

3.2 *Fault interactions*

Migration of normal faulting is common in the Greek extensional province and has led to the initiation of drainage basins in previously subsiding depositional areas, the disruption of pre-existing basins, and drainage capture. In many basins, young normal faults have cut across an older faulted terrain of markedly varying erosional resistance. New drainage systems have developed upon the tilted surfaces produced by successive fault movements during the Quaternary. The flux of sediment inferred to have come out of the new drainage basins is highly non-uniform, illustrating the importance of geological controls in the footwall of the system. The sediment flux is very high where large drainage basins have developed in the unconsolidated deposits of former Neogene extensional basin fills. We illustrate these generalisations by the following three examples.

3.2.1 *Megara basin*

The Neogene Megara basin lies onshore, to the east of the Gulf of Corinth (Fig. 1). It was formed by a major fault bounding its NE margin (Fig. 5), towards which the sediments of that age now mostly dip (Bentham et al. 1991; Roberts and Jackson 1991). This fault is apparently now inactive. The present NW end of the basin is marked by an active fault (Jackson et al. 1982) that cuts the trend

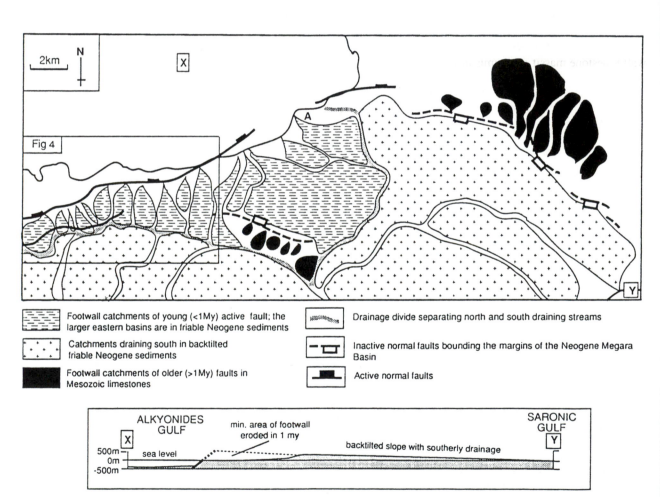

Figure 5. Faulting and drainage catchments in the area of the Skinos-Alepochori active coastal fault (the western part of which is detailed in Fig. 4) and the backtilted, Neogene Megara Basin. A, Alepochori.

of the old fault line obliquely and closely parallels the modern coastline. The c. 350 m of footwall uplift associated with the active fault has reversed the previous NW-draining Neogene drainage system (Bentham et al. 1991) south of the footwall drainage divide, backtilting the former depositional surface to the SE. Extrapolation of uplift rates from a dated marine terrace in the footwall (0.3-0.4 mm a^{-1} since about 100 ka) establishes that the young fault is about 1 Myr old (Leeder et al. 1991; Collier et al. 1992). Drainage basins established in the footwall in this time interval have had their catchment areas strongly controlled by rock type. Thus steep, small drainage basins have developed in Mesozoic limestones, larger basins in serpentinites, and large basins with dendritic drainage have developed in the poorly consolidated Neogene deposits. The drainage divide in the footwall of the modern fault has migrated some 2-3 times further in the last million years or so in the erodible Neogene sediments than in the adjacent, more resistant Mesozoic basement. This high rate of divide retreat (c. 1 cm a^{-1}) has been accomplished mostly by rockfalls and slides at the headcuts of deep gorges.

3.2.2 *Renginion basin*
The Renginion basin of Miocene to Quaternary age (located in Fig. 1) lies 300-400 m above sea level in the footwall of a young coastal fault system (Fig. 6). The Neogene sediments in the basin dip south from the footwall limestone massif of Knimis towards an older fault, probably last active in the Pleistocene (Philip 1974; Mercier 1976), which bounds the footwall mountain of Kalidromon. Unlike the example of the Megara basin discussed previously, we have no information about the age

of initiation of the younger faulting, though young Quaternary activity is implied by river terraces in the gorges where streams cross the fault. The uplifted Neogene sediments are now being eroded rapidly by drainage networks whose trellis-like tributary patterns reflect the influence and structural trend of local rock type. In particular, the central part of the basin is drained by two very large catchments, A and B. The large sediment flux which has come out of these drainage basins in the Holocene is clearly shown by their large coastal deltas (Fig. 6). Sediment flux here reflects both catchment area and sediment flux per unit catchment area, as a function of erodibility of the Neogene sediment. The coastline in front of Knimis has only relatively small fans and deltas and a narrow coastal plain. The tributaries of the large catchments flow axially until they find their escape across the coastal fault system. In the east (B) this occurs at a prominent offset zone between fault segments and in the west (A) through a gorge in less resistant Neogene sediments.

3.2.3 *Western Gulf of Corinth*
The southern coast of the western Gulf of Corinth (Fig. 7) contains a number of right-stepping E-W to ESE-WNW fault segments – the most easterly of which probably moved in a destructive earthquake in 1861 (Bousquet et al. 1983). Two limestone basement ridges occur in the footwalls of the coastal faults. To the south, older and apparently inactive faults are mapped with similar trend. Between the two generations of faults are very thick (> 800 m) Neogene sequences of Gilbert fan-delta deposits which are backtilted towards the south. The bot-

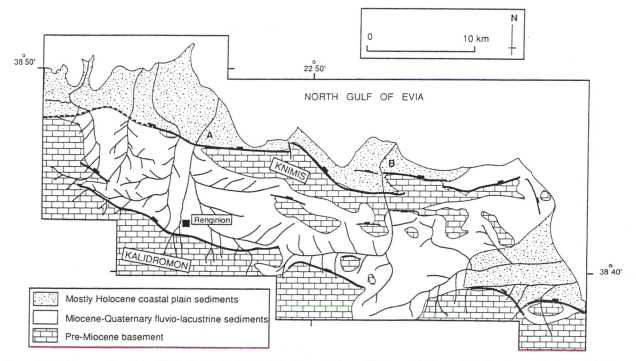

Figure 6. Faulting, geology (modified after Philip 1974; Roberts and Jackson 1991), and drainage (after Leeder and Jackson 1993) in the area of the northern Gulf of Evia (located in Fig. 1) and the Miocene/Quaternary Renginion basin.

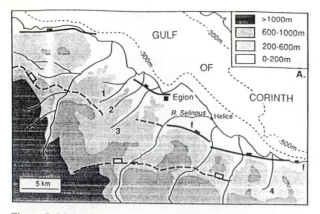
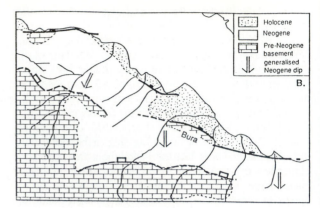

Figure 7. Major fault segments, simplified geology and drainage in the region of Aigion (located in Fig. 1) in the western Gulf of Corinth (after Roberts and Jackson 1991; Leeder and Jackson 1993). Heavy dashed lines denote historically inactive fault segments. See text for discussion of rivers 1-4. Note the generalised Neogene structural dips, not to be confused with steep, basinwards depositional dips in the Neogene syn-rift sediments in the area.

tomsets of these deposits now occur at elevations of up to 1 km (Ori 1989). At present we have no evidence to demonstrate whether these units are lacustrine or marine. If marine, they would leave no doubt that the locus of faulting has shifted northwards and that substantial regional uplift of the northern Peloponnese has occurred in the past few million years (Roberts and Jackson 1991). Estimates of the uplift rate vary from about 0.3 mm a^{-1} in the eastern Gulf of Corinth to over 1.6 mm a^{-1} in the western Gulf of Corinth footwall area (Keraudren and Sorel 1987; Collier 1990; Doutsos and Piper 1990; Collier et al. 1992).

Antecedent drainage basins sourced in pre-Neogene basement have adjusted to the uplift by elongating and downcutting into the newly uplifted downstream parts, with knickpoints gradually propagating upstream. Stream courses have been influenced by local changes in basement lithology and by the lower gradients encountered at the ends of fault segments. Some river courses have found their way between fault offsets, avoiding the resistant limestones in footwall ridges. The River Selinous has been diverted towards the Bura fault by subsidence, the present land surface between the Bura and Aigion faults dipping 3° towards the south (Roberts and Jackson 1991; Fig. 7). Other streams have cut through the more easily erodible Neogene footwall lithologies with no change of course. The antecedent rivers have constructed prominent Holocene fan deltas at the coastline, with large submarine fans offshore (Ferentinos et al. 1988). In addition to the larger antecedent catchments, younger stream systems (1-4 in Fig. 7) have propogated SW from the coastal faults into the easily eroded Neogene footwall deposits. These streams have constructed smaller fan deltas at the coastline.

4 QUATERNARY EROSION RATES AND SEDIMENT FLUXES

The calculation of sediment fluxes from identified drain-

age basins is an important step if we are to understand the impact of tectonic, lithologic, climatic and land-use controls on erosion rates. The basic principles of flux determination, summarised in Figure 8, are that a volume is calculated either of the material eroded from a particular drainage basin or of the sediment accumulated at the depositional (marine) end of the system, over some known period of time. It should be noted, of course, that calculated erosional flux includes chemical denudation, whereas depositional flux equates only to mechanical denudation (i.e. sediment yield). If possible in the latter case, an estimate can be made of material lost to the oceans as dissolved chemical load to obtain total (mechanical plus chemical) denudation rates. In calculating depositional fluxes from marine sediment volumes, a correction must also be made for any difference in the porosity of the sedimentary product compared to the source area lithologies, and for any biogenic (e.g. marine carbonate) input. Importantly, dated stratigraphic surfaces are a prerequisite in the calculation of both erosional and depositional fluxes. To obtain a time-averaged erosion rate for the drainage basin for the period since the dated stratigraphic surface, the catchment area is also needed.

Figure 8 describes the situation in calculating sediment fluxes out of a footwall-sourced drainage basin generated by the growth of a normal fault. If remnants of a pre-faulting planar surface are preserved on the footwall side of the fault, then the flux may be calculated knowing the volume X if the age of the fault can be estimated. Similarly, and again requiring knowledge of the age of the fault, the flux may be estimated from the sedimentary volume Y that progrades (downlaps) onto the pre-faulting surface on the downthrown, hangingwall side of the fault. This configuration offers the simplest situation, without having to make allowances for sediment accumulation (storage) in depocentres between the erosional catchment area and the site of marine deposition.

In Greece, we have the opportunity to constrain flux variation both areally and through time over a range of

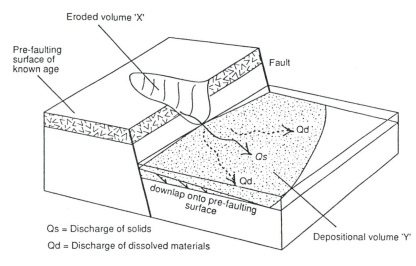

Figure 8. Principles of flux determination. The sediment flux may be calculated either from the eroded volume X or from the deposited volume Y, as indicated, if the age of the pre-faulting surface is known and some partition is made between solid and dissolved discharge (see text for discussion).

geological time scales. It would be useful to be able to compare sediment fluxes from a number of drainage basins from each of the four drainage domains outlined in section 2. This would allow an assessment of the impact of the different geomorphic character and tectonic history of each domain on denudation rates. Unfortunately, the data are unavailable that would allow us to achieve this. However, we are in a position to constrain sediment fluxes from a limited number of catchments from within domain 2, the southern and east-central Greece region characterized by tilt-block controlled drainage patterns.

Stratigraphic marker surfaces of ages between 10^3 and 10^6 years are known within the syn-rift environment of drainage domain 2. Therefore some attempt can be made to compare denudation rates through the Quaternary. Tectonic and/or catchment lithology effects that may act in competition with climatic or land-use (e.g. deforestation) effects can also be compared. Below, we calculate and discuss the implications of sediment fluxes from three drainage basins within domain 2. The results are used to assess the relative impact of tectonic, climatic and other controls upon sediment fluxes. Such quantitative studies are vital if we are to improve our ability to interpret the environmental and tectonic significance of basin-fill sequence geometries elsewhere in the geologic record.

4.1 Erosional flux from the Alepochori/Megara Footwall

The geological setting of the coastal fault at Alepochori has already been outlined in Section 3.2.1 (Fig. 5). The tectonic geomorphology of the area is described in detail by Leeder et al. (1991).

A Pleistocene alluvial plain has been back-tilted to the southeast by the active coastal fault since about 1 Ma. This provides a surface datum from which the volume of materials removed by erosion from footwall cirques may be calculated. The eroded materials are the Plio-Pleistocene sediments of the Megara basin (Bentham et al. 1991), now being re-cycled into the Alkyonides Gulf to the northwest. Leeder et al. (1991) estimate that some

8.9 ($\pm$0.5) km^3 has been eroded from an area of 32.5 km^2 during the last 1 Myr. This produces a mean denudation rate of 270 ($\pm$15) m^3 km^{-2} a^{-1}, which converts to 590 ($\pm$33) $\times$ 10^3 kg km^{-2} a^{-1}, assuming an initial average sediment density of 2200 kg m^{-3}. The mean surface denudation rate is calculated to be 270 ($\pm$15) mm ka^{-1} over the last 1 Myr.

4.2 Depositional fluxes in the Maliakos Gulf

An approximation of depositional fluxes over two intra-Holocene time intervals can be made for the Maliakos Gulf, located in Figure 1. The gulf is fed by the Sperchios River (axial) catchment and by smaller hangingwall and footwall drainage basins to the north and south of the Maliakos Gulf respectively. Archaeological investigations have been combined with a programme of core-drilling across the present coastal plain, and with radiocarbon dating of core samples by Kraft et al. (1987). This has allowed the reconstruction of the shoreline position for the mid-Holocene, at about 4500 years BP, and at 480 BC (of interest as the date of the historic battle at Thermopylae, on the southern edge of the gulf). These shorelines are compared with the present coastline in Figure 9.

The eastern end of the Maliakos Gulf is effectively restricted by the lateral progradation of two deltas, A and B (Fig. 9). Only a neglible veneer of Holocene sediment has 'escaped' to the east of these deltas, as seen on high resolution offshore seismic profiles (Tj. van Andel, pers. comm.). This allows us to treat the Maliakos Gulf to the west of deltas A and B as a closed depositional system (other than for chemical dissolved load).

From the available core data (Kraft et al. 1987) and the present water depth in the Gulf, we estimate that the coastal plain has been advancing since the mid-Holocene into a relatively constant water depth of 25 ($\pm$5) m. This provides us with an average depositional thickness which can be multiplied by the area of coastal plain advance to obtain a depositional volume for each time interval, assuming a constant depositional dip on the delta front. For the interval 4500 years BP to 480 BC (c. 2030 a), the

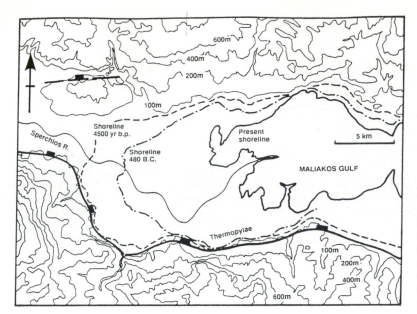

Figure 9. Map of the Gulf of Maliakos (located in Fig. 1) with palaeoshorelines reconstructed for 4500 years BP and 480 BC (the date of the battle at Thermopylae) on the basis of borehole data constrained by radiocarbon ages, after Kraft et al. (1987). Above 200 m the contour interval is 200 m.

coastal plain advanced over some 48 km², depositing a volume of 1.2 (±0.2) km³. Since 480 BC (c. 2470 a), the coastal plain has covered a further 110 km², equivalent to a sediment volume of 2.8 (±0.6) km³.

There are no significant marine carbonate accumulations within the Maliakos Gulf, so any biogenic contribution to the depositional volume can be neglected. However, 20% of each volume is subtracted as a reasonable estimate of the excess porosity in the sediments compared to the basement source lithologies. Source rock-types are dominated by Pelagonian Zone carbonates, but include extensive early Tertiary flysch deposits in the western part of the Sperchios catchment. The total Maliakos Gulf catchment area is calculated to be 1970 km². From these figures we derive average mechanical denudation rates of 240 (±50) mm ka⁻¹ for the interval 4500-2470 years BP and 450 (±90) mm ka⁻¹ since 2470 years BP (480 BC).

The numbers neglect any allowance for sediment accumulation within the axial alluvial plain of the Sperchios basin. They thus provide minimum mechanical denudation rates over the given periods. Holocene soils in the Sperchios basin are however of neglible average thickness (and therefore volume) compared to the sediment volumes in the Maliakos Gulf. Again, we have no direct knowledge of dissolved load flux into the Maliakos Gulf. Given the lithologies being eroded, dissolution is likely to be a significant component of total denudation rates, often exceeding 40% in equivalent drainage basins elsewhere (based on rivers such as the Danube, which drain similar Tethyan carbonate lithologies; see Meybeck 1976).

The available data indicate a major increase in mechanical erosion rates in the late Holocene. Enhanced sediment storage since 480 BC, accommodated by subsidence over the additional area across which the shoreline advanced from 4500-2470 years BP, reinforces this contrast.

4.3 Depositional flux in the Gulf of Argos

The depositional flux from mechanical erosion processes may be estimated for the Gulf of Argos basin over the last c. 12,000 years. During the last glacial period, sea level was as much as 125 m lower than at present. Lowstand deltas and erosional surfaces cutting down to this level are evident in seismic profiles from around the Aegean coast (Aksu et al. 1987; Lykousis and Chronis 1989; van Andel et al. 1990). From about 12,000 years BP to 6000 years BP sea level rose to its present level as the polar ice sheets retreated. During this interval, transgressive marine deposits progressively onlapped coastal margins. Since about 6000 years BP, coastal plains and highstand deltas have prograded seawards at present sea level.

The early Holocene maximum transgressive shoreline at the northern end of the Gulf of Argos has been mapped by Niemi and Finke (1988), using Landsat imagery and subsurface (borehole) investigations. Borehole transects combined with the mapped shoreline allow the volume of the Late Pleistocene-Holocene transgressive to highstand sequence to be estimated for the onshore area. The offshore (Gulf of Argos) transgressive to highstand sequence in the area outlined in Figure 10 is imaged in high resolution, single channel sparker seismic profiles, which are described by van Andel et al. (1990). The base of the sequence is marked by prominent onlapping stratal terminations. An average thickness is estimated from these data to allow calculation of the post-12,000 years BP sediment volume in this part of the Gulf of Argos marine basin.

Our best estimates for the sediment volumes are as follows. Onshore, the transgressive-highstand deposits

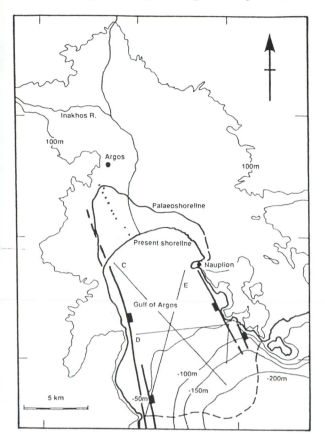

Figure 10. Map of Gulf of Argos (located in Fig. 1) with reconstructed Holocene maximum transgressive shoreline, after Niemi and Finke (1988). Black dots denote the borehole transect published by Niemi and Finke which has been used to constrain the volume of transgressive to highstand sediments (c. 12,000 years BP present) onshore. Lines C, D, and E denote seismic profiles in van Andel et al. (1990) which have been used to calculate the volume of the transgressive to highland sequence in the offshore Gulf of Argos within the area delimited by the heavy dashed line which has received the bulk of the sediment from the Argive catchment.

years, derived from a catchment of 1090 km². After subtracting 20% as an estimate of excess porosity in the sediments compared to the source lithologies, a mechanical denudation rate of 50-90 m³ km⁻² a⁻¹, or 50-90 mm ka⁻¹ is derived.

For a more comprehensive treatment of Late Pleistocene-Holocene depositional fluxes into the Gulf of Argos basin, two factors would again have to be evaluated. Sediment storage by tectonic subsidence (and subsidence driven by sediment loading) on the Argive plain could be constrained by an extensive drilling programme. A limited amount of borehole data is reported by van Andel et al. (1990) which does not suggest a significant component of Holocene sediment storage on the plain, but there is inadequate age control to put further time-lines through the stratigraphy and hence to constrain depositional rates in more detail. Secondly, knowledge of the chemical dissolved load of rivers entering the Gulf of Argos would aid at least modern estimates of denudation rates. No information is available on chemical denudation rates in the past. Climate and precipitation have varied through time (e.g. as summarised by Bull 1991, Fig. 3.3), but there is no direct correlation between variations in mechanical denudation rates through time (even where known) and changes in chemical erosion rates, due to the different thermodynamic rate-limiting processes involved.

5 DISCUSSION OF SEDIMENT FLUX ESTIMATES

The surface denudation rate for the Alepochori footwall cirques (Table 1) can be treated with a high level of confidence, as it is based directly on the volume of the footwall drainage basins. The main uncertainty is in the exact age of the active fault, estimated by extrapolation of uplift rates over the last 100 kyr. Greater uncertainties exist in the calculation of the depositional volumes in the Maliakos Gulf and the Gulf of Argos. Further borehole coverage would be needed to improve the estimates of the thickness of the sedimentary sequences involved. Nevertheless, some of the calculated denudation rates are resolvably different. The increase in erosion rates by a factor of about two for the last 2470 years in the Maliakos catchment area compared to the period 4500-2470 years BP, and the almost order of magnitude contrast in recent denudation rates between this area and the Gulf of Argos catchment, appear robust.

The sediment flux data, such as they are, allow some comment on the competing controls that are inferred to have governed erosion rates and sediment fluxes through time. It is clear that in areas of active faulting, drainage basin areas and relief are strongly influenced by the length scales and vertical displacements of faulting. The relationship between denudation rates and relief is well established (Ahnert 1970). However, as we have seen exemplified by the Skinos footwall drainage basins (Section 3.1), flux contrasts determined by differences in rock type may exceed the influence of relief. The significantly lower denudation rate calculated for the Argive catch-

average 7 (±2) m in thickness over an area of 20 km², equating to a volume of 0.14 (±0.04) km³. Offshore, an average thickness of 8 (±2) m for the same sequence is obtained from seismic data (average 0.01 s two-way-travel-time, converted assuming a velocity of 1600 m/s). The sediments cover an area of 100 km² in the northern part of the gulf (outlined in Fig. 10), giving a volume of 0.8 (±0.2) km³. It is improbable that a significant volume of terrigenous sediment has been transported to the south and southeast of this area. A thin veneer of Holocene marine sediments is imaged in high resolution seismic reflection profiles to the south (van Andel et al. 1993), but this area includes input from catchments to the west and east. We estimate that at most another 20-30% by volume of sediment derived from the main northern Argive catchment may be deposited in this southern area. Combined with the onshore equivalents, a total volume of 0.84-1.5 km³ has thus been deposited in the last 12,000

Table 1. Ranges for average denudation rates from three catchments in the seismically active drainage domain 2, calculated as described in the text, and incorporating error margins as follows: *incorporating errors in depositional volume estimation, as discussed in text, and allowing ±10% error on estimated age of fault; **allowing for errors in depositional volume estimation, as discussed in the text.

Location	Time interval	Surface denudation rate (mm ka^{-1})	Mechanical denudation rate (mm ka^{-1})
Alepochori/Megara footwall	1 Ma to present	220-330*	
Maliakos Gulf catchment	4500-2470 years BP		190-290**
	2470 years BP to present		360-540**
Gulf of Argos catchment	12,000 years BP to present		50-90**

ment, as compared to other areas, similarly reflects lithological controls on mechanical erosion. The Argive catchment area is dominated by mechanically-resistant recrystallized carbonates. This contrasts with the Maliakos catchment, which includes large areas of readily erodible flysch deposits, and the highly erodible Neogene sediments in the Alepochori catchments. However, it may well be that the dominant denudation process affecting the Mesozoic carbonates of the southern Peloponnese during the Quaternary has been chemical dissolution.

It is inappropriate to compare the denudation rates calculated for the Alepochori footwall directly with those for the Maliakos Gulf area. Rates are averaged over very different time scales and it is known that the effects of climate will have changed through time. Runoff and hence denudation rates have varied between the pluvial episode of about 14,000-5000 years BP that affected the whole eastern Mediterranean region (Bull 1991 op. cit.) and the preceding and subsequent periods. There is abundant evidence from the Holocene alluvial fans and fan deltas around Greece, which are now typically undergoing incision, to support the observation that late Holocene climates are generally more arid than in the early Holocene. The Skinos fans outlined in section 3.1 (Fig. 4) illustrate this situation. Discharge reductions and reduced sediment yields have led to incision on the majority of these surfaces, where catchments have not been modified by anthropogenic effects such as deforestation and increased grazing.

The marked increase in the Maliakos Gulf sediment flux rates in historic times is incompatible with an origin in climatic change. No corresponding change in rainfall or increase in hydrograph variability, that might enhance erosivity, is recorded over the relevant time interval in pollen records from central/southern Greece (Wright 1968). Yassoglou and Nobeli (1972) report the results of soil studies in the southwest Peloponnese which indicate a constancy of alfisol-forming conditions for the last few thousand years, since at least the Late Bronze Age. Palynological results from around Greece indicate that since about 2500 BC, human activity has modified the abundance and distribution of tree species by widespread deforestation and cultivation, and increased the extent of grazing, with consequent effects of increased erosion and sedimentation (Davidson 1980). Regionally, the pattern of deforestation varies. In the Messenia region of southwest Peloponnese, pine woods in coastal areas were reduced during the Middle Bronze Age or early Late

Bronze Age, from about 4000 years BP (Wright 1972). Wright describes pollen records from sediments of the Osmanaga Lagoon near Pylos, which indicate that after a period dominated by mixed maquis shrubland (evergreen oaks, *Pistacia* sp., the wild olive *Olea oleaster, Ceratonia siliqua, Phillyrea media*), the area was extensively cultivated for olives from about 1100 to 700 BC. Similarly, a rise of olive pollen is reported in a core from Lake Voulkaria in northwestern Greece, from about 1200 BC. Greig and Turner (1974) contrast the pollen record from Lake Phillipi in northeastern Greece, where a thick oak forest had been established by about 7000 BC. Here, an increase in maquis vegetation from about 2500 BC indicates the spread of grazing. But no major deforestation appears to have occurred before the Medieval period. However, results from Lake Kopais in central Greece, an area 50 km to the southeast of the Maliakos catchment, indicate that the oak forest was much reduced by the Bronze Age at the latest, when population growth was more rapid in central and southern Greece compared to the north. Thus the marked increase in sediment fluxes into the Maliakos Gulf in historic times can be ascribed to anthropogenic effects. We are seeing the product of the intensification of agricultural activities, with goat and sheep grazing of upland slopes, deforestation and cultivation having contributed to greater substrate erodibility and higher erosion rates – despite the relatively dry, Mediterranean late Holocene climate.

The above discussion demonstrates the efficacy of stratigraphically-defined studies of sediment flux variation. The data are not ideal and are limited in number of catchment areas. However, we have demonstrated a methodology for looking at sediment flux variation areally and through time, whether derived from erosional volumes or the volume of depositional sequences. Future work on sediment fluxes from the four distinct drainage domains identified in Greece promises to provide much further information on the tectonic, lithologic, climatic and anthropogenic controls on erosion and sedimentation rates in this seismically active area.

6 ACKNOWLEDGEMENTS

The authors are grateful to the Institute of Geology and Mineral Exploration in Athens for granting permission to work in Greece. RELlC and MRL thank Mobil North Sea Ltd and JAJ thanks Sun Oil for continued financial support of our syn-rift basin studies. We also thank Pete Talling, Jo Wilkin and two anonymous referees for comments on an earlier manuscript.

REFERENCES

Ahnert, F. 1970. Functional relationships between denudation, relief, and uplift in large mid-latitude drainage basins. *American Journal of Science* 268:243-263.

Aksu, A.E., D.J.W. Piper & T. Konuk 1987. Late Quaternary tectonic and sedimentary history of Outer Izmir and Candarli Bays, western Turkey. *Mar. Geol.* 76:89-104.

Anderson, H. & J.A. Jackson 1987. Active tectonics of the Adriatic region. *Geophys. J. R. astr. Soc.* 91:937-983.

Bentham, P., R.E.Ll. Collier, R.L. Gawthorpe, M.R. Leeder, S. Prosser & C. Stark 1991. Tectono-sedimentary development of an extensional basin: the Neogene Megara Basin, Greece. *J. geol. Soc. London* 148:923-934.

Bornovas, J. & Th. Rondogianni-Tsiambaou 1983. *Geological map of Greece, 1:500,000*. Institute of Geology & Mineral Exporation, Athens.

Bousquet, N., J.J. Dufaure & P.Y. Pechoux 1983. Temps historique et evolution des paysages egeen. *Mediterranee* 2:3-10.

Bucknam, R.C. & R.E. Anderson 1979. Estimation of fault-scarp ages from a scarp-height-slope-angle relationship. *Geology* 7:11-14.

Bull, W.B. 1979. Threshold of critical power in streams. *Bull. Geol. Soc. Am.* 90:453-464.

Bull, W.B. 1991. *Geomorphic Responses to Climatic Change*. Oxford University Press.

Carson, M.A. & M.J. Kirkby 1972. *Hillslope Form and Process*. Cambridge University Press.

Collier, R.E.Ll. 1990. Eustatic and tectonic controls upon Quaternary coastal sedimentation in the Corinth Basin, Greece. *J. Geol. Soc. London* 147:301-314.

Collier, R.E.Ll., M.R. Leeder, P.J. Rowe & T.C. Atkinson 1992. Rates of tectonic uplift in the Corinth and Megara Basins, central Greece. *Tectonics*, 11:1159-1167.

Davidson, D.A. 1980. Erosion in Greece during the first and second millenia BC. In R.A. Cullingford, D.A. Davidson & J. Lewin (eds), *Timescales in Geomorphology*:143-158. Chichester: John Wiley & Sons.

Doutsos, T. & D.J.W. Piper 1990. Sedimentary and morphological evolution of the Quaternary, eastern Corinth rift, Greece: first stages of continental rifting. *Bull. Geol. Soc. Am.* 102:812-829.

Ferentinos, G., G. Papatheodorou & M.B. Collins 1988. Sediment transport processes on a submarine escarpment of an active asymmetric graben, Gulf of Corinth, Greece. *Mar. Geol.* 83:43-61.

Greig, J.R.A. & J. Turner 1974. Some pollen diagrams from Greece and their archaeological significance. *J. Arch. Sci.* 1:177-194.

Hanks, T.C., R.C. Bucknam, K.R. Lajoie & R.E. Wallace 1984. Modification of wave-cut and faulting-controlled landforms. *J. Geophys. Res.* 89:5,771-5,790.

Jackson, J.A. 1987. Active normal faulting and crustal extension. In M.P. Coward, J.F. Dewey & P.L. Hancock (eds), *Continental Extensional Tectonics*:3-17. Geological Society of London Special Publication no. 28.

Jackson, J.A., J. Gagnepain, G. Houseman, G.C.P. King, P. Papadimitriou, C. Soufleris & J. Virieux 1982. Seismicity, normal faulting, and the geomorphological development of the Gulf of Corinth (Greece): the Corinth earthquakes of February and March 1981. *Earth & Planet. Sci. Lett.* 57:377-397.

Jackson, J.A. & N.J. White 1989. Normal faulting in the upper continental crust: observations from regions of active extension. *J. struct. Geol.* 11:15-36.

Jacobshagen, V., S. Durr, F. Kockel, K.O. Kopp & K.G. Kowalczyk 1978. Structure and geodynamic evolution of the Aegean region. In H. Cloos, D. Roeder & K. Schmidt (eds), *Alps, Apennines, Hellenides*: 537-564. Inter Union Commission on Geodynamics Science Report 28. Stuttgart: E. Schweizerbart'sche Verlagsbuchhandlung.

Keraudren, B. & D. Sorel 1987. The terraces of Corinth (Greece) - a detailed record of eustatic sea-level variations during the last 500,000 years. *Mar. Geol.* 77:99-107.

Kissel, C. & C. Laj 1988. The Tertiary geodynamical evolution of the Aegean arc: a palaeomagnetic reconstruction. *Tectonophys.* 146:183-201.

Kraft, J.C., G.Jr. Rapp, G.J. Szemler, C. Tziavos & E.W. Kase 1987. The pass at Thermopylae, Greece. *J. Field Archaeology* 14:181-198.

Leeder, M.R., M.J. Seger & C.P. Stark 1991. Sedimentation and tectonic geomorphology adjacent to major active and inactive normal faults, southern Greece. *J. Geol. Soc. London* 148:331-343.

Leeder, M.R. & J.A. Jackson, 1993. The interaction between normal faulting and drainage in active extensional basins, with examples from the western United States and central Greece. Basin Research, 5:79-101.

Lykousis, V. & G. Chronis 1989. Mechanisms of sediment transport and deposition: Sediment sequences and accumulation during the Holocene on the Thermaikos plateau, the continental slope and basin (Sporadhes Basin), northwestern Aegean. *Mar. Geol.* 87:15-26.

Lyon-Caen, H., R. Armijo, J. Drakopoulos, J. Baskoutass, N. Delibassis, R. Gaulon, V. Kouskouna, J. Latoussakis, K. Makropoulos, P. Papadimitriou, D. Papanastassiou & G. Pedotti 1988. The 1986 Kalamata (south Peloponessus) earthquake: detailed study of a normal fault and tectonic implications. *J. geophys. Res.* 93:14,967-15,000.

Makris, J. 1977. *Geophysical Investigations of the Hellenides. Hamburger Geophysikalische Einzelschriften*. Hamburg: G.M.L. Wittenborn Sohne.

Mercier, J.L. 1976. La néotectonique, ses methodes et ses buts. Un example: l'arc egéen (Mediterrane orientale). *Rev. Geol. Dyn. Geog. Phys.* 18:323-346.

Mercier, J.L., E. Carey, N. Philip & D. Sorel 1976. La néotectonique plio-quaternaire de l'arc égéen externe et de la mer Egée et ses relations avec la séismicité. *Bull. Géol. Soc. de la France* 18:355-372.

Mercier, J.L., N. Delibassis, A. Gauthier, J.-J. Jarrige, F. Lemeille, H. Philip, M. Sebrier & D. Sorel 1979. La néotectonique de l'arc Egéen. *Rev. Géol. Dyn. Geog. Phys.* 21:67-92.

Mercier, J.L., P. Vergely & K. Simeakis 1989. Extensional tectonic regimes in the Aegean basins during the Cenozoic. *Basin Res.* 2:49-71.

Meybeck, M. 1976. Total mineral dissolved transport by world major rivers. *Hydrological Sciences Bulletin* 21:265-284.

Niemi, T.M. & E.A.W. Finke 1988. The application of remote sensing to coastline reconstruction in the Argive Plain, Greece. In A.R. Hands & D.R. Walker (eds), *Archaeology of Coastal Changes*:119-136. Oxford: B.A.R.

Ori, G.G. 1989. Geologic history of the extensional basin of the Gulf of Corinth (?Miocene-Pleistocene), Greece. *Geology* 17:918-921.

Papanikolaou, D.J., E. Lekkas, I. Mariolakis & R. Mirkou 1988. Geodynamic evolution of the Mesohellenic basin. *Bull.*

Geol. Soc. Greece 20:17-36.

Pavlides, S.B. & A.A. Kilias 1987. Neotectonic and active faults along the Serbomacedonian zone (SE Chalkidiki, northern Greece). *Annales Tectonicae* 1:97-104.

Philip, H. 1974. Etude neotectonique des rivages Egeens en Locride et Eubee nord occidentale (Grece). Thesis, Universite des Science et Techniques du Languedoc, Montpellier.

Roberts, S. & J.A. Jackson 1991. Active normal faulting in central Greece: an overview. In A.M. Roberts, G. Fielding & B. Freeman (eds), *The Geometry of Normal Faults*:125-142. Special Publication of the Geological Society of London no. 56.

Soufleris, C., J.A. Jackson, G.C.P. King, C.P. Spencer & C.H. Scholz 1982. The 1978 earthquake sequence near Thessaloniki (northern Greece). *Geophys. J. R. astr. Soc.* 68:429-458.

Stark, C.P. 1991. An invasion percolation model of drainage network evolution. *Nature* 352:423-425.

Underhill, J.R. 1989. Late Cenozoic deformation of the Hellenide foreland, western Greece. *Bull. Geol. Soc. Am.* 101:613-634.

van Andel, Tj. H., E. Zangger & C. Perissoratis 1990. Quaternary transgressive-regressive cycles in the Gulf of Argos, Greece. *Quaternary Res.* 34:317-329.

van Andel, Tj. H., C. Perissoratis & T. Rondoyanni 1993. Quaternary tectonics of the Argolikos Gulf and adjacent basins, Greece. *J. geol. Soc. London*, 150:529-539.

Wallace, R.E. 1978. Geometry and rates of changes of fault-generated range fronts, north-central Nevada. *US Geol. Surv. Journal of Research* 6:637-650.

Wallace, R.E. 1979. Map of young fault scarps related to earthquakes in north-central Nevada. *US Geol. Surv. Open-File Report* 79:1554.

Wright, H.E. 1968. Climatic change in Mycenaen Greece. *Antiquity* 42:123-127.

Wright, H.E. 1972. Vegetation history. In W.A. McDonald & G.R. Rapp (eds), *The Minnesota Messenia Expedition: Reconstructing a Bronze Age Regional Environment*:188-199. Minneapolis: University of Minnesota Press.

Yassoglou, N.J. & C. Nobeli 1972. Soil studies. In W.A. McDonald & G.R. Rapp (eds), *The Minnesota Messenia Expedition: Reconstructing a Bronze Age Regional Environment*:171-176. Minneapolis: University of Minnesota Press.

CHAPTER 4

River response to Quaternary tectonics with examples from northwestern Anatolia, Turkey

CATHERINE KUZUCUOGLU
Laboratoire de Géographie Physique, URA-DO141, Meudon, France

ABSTRACT: In common with much of the Aegean part of northwestern Anatolia, landform development in the Ayvalik-Kozak region has been strongly affected by Mio-Pliocene volcanic and tectonic events. Quaternary tectonic activity has produced regional uplift and subsidence of adjacent areas.

Continental formations which accumulated in the subsiding western margins record most of the volcanic and tectonic events, allowing a reconstruction of the relative chronology of Tertiary and Quaternary tectonic activity. Volcanic effusions during the Miocene and Pliocene produced a mountainous belt around the older granodioritic and metamorphic core of the Kozak Mountains. As a result, the drainage network in the western part of the Kozak Mountains was totally modified.

During the Quaternary, continuous uplifting of the Kozak Mountains and surrounding areas, and the recent sinking of rift valleys now partly submerged by the Aegean Sea, led both to the deepening of entrenched meanders of the valleys within the mountains and to relief inversion in some valleys in the sedimentary margins of the mountains. In this latter case, basalt flows invading Quaternary valleys have sealed off a palaeotopography which is quite different from the present landscape. Field observations and geological mapping have demonstrated the strong influence of recent tectonic sinking of the marine pass between the Altinova Plain and the Aegean island of Lesbos upon the drainage pattern.

KEYWORDS: Northwestern Anatolia, volcanism, Quaternary tectonics, basaltic flow, valley inversion, drainage pattern, river capture, entrenched meanders.

1 INTRODUCTION

Over the last two decades a great deal of geological research has been undertaken in the northwestern part of Mediterranean Turkey. Several authors have attempted to establish the chronology of geological, tectonic and volcanic events in western Anatolia during the Tertiary and the Quaternary (see Fytikas et al. 1984; Ercan et al. 1985; Ercan et al. 1986; Akyürek 1989; Yilmaz 1989; McKenzie and Yilmaz 1991). Andesites are widespread across this area and radiometric dating of these materials has given Miocene to Pliocene ages (see Borsi et al. 1972; Bellon et al. 1979; Innocenti et al. 1982; Yilmaz 1989). Most of this geological research has focused on the Tertiary Period and the Quaternary history of the region has not been extensively documented, although some geomorphological studies have been carried out (see Erol 1980, 1981; Kuzucuoglu 1980, 1982; Kozan et al. 1982).

2 STUDY AREA AND REGIONAL TECTONIC SETTING

The study area is located south of the Biga Peninsula to the north of Izmir in the northern part of Mediterranean Turkey (Fig. 1). The mountain range of Kaz dag lies immediately to the north and the island of Midilli (Lesbos, Greece) to the west. The main towns of the region, some of which are ancient Greek cities, are located near the coast (e.g. Ayvalik) and on the edge of luxurious river valleys (Edremit/Adramyttion, Bergama/Pergamon) (Dörpfeld 1928). These wide and flat valleys are sunken Quaternary rifts (Havran River, Gediz/Kaikos River) (McKenzie 1972).

Throughout the Quaternary Period, the whole Aegean region has been characterized by widespread subsidence and large areas are now partly submerged by the Aegean Sea. The landforms associated with the present coastline of the Edremit Gulf on the northern shore of the study area (Fig. 1) are, for example, typical of a recently submerged topography (Kuzucuoglu 1980).

The landforms of this region associated with Quaternary tectonic activity are related to two simultaneous events: (1) the sinking of the bordering rifts (the Edremit Gulf to the north, the Lesbos pass to the southwest, and the Gediz River valley to the south), and (2) the uplift of the Kozak Mountains which constitute the heart of the study area and are well defined by rift faults (Fig. 1).

In addition, recent basaltic lava has locally buried palaeolandforms and this flow represents one of the last volcanic lava emissions in the region. Volcanism produced huge amounts of andesitic-type lava during the Miocene (Borsi et al. 1972; Benda et al. 1974) and Pliocene (Ercan et al. 1986). It is only very recently that

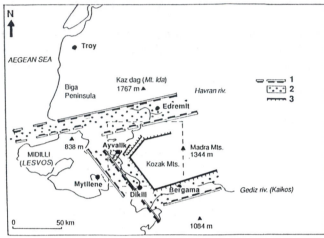

Figure 1. Location of the study area. Lower map shows the major fault lines of the area around Ayvalik. Legend for lower diagram: (1) Main Quaternary trough fault lines; (2) Quaternary rift valleys (either invaded by sea or corresponding to main coastal and river plains); (3) Main fault lines bordering the uplifted Kozak Mountains and the Quaternary sunken valleys within the Tertiary basins.

basalts have been reported from the region of Ayvalik-Kozak (see Kuzucuoglu 1980; Ercan et al. 1986; Akyürek 1989) although the presence of basaltic flows in Lesbos has been known for some time (Präger 1965; Peccerillo and Villari 1980).

3 DRAINAGE BASIN CHANGES DURING THE MIOCENE-PLIOCENE

3.1 The pre-Miocene volcanism basal conglomerate

The stratigraphic sequences in the western sedimentary margin of the Kozak Mountains begin with a series of grey sandstones and marls alternating with beds of a very hard, dark, conglomerate which outcrop at Kurutepe (Fig. 2). Three elongated hills correspond locally to the outcrops of this conglomerate (Fig. 3) and observations of its structure have demonstrated that it contains rounded gravels deposited in a fluvial environment rich in well-sorted sand.

These sandstones and conglomerates are mainly composed of metamorphic material mixed with a few granitic elements; these latter elements are derived from the Eocene/Oligocene Kozak granodioritic intrusion which began to be eroded at this time. Almost no volcanic rock is present, with the exception of some rare rounded and weathered rhyolitic gravels. This deposit is of great stratigraphic significance as it demonstrates that this part of the

lacustrine sequence is older than the regional andesitic volcanic activity (Fig. 2).

At the time of the deposition of these gravels and sand (probably at the beginning of the Miocene), rivers were eroding both the granodiorite and the surrounding metamorphic schists. These rivers were flowing westwards, down to the shores of the western lakes where they discharged their sediment loads.

3.2 The Early-Miocene deposits

Potassium-Argon (K/Ar) ages of volcanic products obtained by Borsi et al. (1972) and Benda et al. (1974) indicate a Miocene age for andesites from areas near Bergama and Ayvalik. When these Miocene andesitic emissions started, erosion also began on the volcanic structures. Thus, at the base of the sequence following the sedimentation of the Kurutepe Formation, another fluvial conglomerate (the Türkün-burnu Formation) may be observed (Figs 2 and 3). This conglomerate is almost always in faulted contact with either the previous sequence, or with the younger lava flows. It contains granodioritic and metamorphic gravels together with andesitic elements.

Deposited at approximately the same time, the numerous river gravels included within the volcano-sedimentary series (Mount Ahlat) are also similar in composition with several lithologies, including granodiorite, represented.

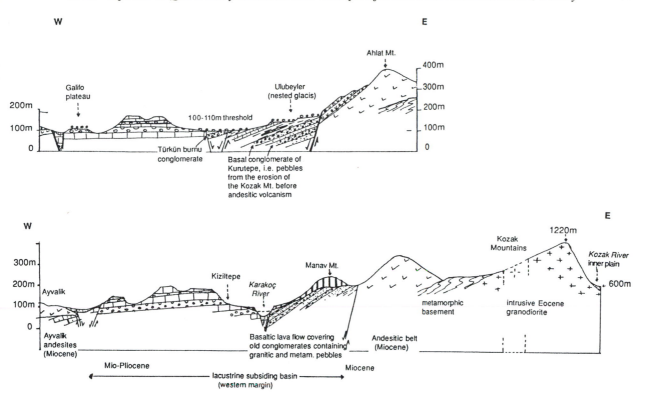

Figure 2. Schematic sections showing the stratigraphic relationships in the western margins of the Kozak Mountains. Upper section shows the northern part of the Karakoç River depression and the lower section shows the southern part (distances not to scale).

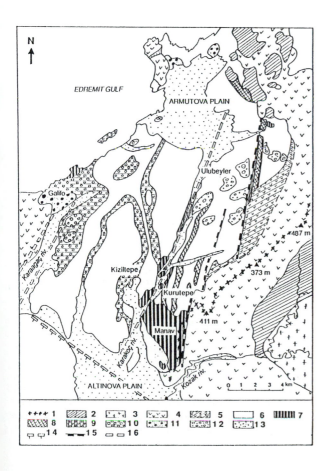

Figure 3. The major continental formations related to the uplift of the Kozak Mountains and the main fault lines. Legend: (1) Actual drainage divide separating the inner Kozak Mountains from their western margins. *Main rock outcrops:-* (2) Metamorphic rocks; (3) Granodiorite; (4) Mio-Pliocene lava flows; (5) Miocene volcano-sedimentary layers; (6) Mio-Pliocene lacustrine limestone; (7) Plio-Quaternary (?) basalts. *Main continental formations-* Mio-Pliocene: (8) Grey conglomerate (Kurutepe) previous to andesitic flows; (9) Yellow conglomerate (*Türkün burnu*), prior to the isolation of the granodioritic and metamorphic heart of the Kozak Mountains; (10) Reddish conglomerates and coarse formations providing evidence for the isolation of the Kozak Mountains and for destruction of the andesitic belt during the Pliocene (*Kiziltepe*). *Quaternary:-* (11) *Galifo* Formation with basaltic elements; (12) *Ulubeyler* formations over stepped floors at the foot of the main fault line; (13) Holocene alluvium. *Quaternary tectonics-* (14) Trough lines of rift valleys related to the Aegean subsidence; (15) Main Quaternary fault lines bordering the uplifted mountains; (16) Faults inducing the subsidence of fault line valleys.

The petrographic composition of both deposits indicates that during the first stage of the Miocene andesitic volcanic activity, rivers were still flowing from the heart of the Kozak Mountains, transporting gravels to the western lacustrine margins.

3.3 *Drainage basin changes related to volcanic activity*

During the Miocene and Pliocene, the volcanic belt rose so high with the growing mass of andesitic emissions, that it presented a barrier to the rivers flowing from the east. All the rivers draining the western part of the Kozak Mountains were then deflected along the foot of this belt to reconnect with the Kozak River which was already entrenched into the granodiorite and the metamorphic basement (Fig. 4). The same phenomenon can be observed at the southern border of the intrusion, where the Miocene andesitic mountains bar the southward access to the Gediz River valley.

During the periods of relative volcanic calm, western streams, now solely confined to the volcanic landforms, deposited several metres of red-coloured continental sediments into the western lakes (outcropping at Kiziltepe, see Figs 2 and 3). These beds of rounded volcanic gravels alternate with beds of andesitic gravels and well-sorted sands. Several occurrences of these continental deposits can be recognized within the Miocene and Pliocene sequences. Their extension progressed in conjunction with the westward migration of the Pliocene lakes.

The clasts contained in these formations are mainly andesitic. The most recent beds also contain limestones and sandstones eroded from the older Miocene lacustrine beds which had been uplifted to the east. No granodioritic or metamorphic materials have been observed in any of these deposits.

The petrographic composition of these reddish formations indicate the total isolation of the Kozak Mountains from their western margins (Fig. 4). This drainage system has been in existence since Mio-Pliocene times as the present watersheds still follow the contact lines between the metamorphic outcrops and the surrounding andesitic uplands.

4 THE IMPACT OF QUATERNARY UPLIFT

4.1 *The Kozak inner plain*

The presence of an extensive plain (15 × 10 km) at an altitude of 525 to 600 m within the granodioritic intrusion in the very heart of the high (1200-1340 m) mountains can be explained by a difference in the petrographic nature of the granodiorite and as a response to regional uplift.

Examination of rock samples from the inner granodiorite reveals greater susceptibility to weathering and erosion due to the fact that it is the heart of the intrusion which has been incised and widened by fluvial erosion (Kuzucuoglu 1982). However, differences in petrography cannot fully explain all of the huge slope dominating the

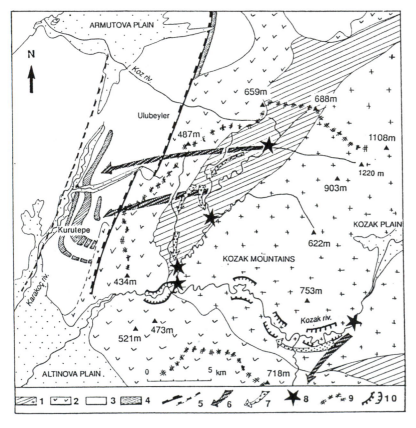

Figure 4. Catchment changes in the western part of the Kozak Mountains following Miocene and Pliocene andesitic activity. *Main rock outcrops:-* (1) Metamorphhic rocks and granodiorite; (2) Mio-Pliocene lava flows; (3) Mio-Pliocene sedimentary layers deposited in the sinking western margins; (4) Miocene conglomerates including metamorphic and granodioritic gravels, showing drainage from the heart of the mountains towards the western area; (5) Main fault lines since Pliocene river responses to volcanic and tectonic activity in the area; (6) Direction of Miocene water flows and sediment distribution; (7) Changes in drainage pattern during the Mio-Pliocene, due to lava flows interrupting previous drainage; (8) Captures due to changes in drainage pattern during the Mio-Pliocene; (9) Actual drainage basin divide separating the inner Kozak Mountains and their western margins; (10) Entrenched river course of the Kozak River, with gorges due to the antecedence of its meanders before the Quaternary uplift of the Kozak Mountains.

inner plain. As no fault line appears on the contact line between the plain and the dominating uplands, the magnitude of the surrounding slopes (up to 700 m on the northern side of the plain) may also be the result of the regional uplift of the Kozak Mountains during the Quaternary.

In the upstream part of the Kozak River, some minor recent changes in the drainage network of the plain may be noticed. At the easternmost part of the plain, a small torrent flows in a minor channel that appears to be too wide for the volume of water flowing in it (Fig. 6). This underfit stream could be explained by the fact that the confluence with the Gökmen River, a much longer tributary whose source is at 1300 m, may have been displaced downstream. Could this displacement be due to local tectonic movement? If the northern slope dominating the Kozak inner plain has recently been subject to a local uplift, a left bank minor tributary could well have captured the Gökmen River, thus displacing its junction downstream (Fig. 6). In this case, the Gökmen River must be considered as having been previously the upper part of the Kozak River.

4.2 *The entrenched meanders of the Kozak River*

The entrenchment of meanders of the Kozak River in its middle course, downstream from the Kozak inner plain,

in the western part of the granodioritic intrusion (Fig. 5), and the horizontal steps in the profile of the Kozak River (Fig. 7), are geomorphological indicators which record the nature of the regional uplift. Several gorges correspond to steps in the river's long profile. These steps can be interpreted as a response to continuous uplift during the Quaternary, since petrographic analyses of the granodiorite have shown no sign of marked spatial variation in rock resistance.

The longitudinal profiles of the Kozak River tributaries, in its middle course, also support the existence of the captures suggested above. The captures occurred at the Miocene-Pliocene transition (depending on the age of the main andesitic lava) with the building up of the western volcanic belt. All the channels of the right bank tributaries, diverted by these volcanic uplands, are more deeply incised than the Kozak River channel itself. This can be explained by the fact that these tributaries are eroding the less resistant metamorphic (schist-like) basement. However, the downstream part of the profile of the main right bank tributary, the Bakirlik River, contains a step which is similar to that on the profile of the Kozak River. This step is close to where the Kozak River leaves the granodioritic intrusion and it could be interpreted as the result of differential erosion. However, the existence of an active fault line distinct from the contact line between the granodiorite and the andesites can be suggested, since

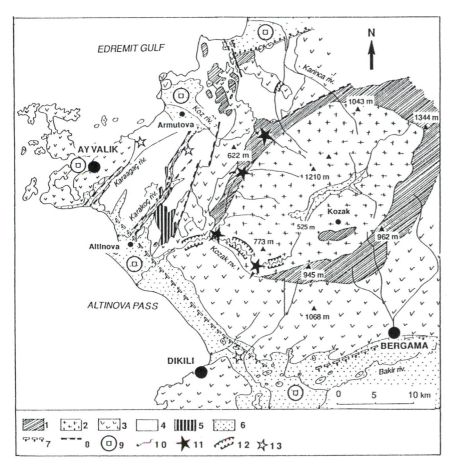

Figure 5. River response to Quaternary tectonics. *Main rock outcrops:*- (1) Metamorphic rocks (Primary age); (2) Granodiorite (Tertiary age); (3) Mio-Pliocene volcanic rocks; (4) Mio-Pliocene lacustrine and continental deposits; (5) Plio-Quaternary (?) basalt; (6) Holocene alluvium. *Tectonic environment:*- (7) Trough lines of rift valleys; (8) Main Quaternary fault lines; (9) Sinking valleys and plains. *River responses to tectonics:*- (10) Main river courses; (11) Captures (Tertiary drainage basin changes) with ancient direction of drainage; (12) Quaternary gorge (antecedent river with entrenched meanders); (13) Thresholds (showing Quaternary inversion in the direction of river courses).

50   Catherine Kuzucuoglu

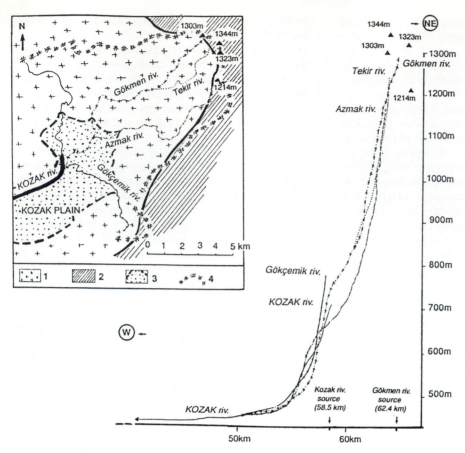

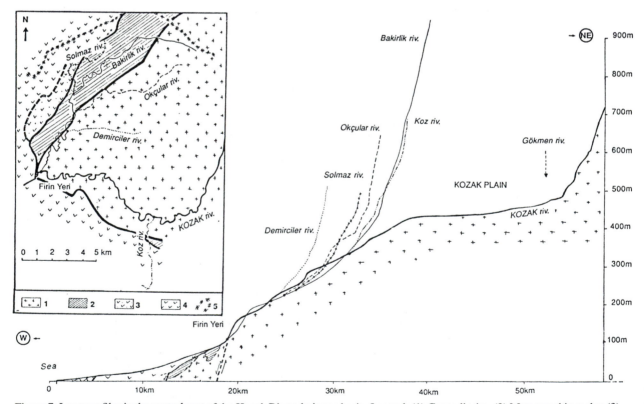

Figure 6. Long profiles of the Kozak River and its main tributaries upstream of the Kozak inner plain. Legend: (1) Granodiorite; (2) Metamorphic rocks; (3) Approximate limit of Kozak inner plain and of its alluvial fill; (4) Kozak River watershed.

Figure 7. Long profiles in the central part of the Kozak River drainage basin. Legend: (1) Granodiorite; (2) Metamorphic rocks; (3) Trachytes in basement; (4) Andesites; (5) Kozak River watershed.

the step is observed at least 1km upstream of the granodiorite-andesite contact (Fig. 7).

4.3 *The stepped floors of the emerged western basin*

The western border of the andesitic belt is marked by a series of continuous fault lines which have been active since the Tertiary. To the west is the subsiding area, filled with lacustrine and continental sediments; to the east are the rising uplands, composed of Tertiary granodiorite intrusive into an older epimetamorphic basement and partly covered by the thick lava flows of the Mio-Pliocene andesitic belt (Fig. 3). At the foot of this boundary a depression has been shaped by subaerial erosion of the lacustrine sediments. The present drainage network erodes this floor – this drainage is mainly due to the Karakoç River, with its source near the Armutova Plain, but flowing north-south towards to the Altinova Plain, so draining most of this central plain.

Several landforms in the upstream part of this depression provide evidence of stages in the Quaternary lowering of the basin and of its more recent partial destruction. At least two nested levels ('glacis') can be distinguished, which probably result from fault line activity during the Quaternary (Kuzucuoglu 1980). They are particularly well preserved near the village of Ulubeyler (Fig. 4). These glacis are covered with red detrital sediments containing gravels (including granitic ones) from the uplifted volcano-sedimentary layers (Mount Ahlat) east of the fault line. The clay content and also the black iron- and manganese-rich film covering the gravels are similar to those observed in other Quaternary detrital sediments preserved on the top surface of the westernmost plateau (Galifo plateau).

The lowest step (100-110 m) in the floor of the basin corresponds to the topographic threshold which separates the Karakoç River watershed (flowing southwards to the Altinova Plain) from the Kanli River watershed (flowing northwards to the Armutova Plain) (Figs 3 and 5). After the establishment of this level, the local rivers incised their valleys and both valley systems correspond to a fault line.

5 THE IMPACT OF QUATERNARY SUBSIDENCE

5.1 *Ancient uplands and their river drainage around Ayvalik and north of the Armutova Plain.*

Early Quaternary drainage in the area around Ayvalik was probably quite different from the present day. On the surface of the Galifo plateau, east of Ayvalik, a detrital cover can be observed which is somewhat similar to that covering the nested levels of the Karakoç River depression (Figs 2 and 3). A similar detrital deposit may also be found covering the floor of a dry valley between two hills of limestone overlooking the north of the Armutova Plain. The black coloured gravels, incorporated into a very dark brown and reddish sandy matrix come from

various basement rocks including granodiorite, metamorphic elements and basalt.

The basaltic gravels may derive from the erosion of a basalt flow which occurred 1 km to the north of the Galifo plateau (Fig. 3). Upland sources for the granitic and metamorphic gravels may have been located some distance to the east. However, these are presently located outside the river systems of the plateau surface being considered. Another possibility is that uplands might also have been located to the west where the sea is now. The disappearance of part of the early Quaternary watershed of the plateau surface would have to be related to the local subsidence of the Edremit Gulf.

5.2 *The recent sinking of the Altinova Plain*

The linear erosion, which destroyed parts of the lowest floor of the Ulubeyler stepped floors in the Karakoç River depression, has revealed a palaeotopography preserved beneath a basalt flow (Figs 2 and 3).

North-south geological sections through the Karakoç River valley in the western sedimentary margins of the Kozak Mountains show that the slope of the palaeotopography buried under the basalt diminishes progressively towards the north. In contrast, the present direction of the Karakoç River is the reverse of the ancient river since it now flows southwards towards the Altinova Plain. The altitudes of the plain preserved under the basalt (established from field mapping) provide valuable information about conditions prior to and following the basalt flow.

From an elevation of 80 m in the Karakoç River valley at the foot of Mount Manav (the summit of the basaltic outcrop), to the lowest remnants sited at 50 m, the altitudes diminish progressively towards the north. There must have been a decisive topographical change due to tectonic movement within the depression (Fig. 8). The northern part of the depression was uplifted while the southern part subsided, attracting southwards the rivers draining the depression.

The whole of the Karakoç River valley, continued by the Kanli River valley, follows a fault line orientated north-south. This fault line disrupts the central depression in the same way as the parallel main fault line limits the volcanic belt to the east. The sinking of this rift is probably more intense in the south than in the north, causing the river drainage to be oriented southwards, instead of the previous northward direction.

6 CONCLUSIONS

The main landforms related to Quaternary tectonics around the Aegean Sea may be classified into two types: (1) uplands having risen along active fault lines bordering rift valleys and (2) lowlands which have been partly submerged in the subsiding basins at the foot of the rising uplands.

In the Ayvalik-Kozak region, limited by the two west-

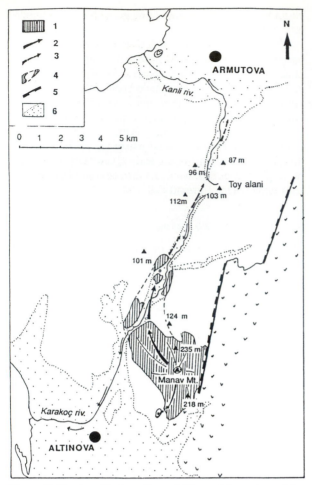

Figure 8. The basaltic lava flow of Manav Mountain and its geomorphological significance. Legend: (1) Basalt; (2) Direction followed by the basalt flow; (3) Ancient course of the palaeovalley buried under the basalt; (4) Limits of the present day outcrop of basalt; (5) Main Pliocene fault line (also cutting the basalt flow); (6) Geological limits between Tertiary basement and the alluvial fill of valleys and coastal plains.

east oriented rifts of the Edremit Gulf and Gediz River valley, river network development has been strongly influenced by tectonics since the Miocene. In this region Quaternary tectonic activity has not produced landforms as striking as in other parts of the Aegean (for example on some mountainous shorelines of Greece or along the Büyük Menderes River rift valley) and the magnitude of the movements has remained rather modest when considering the extended duration of activity.

Inversion (in the Karakoç River basin), captures (in the Kozak inner plain and around the andesitic belt) and entrenchments (in the Kozak intrusion) may be observed within both the uplifted Kozak Mountains and the subsiding lacustrine margins. Field studies show that the rhythms and characteristics of the regional uplift and of the faulting activity are best illustrated by nested landforms preserved in the subsiding margins.

More locally, the relations between sequences of volcanic emissions and river incision show that the subsi-

dence of the Altinova Plain has been taking place for a longer period of time than that of the Armutova Plain.

Thus, in the study area, Quaternary tectonics have had a less important impact on the landscape than the Mio-Pliocene tectonics. Nevertheless, Quaternary tectonics are responsible for the continuation of regional uplift of mountains and sinking of rifts, as illustrated by their effects on river drainage patterns.

7 ACKNOWLEDGEMENTS

Most of the results presented here formed part of my Doctoral thesis, the fieldwork for which I was able to undertake with the generous help of M.T.A.E. (Maden Tetkik Arama Enstitüsü, Ankara) and in collaboration with Professor O. Erol. The author would also like to thank Mrs A. Bigonneau who kindly revised the English and Professor John Lewin and Dr Jamie Woodward for their helpful comments on the manuscript.

REFERENCES

Akyürek, B. 1989. *Geological map of Ayvalik (1:100,000).* Türkiye Jeoloji Haritalari Serisi, Ayvalik G3 Paftasi. Ankara: MTA (in Turkish).

Bellon, H., J.-E. Jarriage & D. Sorel 1979. Les activités magmatiques égéennes de l'Oligocène à nos jours et leurs cadres géodynamiques. Données nouvelles et synthèse. *Rev. Géol. Dyn. Géogr. Phys.* 21(1):41-55.

Benda, L., F. Innocenti, R. Mazzuoli, F. Radicati & P. Stefens 1974. Stratigraphic and radiometric data on the Neogene in Northwest Turkey. *Z. Deutsch. Geol. Ges.* 125(2):183-193.

Borsi, S., G. Ferrara, F. Innocenti & R. Mazzuoli 1972. Petrology and geochronology of recent volcanism in the Eastern Aegean Sea (W. Anatolia and Lesvos Island). *Bull. Volc.* 36:473-476.

Dörpfeld, W. 1928. Strabon und die Küste von Pergamon. *Mitt. des deutsch. Archäol. Inst. Athen. Abteil.* LIII:117-159.

Ercan, T., M. Satir, M. Kreuzer, A. Türkecan, E. Günay, A. Cevikbas, M. Ates & B. Can 1985. Bati Anadolu Senozoik volkanitlerine ait yeni kimyasal, izotopik ve radyometrik verilerin yorumu (Recent results about the chemical, isotopic and radiometric characteristics of Cenozoic volcanic products in Western Anatolia). *Türk. Jeol. Kur. Bült.* 28:121-136 (In Turkish).

Ercan, T. 1986. Ayvalik Çevresinin Jeolojisi ve Volkanik Kayaçlarin Petrolojisi (The geology of the Ayvalik area and the petrology of magmatic rocks). *Jeol. Müh.* 27:19-30 (In Turkish).

Erol, O. 1980. The Neogene and Quaternary erosion cycles of Turkey in relation to the erosional surfaces and their correlated sediments. *Bull. Geom.* 9(8):1-40 (In Turkish).

Erol, O. 1981. Neotectonic and geomorphological evolution of Turkey. *Z. Geomorph.* 40:193-211.

Fytikas, M., F. Innocenti, P. Manetti, R. Mazzuoli, A. Peccerillo & L. Villari 1984. Tertiary to Quaternary evolution of volcanism in the Aegean region. In: J.E. Dixon & A.H.F. Robertson (eds.), *The geological evolution of the Mediterranean.* Geological Society Special Publication 17: 687-699.

Innocenti, F., P. Manetti, R. Mazzuoli, G. Pasquaré & L. Villari 1982. Neogene and Quaternary volcanism in Anatolia and northwestern Iran. In: R.S. Thorpe (ed.) *Orogenic Andesites and related rocks*. New-York: John Wiley and Sons 327-349.

Kozan, T. A., F. Ogdüm, E. Bozbay, A. Bircan, M. Keçer, K. Tüfekçi, A. Durukal, S. Durukal, S. Ozaner & M. Herece 1982. *Burhaniye – Menemen arasi kiyi bölgesinin jeomorfolojisi (Geomorphology of the coastal area between Balikesir and Izmir*. MTA rap. 7287. Ankara, MTA (In Turkish).

Kuzucuoglu, C. 1980. Le massif de Kozak et ses bordures (Anatolie occidentale). Etude géomorphologique (2 Volumes and 5 maps) Thèse de Doctorat. Paris: University of Paris I Sorbonne. (Unpublished).

Kuzucuoglu, C. 1982. L'origine des alvéoles en milieu cristallin. L'exemple du massif de Kozak (Turquie). *Physio-Géo.* 4:1-23.

McKenzie, D. 1972. Active tectonics of the Mediterranean region. *Geophys. J.R. Astron. Soc.* 30:109-185.

McKenzie, D. & Y. Yilmaz 1991. Deformation and volcanism in western Turkey and the Aegean. *Bull. Techn. Univ. Istanbul.* 44(3-4):345-373.

Peccerillo, A. & C. Villari 1980. Neogene volcanism of the northern and central Aegean region. *Ann. Géogr. Pays Hell.* 30:106-129.

Präger, M. 1965. Présentation d'une esquisse géologique des terrains volcaniques de l'île de Lesbos (Note préliminaire). *Ann. Géogr. Pays Hell.* 16:512-527.

Yilmaz, Y. 1989. An approach to the origin of young volcanic rocks of western Turkey. In A.M.C. Sengör (ed.) *Tectonic evolution of the Tethyan Region*. Kluwer Academic Publishers 159-189.

CHAPTER 5

The influence of Quaternary tectonics on river capture and drainage patterns in the Huércal-Overa basin, southeastern Spain

E. WENZENS and G. WENZENS
Heinrich-Heine-Universitat Düsseldorf, Geographisches Institut, Düsseldorf, Germany

ABSTRACT: Quaternary relief development in the Huércal-Overa basin has been strongly influenced by tectonic disturbances and river captures, which have had contrasting effects on different parts of the basin. Different heights and degrees of dissection of the oldest glacis remnants in the western and eastern part of the basin can be attributed to the uplift of the eastern margin of the basin. A tectonic disturbance trending NNE-SSW, affected the eastern margin of the basin and caused intense erosion there. The southwestward continuation of the same fault crosses the central part of the basin. The eastern central part of the basin, unaffected by tectonics, mainly consists of one mid-Quaternary glacis, whereas in the west three separate mid-Quaternary and one late Quaternary glacis can be identified. In addition, the incision of the Rambla Saltador in the Sierra de Almagro, which took place in the late mid-Quaternary, has further modified the morphogenesis. In the eastern part of the basin, headward erosion caused by capture occurs only in the form of shallow channels. In the west however, in the drainage of the Rambla Guzmaina, increased erosional activity is particularly evident north of La Parata. Here, the incision caused by the Rambla Guzmaina amounts to more than 50 m. The recent valley floor of the neighbouring Rambla Almajalejo is only 4 m below the water divide. A considerable amount of erosion in the Rambla Guzmaina took place in late Quaternary times.

KEYWORDS: Pliocene and Quaternary drainage patterns, Quaternary tectonics, glacis terraces, Rio Almanzora, Huércal-Overa basin, river capture, incision.

1 INTRODUCTION

1.1 *The study area*

The research area is part of the eastern Betic Cordillera, which is characterized by an alternation between Neogene sedimentary basins and surrounding mountain chains (Fig. 1). There is a longitudinal valley running from west to east bounded by the Sierra de los Filabres (2168 m) and the Sierra de las Estancias (1722 m). In the east this longitudinal valley widens towards the Huércal-Overa basin. The basin itself is bordered by the Sierra de las Estancias and the Sierra de Enmedio in the north and by the Sierra de los Filabres and the Sierra de Almagro in the south (Fig. 1). In the east the basin slopes gently towards the Pulpi basin running from north to south.

The Rio Almanzora is the main river, flowing at the foot of the northern flank of the Sierra de los Filabres (Fig. 1). The Sierra de Almagro has a dominant position as it marks the boundary between the Vera basin to the south, the Huércal-Overa basin to the north, and the Pulpi basin to the east (Fig. 2). It also forms a barrier across the valley of the Rio Almanzora, which the river crosses in a deeply incised valley. Just before its mouth into the Mediterranean Sea, the Rio Almanzora receives its last and most important tributary, the Rio Canalejos. This river system drains the whole of the Pulpi basin and, with

the Rambla de las Norias, the northern part of the Huércal-Overa basin.

1.2 *Climate and river regime*

The climate of the Huércal-Overa basin is semi-arid with a precipitation of 250-350 mm at elevations of 300 to 800 m, and rising to 500 mm at elevations exceeding 800 m in the surrounding mountains. With the exception of the perennial reaches of the Rio Almanzora at the southern margin of the basin, all other streams in the basin are ephemeral. For this reason they are called torrentes or ramblas.

2 GEOLOGICAL-TECTONIC STRUCTURES

The geological-tectonic framework is shown in Figure 1 and is defined by a system of interconnected intramontane Neogene basins with the surrounding mountains consisting of variously metamorphosed rocks of the Nevado-Filabride and Alpujarride complexes (Montenat et al. 1990). In Tortonian times the Huércal-Overa basin was filled initially with terrestrial and later with marine sediments. During the Messinian the sedimentation of marl continued for a time, but with the regression of the sea the marls pass into continental facies at the northern

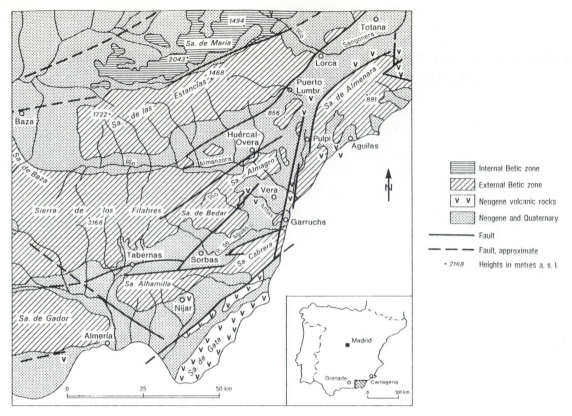

Figure 1. Geological-tectonic structures of the eastern Betic Cordillera (based on Briend 1981).

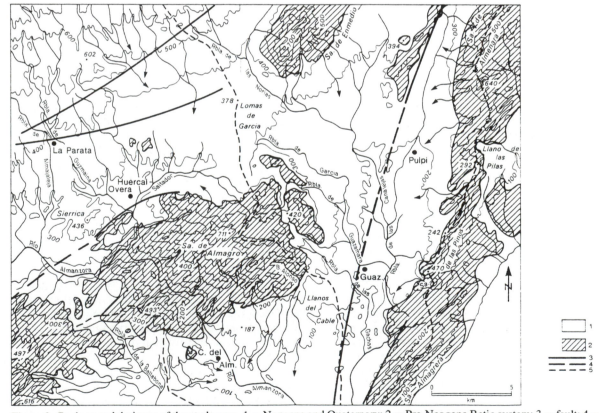

Figure 2. Geology and drainage of the study area. 1 = Neogene and Quaternary; 2 = Pre-Neogene Betic system; 3 = fault; 4 = fault, approximate; 5 = boundary line of the Rio Almanzora catchment.

and southern margin of the basin. Briend (1981) has reconstructed the position of the former coastline from the distribution of the Messinian sediments (Fig. 3), with a bay between the Sierra de Enmedio and the Sierra de Almagro opening towards the Pulpi basin. In the neighbouring Vera and Pulpi basins the Messinian deposits are covered by Pliocene marine sediments. This Pliocene transgression did not extend to the Huércal-Overa basin

(Briend 1981) where all the post-Messinian sediments are of continental origin.

The palaeogeographical development of the study area has been determined mainly by the activity of two important fault systems (Figs 2 and 3). Thus 'the Huércal-Overa basin is located at the compressional ending of the sinistral Alhama de Murcia fault' (Montenat et al. 1987). Another disturbance representing the northern boundary

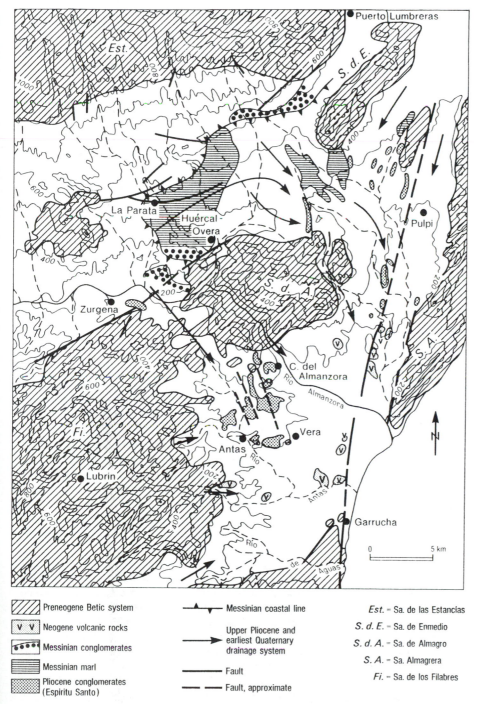

Figure 3. Upper Pliocene and earliest Quaternary drainage of the Huércal-Overa basin (based in part on Briend 1981; Veeken 1983; Völk 1967).

Table 1. Proposed correlation of river capture, fluvial erosion and glacis sediment accumulation with tectonics in the Huércal-Overa basin.

period	m.y. B.P.	tectonics	western part of the basin river capture/ drainage pattern	western part of the basin processes and forms	tectonics	eastern part of the basin river capture/ drainage pattern	eastern part of the basin processes and forms
Quaternary — Upper			Rbla. Guzmaina	intense erosion two river terraces		channels divert the drainage towards the south	below 280 m a. s. l.: erosion two river terraces
	0.125		Rbla. Almajalejo	accumulation: glacis			
Quaternary — Middle			capture of Rbla. Saltador	erosion; accumulation of glacis 3; erosion; accumulation of glacis 2; erosion; accumulation of glacis 1		capture of Rbla. Saltador	accumulation of "main glacis"; erosion in the area of Huércal-Overa; accumulation of glacis 1
	0.7						
Quaternary — Early	0.9		capture of Río Almanzora	erosion			erosion
Quaternary — Earliest			Río Almanzora flows west of Sierra de Almagro	accumulation of one or two glacis		drainage to the Pulpi basin towards the east: in the north and in the south of Loma Garcia	accumulation of glacis; erosion; accumulation of glacis (travertine)
	1.9						
Pliocene — Upper				erosion; accumulation of fluvial sediments (southeast of Zurgena)		drainage to the Pulpi basin towards the east	erosion; accumulation of fluvial sediments (Loma Garcia)

fault of the Sierra de Almagro runs almost parallel to it. The importance of Quaternary tectonics in the Huércal-Overa basin was pointed out by Briend (1981) and he mapped several faults mainly trending ENE-WSW.

In the following exposition the drainage of the Huércal-Overa basin since the Upper Pliocene will be reconstructed, the main causes of change in the drainage patterns will be demonstrated and the effects on morphogenesis will be considered (Table 1). Drainage development during the Villafranchian/earliest Quaternary will be dealt with only briefly. The study will focus on drainage development since the mid-Quaternary.

3 THE UPPER PLIOCENE AND EARLIEST QUATERNARY DRAINAGE PATTERN

The reconstruction of the fluvial drainage system in the research area can usefully begin with the palaeogeographical setting of the Upper Pliocene (Fig. 3). The regression from the Pulpi and Vera basins started during the Pliocene and is represented by coastal plain sediments, the so-called Espiritu-Santo-Formation (Völk 1967; Veeken 1983) whereas at the same time fluvial sediments were deposited in the Huércal-Overa basin. Southeast of Zurgena, hills extending up to 335 m are composed of

gravelly sediments with a strikingly large number of subrounded coarse blocks. They appear to correspond to this epoch and suggest that at that time the former Rio Almanzora flowed through a narrow gap between the spurs of the Sierra de los Filabres and the Sierra de Almagro. The drainage of the southern Huércal-Overa basin took place through this narrow gap towards the bay – at that time situated to the northwest of Vera.

North of the Sierra de Almagro there are also hills (Lomas de Garcia, Fig. 2), c. 1 km wide, 5 km long, and c. 30-40 m high, consisting of Upper Pliocene fluvial sediments. These show that the former drainage of the middle and eastern part of the Huércal-Overa basin was directed towards the Pulpi basin.

In another tectonic phase at the end of the Pliocene the sea not only retreated completely from the Vera and Pulpi basins, but the tilted Lomas de Garcia also formed a barrier to eastward drainage. After this, the rivers of the central and eastern part of the Huércal-Overa basin became tributaries to the Rambla de las Norias in the north and to the Rambla de las Gachas in the south of the Lomas de Garcia (Fig. 2).

At this time extensive deposition occurred over the eroded surface of the basins forming the first glacis (Wenzens 1992a). The term glacis here means erosional surfaces cut across non resistant Neogene sediments, later mantled by a veneer of, at most, a few metres of deposits. Such sedimentary sequences may consist of several different aggradation phases. They reflect the strongly alternating conditions of erosion and accumulation which were also interrupted by soil formation processes. This earliest Quaternary phase of accumulation came to an end with the renewed tectonic activity along the northern boundary fault of the Sierra de Almagro. The uplift of the Sierra de Almagro favoured the headward erosion of a river in the northern Vera basin so that this river dissected the Sierra de Almagro and captured the Rio Almanzora east of Zurgena. Harvey (1987) suggests early superposition and then antecedence as uplift proceeds. But neither remnants of fluvial terraces nor gravels of this period are present. The capture of the Rio Almanzora initiated intense erosion of the Villafranchian/earliest Quaternary glacis in the Vera basin, and its surface was lowered by 100 m in the present drainage area of the Rio Almanzora (Wenzens 1992b). The Upper Pliocene to earliest Quaternary valley of the Rio Almanzora west of the Sierra de Almagro has not been subjected to intense erosion since the capture and it is now a tributary of the Rio Antas (Fig. 3).

The genesis of a major transverse valley in the early Quaternary has had very different effects on the further morphogenesis of the Vera, Pulpi, and Huércal-Overa

basins. In the following discussion the extent to which the glacis and river development in the Huércal-Overa basin were influenced by capture and Quaternary tectonics will be demonstrated.

4 QUATERNARY RIVER AND GLACIS DEVELOPMENT IN THE HUÉRCAL-OVERA BASIN: THE INFLUENCE OF TECTONICS AND CAPTURE

4.1 *Earliest to mid-Quaternary relief development*

Isolated relics within the northeastern margin of the basin at an altitude of between 500 and 900 m indicate the oldest formation of glacis in the Huércal-Overa basin (Fig. 4). North of the abandoned village of Abejuela (Sheet 974, Vélez Rubio) at an altitude of 850 m the uppermost erosion surface is covered by travertine approximately 5 m thick, truncating the red conglomerates of the Lower Tortonian. The relief is strongly dissected within the margin of the basin at 500-800 m above sea level. Since the formation of this glacis level the Rambla de Abejuela has incised 150 m. Both the high position above the recent valley floor and the karst pockets in the travertine filled with reddish soil relics probably indicate at least an early Quaternary age.

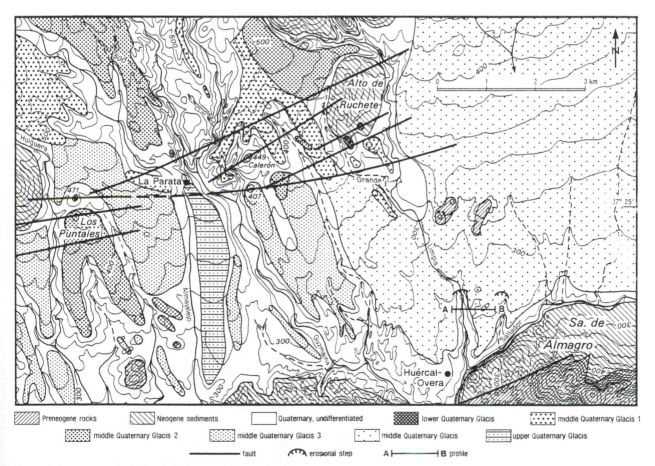

Figure 4. Quaternary glacis-levels in the central part of the Huércal-Overa basin.

Figure 5. Fluvial erosion of the tilted Early Quaternary glacis south of Alto de Ruchete. The background illustrates the non-dissected eastern part of Huércal-Overa basin consisting of a mid-Quaternary glacis.

Also near Abejuela there is another remnant of an early Quaternary glacis at an altitude of 800 m – i.e. 100 m above the modern valley floor. It can be correlated with a sedimentary cap 2 km to the south at an elevation of 713 m, i.e. 110 m above the recent valley bottom. This mainly consists of coarse gravels. On the shoulders of the sharply incised valley near Abejuela there is a glacis surface on both sides of the torrente at an altitude of 760 m. It is 50-60 m above the modern valley floor and is probably early mid-Quaternary in age.

Below approximately the 500 m contour, the central Huércal-Overa basin is hardly dissected in its eastern part, consisting of a glacis surface gently sloping towards the southeast (Fig. 4). The sediments associated with this surface are probably related to mid-Quaternary times (see below); exact dating however is not yet possible (Fig. 5).

Completely different relief conditions can be found in the western part of the basin. A well preserved glacis surface begins at the margin of the basin at an altitude of 800 m. Several spurs continue southwards where they terminate at a height of about 660 m. North of La Hoya, they consist of deposits up to 30 m thick. The isolated remnants at Pico Agujereado (644 m) and Cerro Gordo (625 m) belong to this level. However, the gravel rich capping on top of the Gatero (702 m) is probably a remnant of an older glacis. Although the incision of the torrentes into the glacis remnants is only 80-100 m or 60-70 m in this area, they are accorded an early Quaternary age.

The different altitude and degree of dissection of the oldest glacis relics in the western and eastern part of the Huércal-Overa basin can be explained by the increased uplift of the eastern margin of the basin. Thus the distance between the contour lines at 800 m and 500 m is around 4 km here, whereas it is 6-7 km in the western part. The cause of this is the significant Alhama de Murcia fault. It affects the eastern margin of the Huércal-Overa basin and initiated strong uplift causing vertical erosion. The fault continues southwestwards in the central part of the basin where it splits into a number of disturbances (Fig. 4). The network of faults is very dense between Los Puntales and Alto de Ruchete and concentrated in an area 2-5 km wide (Briend 1981).

Accordingly, the formation of the Quaternary glacis sequences is more influenced by tectonics here although the effects of river captures on the general relief development are also evident (Table 1). This can be further illustrated by an area particularly influenced by tectonics (Fig. 4). Whereas the eastern part of the Huércal-Overa basin consists chiefly of one mid-Quaternary glacis, (the 'main glacis') the western part has three mid-Quaternary and one late Quaternary glacis, together with isolated remnants of early Quaternary glacis. At least two river terraces are also present, though not illustrated in Fig. 4. The glacis sequences have been assigned relative ages on the basis of lithostratigraphy, altitudinal data and degree of soil development.

The mapped early Quaternary glacis remnants are isolated accumulations, related to faults (north of Los Puntales, north of La Parata, south of Alto de Ruchete, Fig. 4). Tectonic disturbances have been active in this area at various periods. The Calerón ridge (449 m) consists of conglomeratic layers, steeply tilted towards the south. They comprise well-rounded Messinian beach pebbles, deposited as deltaic sediments in the Messinian coastal area (see Fig. 3). The other Neogene sediments also dip steeply within a fault zone which stretches from Alto de Ruchete via Calerón to the north of Puntales (Fig. 4). The tilting must have taken place between Messinian and earliest Quaternary times as they are truncated by early Quaternary glacis deposits north of Los Puntales and south of Alto de Ruchete.

A second tectonic phase took place in the late mid-Quaternary. In the area of Los Puntales (461 m) and Las Zorreras (407 m – south of Calerón) the mid-Quaternary glacis 2 was disturbed. Near La Parata (Fig. 6) its base dips steeply and is covered by sediments of the late Quaternary glacis. Also south of Alto de Ruchete the mid-Quaternary glacis ends at the same fault and the 'main glacis' follows (Fig. 5).

4.2 *Morphological consequences of the mid-Quaternary valley formation of the Rambla Saltador*

Apart from the disturbances of the Quaternary glacis, there is a morphologically distinctive feature of great significance north of La Parata (Fig. 6). Here, the two Almajalejo and Guzmaina torrentes are less than 100 m apart (Fig. 7). The water divide here is formed by remnants of glacis of mid- (428 m) and early Quaternary (422 m) ages. The main fault runs between the two glacis relics. Here, the altitude of the water divide is only 394 m. The modern valley floor of the Rambla Almajalejo is only 4 m deeper, but the Rambla Guzmaina has incised more than 50 m. Only road construction has prevented the Rambla Almajalejo from being captured.

What are the causes of the very different degrees of erosion of the two torrentes, which have almost equally large drainage areas? The Rambla Almajalejo reaches the Rio Almanzora directly before its entry into the Sierra de Almagro. The Rambla Guzmaina, however, flows into the Rambla Saltador, which in turn is deeply incised into the Sierra de Almagro and reaches the Almanzora River within the mountainous area. The gorge of the Rambla Saltador, however, is younger than the valley of the Rio Almanzora, which developed in the early Quaternary (Wenzens 1992b).

The only evidence for a mid-Quaternary age for the Saltador gorge is found east of Huércal-Overa (Figs 5 and 8). Here the 'main glacis' is overlooked by a few remnants of the mid-Quaternary glacis level one whose base does not slope towards the Rambla Saltador (Fig. 8). The base of the 'main glacis' consists of a massive conglome-

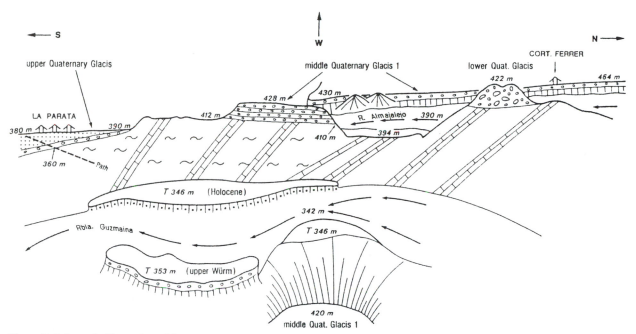

Figure 6. Schematic illustration of the terrace- and glacis-levels north of La Parata (see Fig. 4).

Figure 7. The future capture of the Rio Almajalejo (right side) by the Rio Guzmaina (left side) along the mid-Quaternary fault (foreground).

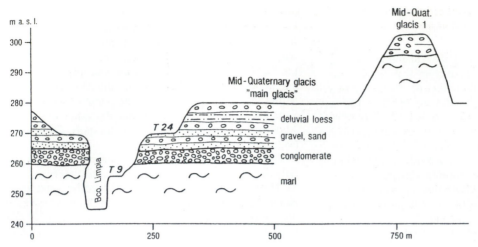

Figure 8. West-east cross section across the Barranco Limpia northeast of Huércal-Overa (location of section shown in Fig. 4).

ratic layer, clearly sloping towards the south, so that its formation must already have been related to the main stream, the Rambla Saltador. In the 20 m thick sediments, loess deposition has alternated with periods of soil formation particularly in the upper layers (Fig. 8), and the sedimentation of the 'main glacis' may include several glacial epochs (cf. Wenzens 1992). The Barranco Limpia – a tributary of the Rambla Saltador – has incised 34 m in the 'main glacis', and thus forms two narrow erosional terraces at 9 m and 24 m above the modern valley floor (Fig. 8). It is not certain whether the two terraces each represent a glacial period.

Only a few hundred metres north of the cross section illustrated in Fig. 8 the deep incision of the Barranco Limpia comes to an abrupt end (Fig. 4). At this point the conglomeratic base forms an erosional step, 8 m in thickness. Upstream the torrente is only incised about 2-3 m into the 'main glacis'. Some tributaries of the Rambla Saltador, flowing from further east, have similar erosional steps.

The headward erosion caused by the Rambla Saltador is continued only in the form of flat channels in this part of the basin. As a result, the torrentes originally draining eastwards to the Rambla de las Gachas were diverted to the Rambla de Saltador towards the south. The Rambla de las Norias draining the northwestern part of the Huércal-Overa basin to the Pulpi basin (Fig. 4) is an exception. This torrente follows the regional slope that has existed since Pliocene times – its course has not been changed either by Quaternary tectonics or by river capture. The dissection of the surface, slightly sloping towards the southeast, nowhere reaches more than 5 m and Quaternary terrace development has therefore not taken place.

The effects of the headward erosion caused by the Rambla Saltador are very different in the western part of the basin. As in the lower course of the Rambla Guzmaina the conglomeratic base of the 'main glacis' is not present, and the river has here incised into the weak Neogene marls. Thus the headward erosion could advance to the margin of the basin in a relatively short time. The enormous erosional activity concentrated in the

drainage area of the Rambla Guzmaina can be attributed to the interaction of the headward erosion, initiated by the major trunk stream, the Rambla Saltador, and the mid-Quaternary tectonics.

Although the disturbances in the western part of the basin have activated general erosion in this area since the mid-Quaternary – as the dissection of the glacis sequences in the area of Los Puntales and Alto de Ruchete shows – it has not advanced towards the trunk stream. In the west this was due to the Rio Almanzora having only moderately incised since the mid-Quaternary. In the east, headward erosion caused by the receiving stream, the Rambla de Saltador, and dissection initiated by Quaternary tectonics, could not link up.

The intense dissection caused by the Rambla Guzmaina took place largely in the late Quaternary. Whereas the Rambla Almajalejo has only incised 2-3 m into the Würm glacis south of La Parata, the Rambla Guzmaina has incised about 50 m and produced a terrace of Late Würm age at a height of 13 m above the valley floor. There is also a Holocene terrace at 3-4 m above sea level. Intense dissection is still continuing in this area.

5 CONCLUSIONS

Many geomorphologists presume that in the basin topography in southeastern Spain a regular arrangement of glacis sequences can be identified (Table 2). In the basins between Valencia and Cartagena Dumas (1977) has mapped six levels and dated at least four of them as Riss or younger. According to Cuenca Paya and Walker (1985) the valleys in the area between Murcia and Alicante show four terrace levels at heights of 120 m, 75 m, 35 m, and 10 m above the modern valley floor, the two youngest of which are dated to the late Quaternary. In the Sorbas basin Harvey and Wells (1987) have investigated the fluvial terraces of the Feos and Aguas river system and recognized four main terrace levels. In the neighbouring basin of Vera in the catchment area of the Rio Almanzora, the author has found one Villafranchian glacis and one fluvial

Table 2. Glacis sequences and river terraces in some basins of southeast Spain.

author + region chrono-logy	DUMAS (1977) Valencia-Cartagena	CUENCA PAYA/ WALKER (1987) Alicante-Murcia	HARVEY/ WELLS (1987) Sorbas basin	WENZENS (1991, 1992 a+b) Vera basin	WENZENS Huércal-Overa basin	
					western part	eastern part
Earliest	glacis Villafranca	glacis D (120 m) Pliocene to	terrace A Late Pliocene to	glacis and terrace Villafranca	glacis Pliocene to Earliest Quat.	
Early	glacis	Early Quat.	Early Quat.		glacis	
Middle	glacis Riss	glacis C (75 m)	terrace B ?	1-3 glacis 2 terraces	glacis 1 2 3	glacis 1 "main glacis"
Upper	glacis Lower Würm glacis Upper Würm	glacis B (35 m) Lower Würm glacis A (10 m) Upper Würm to recent	terrace C Lower Würm terrace D Holocene	terrace Würm terrace Holocene	glacis and terrace Würm terrace Holocene	2 terraces Upper Quat.

terrace, up to three mid-Quaternary glacis and two fluvial terraces, and one Würm and one Holocene terrace (Wenzens 1991; 1992a, b).

The relief development of the Huércal-Overa basin differs strongly in its western and eastern part, and this explains the difficulty in comparing the number and the age of the glacis in the various basins of the Betic Cordillera. There is evidence that phases of accumulation and erosion are less dependent on the climatic fluctuations of the Quaternary than on local tectonics. Each river may have formed a varying number of glacis and river terraces due to both the specific tectonic influences in the drainage area and to changes in the receiving stream. The development of the valleys of the Rambla Almajalejo, the Rambla Guzmaina, and the Rambla de las Norias indicates that the river erosional activity and the number of glacis sequences and fluvial terraces within one basin may vary widely.

REFERENCES

Briend, M. 1981. Evolution Morpho-Tectonique du Bassin Néogène de Huércal Overa (cordillères bétiques orientales-espagne). Documents et Travaux de l'Institut Géologique Albert de Lapparent No. 4: 208p.

Cuenca Paya, A. & M. Walker 1986. Palaeoclimatological oscillations in continental Upper Pleistocene and Holocene formations in Alicante and Murcia. In F. López-Vera (ed.), *Quaternary Climate in Western Mediterranean*: 353-363. Madrid.

Dumas, B. 1977. Le Levant espagnol. La genèse du relief. Thèse d'Etat, Centre National de la Recherche Scientifique: 520p. Paris.

Harvey, A. 1987. Patterns of Quaternary aggradational and dissectional landform development in the Almeria region, southeast Spain: a dry-region, tectonically active landscape. *Die Erde* 118: 193-215.

Harvey, A. & S. Wells 1987. Response of Quaternary fluvial system to differential epirogenic uplift: Aguas and Feos river system, south-east Spain. *Geology* 15: 689-693.

Montenat, C., P. Ott d'Estevou & P. Masse 1987. Tectonic sedimentary characters of the Betic Neogene basin evolving in a crustal transcurrent shear zone (SE Spain). *Bulletin Centres Rech. Explor.-Prod. Elf Aquitaine* 11: 1-22.

Montenat, C., P. Ott d'Estevou, J. Rodriguez Fernández & C. Sanz de Galdeano 1990. Geodynamic evolution of the betic neogene intramontane basins (S and SE Spain). *Paleontologia y Evolució Memória Especial* 2: 5-59.

Veeken, P. 1983. Stratigraphy of the Neogene-Quaternary Pulpi basin, provinces Murcia and Almeria (SE Spain). *Geologie en Mijnbouw* 62: 255-265.

Völk, H. 1967. Zur Geologie und Stratigraphie des Neogenbeckens von Vera, Südost-Spanien. Unpublished thesis, University of Amsterdam.

Wenzens, G. 1991. Die mittelquartäre Reliefentwicklung am Unterlauf des Rio Almanzora (Südostspanien). *Freiburger Geographische Hefte* 33: 185-197.

Wenzens, G. 1992a. The influence of tectonics and climate on the villafranchian morphogenesis in semiarid southeastern Spain. *Zeitschrift für Geomorphologie N.F. Suppl.-Bd.* 84: 173-184.

Wenzens, G. 1992b. Mittelquartäre Klimaverhältnisse und Reliefentwicklung im semiariden Becken von Vera (Südostspanien). *Eiszeitalter und Gegenwart* 42: 121-133.

CHAPTER 6

Controls on drainage evolution in the Sorbas basin, southeast Spain

A.E. MATHER
Department of Geographical Sciences, University of Plymouth, Drake Circus, Plymouth, Devon, UK
A.M. HARVEY
Department of Geography, Roxby Building, University of Liverpool, Liverpool, UK

ABSTRACT: The Sorbas basin lies in the Betic Cordillera of southeast Spain, and is one of a series of east-west orientated sedimentary basins defined within the left-lateral shear zone which characterises the regional structural framework. Since the origin of the basins in the mid-Miocene, the area has been subjected to N-S compression as a result of interaction between the African and Iberian plates. The basement is currently undergoing epeirogenic uplift as a result of this regional compression and isostatic uplift associated with the Oligocene nappe emplacement.

Until the Late Messinian the basin fill was dominantly marine. In the Pliocene a number of marginal, prograding alluvial fans established a basinally convergent drainage network which exited the basin to the south. Inversion of the basin in the Pliocene stimulated dissection and superimposition of this drainage system. This is recorded by a series of terrace gravels which lie unconformably on the Pliocene deposits, post-dating the main phase of deformation. Continued activity during the Quaternary further induced localised uplift, faulting and halokinesis. In addition, accelerated headward erosion, stimulated by the regional differential uplift led to extensive modification by river capture of the superimposed network.

These modifications are reflected in the present drainage network by 4 main genetic river types: Type 1 original consequent drainage; Type 2 transverse drainage; Type 3 beheaded original consequents and Type 4 aggressive subsequents.

Despite the intervention of coeval tectonics, the principal modification of the drainage network has been indirectly related to the tectonics, in the form of river capture. Much of the drainage still reflects the original Pliocene topography from which it was inherited.

KEYWORDS: Betics, capture, antecedence, tectonics.

1 INTRODUCTION

In the simplest case (i.e. stable, uniform conditions) drainage networks developed on an exposed surface will typically show a progressive evolution from initiation, expansion to a maximum drainage density and then reduction, with a decrease in stream density as a result of progressive erosion and lowering of drainage divides. Experimental work has indicated that drainage networks are ultimately fairly stable in nature with the final pattern closely resembling the initial pattern of drainage development (Schumm et al. 1987). Thus the final pattern will reflect the initial topography on which the surface developed. This is assuming no external influences from tectonics, climate or base level changes. In the long term, however, these latter variables are rarely passive and the drainage system will tend to adjust to these external changes.

Rarely is it possible to examine drainage evolution over long time scales in detail, due to lack of preservation of evidence. The Sorbas basin in southeast Spain, however, offers excellent exposure through a Neogene sedimentary basin and enables the evolution of a drainage system to be traced from its initiation in the Upper Messinian through to the present day. It is the aim of this paper to examine the evolution of the Sorbas basin fluvial system over this time period (some 5 million years) and evaluate the evolving roles of relative sea level change, tectonics, climate and topography on its development. An appreciation of the Messinian and Pliocene drainage development gives a clearer understanding of the factors that have determined the Quaternary to recent drainage patterns.

2 REGIONAL SETTING

The Sorbas basin is situated in the Betic Cordilleras of southeast Spain (Fig. 1). Since the origin of the Cordillera in the mid-Tertiary (Bourrouilh and Gorsline 1979) a general north-south compressive regime has existed (Hall 1983; Sanz de Galdeano 1990; Ott d'Estevou 1980; Montenat et al. 1987; Coppier et al. 1989). Much of this movement is taken up along major left-lateral strike-slip faults (Weijermars 1991; Bousquet 1979) which form a wide left-lateral shear zone (Montenat et al. 1987). This faulting in part defines the Neogene sedimentary basins of the Betics, generating localised areas of extension and

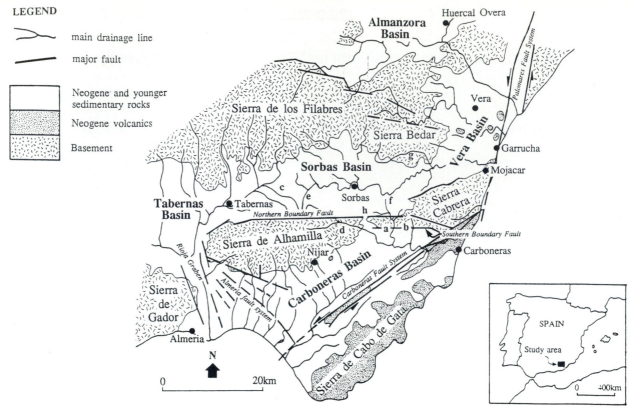

Figure 1. Simplified sketchmap of the Almeria region, southeast Spain showing major tectonic units and principal lines of drainage (modified from Harvey 1990). Letters a to h indicate locations discussed in the text; a: Rambla de los Feos; b: Rambla Gafares; c: Rambla de los Molinos; d: Rambla Lucainena/Alias; e: Rambla de Moraila; f: Rio Aguas capture site; g: Rio Jauto; h: Cantona.

compression (Hall 1983). The Sorbas basin was inverted in the Late Pliocene and subjected to compression, uplift and later dissection. The region is currently undergoing epeirogenic uplift as a result of nappe emplacement and the compressional regional tectonics (Weijermars 1985a, b; Platt 1982; Hall 1983).

The basin is defined by the Sierra de los Filabres in the north and the Sierras Cabrera and Alhamilla in the south (Fig. 1). The eastern and western margins are poorly defined structural highs. The main thickness of the sedimentary basin fill comprises Tortonian turbidites of the Chozas Formation (Fig. 2). These deposits were unconformably overlain by transgressive shallow water calcarenites of the Azagador Member in the Messinian. Reefs of the Cantera Member became established on basement highs at about this time. As sea level declined in response to the Messinian salinity crisis these reefs developed in the deeper, more central basin locations (Ott d'Estevou 1980). Thick sequences of gypsum were subsequently deposited in more central basin locations. In the Late Messinian, after re-establishment of marine connections with the Atlantic, less saline conditions returned and basinal marls and marginal coastal sequences of the Sorbas Member were deposited.

With continued uplift of the basin during the Messinian and Pliocene, marine conditions withdrew. Subaerial drainage was initiated and the subaerial sediments, which form the focus of this paper, were deposited. These sediments comprise the coastal plain sequence of the Zorreras Member (Fig. 2) deposited in the basin centre and alluvial fans of the Moras Member (Fig. 2) at the basin margin. A weakly convergent drainage pattern was thus developed which exited from the south of the basin into the Almeria/Carboneras basin via a low in the Sierras Alhamilla and Cabrera (a in Fig. 1). After a brief marine incursion in the Lower Pliocene the fluvial system became re-established, developed and expanded into an extensive braidplain which drained into the Carboneras basin to the south during the Late Pliocene to early Quaternary.

Continued differential, regional uplift stimulated incision and superimposition of the established drainage network during the Quaternary. This network became progressively modified by captures (Harvey and Wells 1987; Mather 1993a). The main river system now drains axially into the Vera basin to the east (Fig. 1).

3 TEMPORAL EVOLUTION OF THE FLUVIAL SYSTEM

3.1 Inauguration

Information about the nature of the original drainage system which developed prior to uplift and dissection is contained in the sediments of the Cariatiz (Messinian and

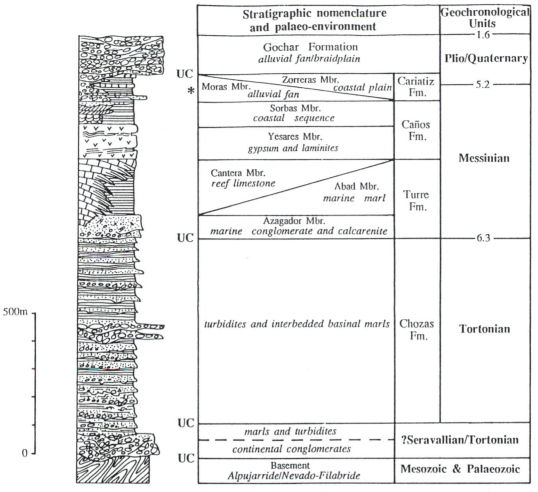

Stratigraphic nomenclature and palaeo-environment			Geochronological Units
			— 1.6 —
Gochar Formation *alluvial fan/braidplain*			Plio/Quaternary
Moras Mbr. *alluvial fan*	Zorreras Mbr. *coastal plain*	Cariatiz Fm.	— 5.2 —
Sorbas Mbr. *coastal sequence*		Caños Fm.	Messinian
Yesares Mbr. *gypsum and laminites*			
Cantera Mbr. *reef limestone*	Abad Mbr. *marine marl*	Turre Fm.	
Azagador Mbr. *marine conglomerate and calcarenite*			— 6.3 —
turbidites and interbedded basinal marls		Chozas Fm.	Tortonian
marls and turbidites *continental conglomerates*			?Seravallian/Tortonian
Basement *Alpujarride/Nevado-Filabride*			Mesozoic & Palaeozoic

Figure 2. Neogene stratigraphy of the Sorbas basin. Nomenclature from Dronkert (1976) modified (asterisked) by Mather (1991). Time divisions as used in Haq et al. (1987). Note that age divisions of the sediments are not based on radiometric data and are mainly relative. UC indicates principal unconformities.

Lower Pliocene) and Gochar (Pliocene and early Quaternary) Formation deposits (Fig. 2). As marine conditions withdrew from the Sorbas basin in the Late Messinian, so the primary fluvial drainage of the Cariatiz Formation was established.

At the northern basin margin, conglomerates of the Moras Member were deposited by a series of alluvial fans. These deposits suggest a dominance of unconfined flood events with rare bar and channel deposits in poorly defined, wide, shallow channels. Clast assemblages suggest at least two source areas – the Sierra de los Filabres and Sierra Bedar to the NE (Fig. 1). The fans were developed over a topographic step defined by a palaeo (Messinian) reef cliff-line developed along a structural lineament (Mather 1991; Mather and Westhead 1993). Distally the fans interfinger with the coastal plain deposits of the Zorreras Member.

The Zorreras Member is dominated by finer, silt and sand grade material, with only rare conglomerate filled channels towards the western margin of the basin, which drained from west to east along the axis of the basin. The

Zorreras deposits dominantly represent deposition on a low gradient coastal plain (Mather 1991). Microfossils indicate a marine connection in the south (Ott d'Estevou 1980). The unconfined nature of the fluvial sediments and lack of sedimentary structures indicates rapid deposition in a fairly flashy regime with infrequent, high discharge events. Poorly developed palaeosols within the sequence are typically well drained calcretes.

The Cariatiz Formation is punctuated by the presence of three extensive marker beds which affect both the Zorreras and Moras Members. These are two lacustrine carbonate beds and a yellow marine band which caps the sequence.

The lacustrine beds reflect deposition in a warm, shallow, unstratified, well oxygenated brackish water lake which extended across the basin on two occasions (Mather 1991). Facies variations and microfossils indicate the lake was connected to the sea in the south, which buffered the lake levels against seasonal fluctuations (Mather 1991).

Facies distributions within the yellow marine band

which caps the Cariatiz Formation again indicate open marine conditions in the south of the area (Mather 1991). Towards the basin margins the incursion drowned the fan topography, developing gravel beaches and providing niches for the development of small coastal lagoons.

Both the marine and lacustrine invasions into the basin were typically restricted to the area south of the structural lineament which defined the topographic high along the northern margin of the basin.

3.2 Expansion

After the retreat of the end Cariatiz Sea towards the south, as a result of uplift of the basin, a convergent drainage pattern similar to that of the earlier, Cariatiz Formation was re-established. Initially, small alluvial fans developed along the margins of the basin. With continued development of the system the fluvial conglomerates prograded from the basin margins over the entire basin, producing the Gochar Formation conglomerates and sandstones.

On the basis of clast assemblages and palaeocurrents it is possible to distinguish four drainage sub-systems within the Gochar Formation conglomerates (Fig. 3). These comprise two feeder systems from the north (Gochar and Marchalico systems), one from the south (Mocatán System) and one which longitudinally drains the centre of the basin from west to east (the Los Lobos System) (Fig. 3a). The latter, main drainage line exited the basin to the south to feed a fan delta in the Carboneras basin (Mather 1993b).

Early Gochar
Initially sedimentation was dominated by sands and silts, and rapidly succeeded by channel fills of mass flow and fluvial origin at the basin margins. In the central part of the basin the sands and silts fed into pre-existing topographic lows which were developed around the Sorbas palaeohigh (Fig. 4a). The latter was an area of non-deposition, the exposed sediments of the marine band being subject to pedogenesis at this time (Mather 1991). Subaqueous channel fills and shell fragments indicate that the topographic lows were occupied by small fan deltas, being fed by the then dominant, Gochar System. Palaeocurrents were deflected around the Sorbas palaeo-high and then exited to the south of the basin into the palaeo-Mediterranean which at this time occupied the Almeria/Carboneras basin (Fig. 4a).

Along the southern margin of the basin evidence is more sparse, but the dominance of sandstone in the lower parts suggests a similar depositional setting. Basal sandstones contain abundant foraminifera and ostracods (Mather 1991), implying that here too a small lake system may have existed. The standing body of water into which the sediments prograded was fairly shallow, being periodically exposed to weak pedogenesis. The latter were dominated by the development of pseudogleys, indicating variable water table levels and periodic waterlogging (Mather 1991).

Middle Gochar
The Middle Gochar is marked by the pronounced progradation of the Gochar fluvial systems (Fig. 4b). In most of the fluvial systems this is characterised by a coarsening upwards trend. Most of the marginal feeding systems were dominated by the movement of sediment as gravel sheets, implying slope gradients of around 5% (Bluck 1987). In more central parts of the basin a braided system developed containing abundant unit bars of lateral, longitudinal and subordinate transverse type (Mather 1991). This dominance of unit bars over gravel sheets suggests a lower gradient, probably in the region of 1% (Bluck 1987).

In the Marchalico System to the NE of the basin (Fig. 3), two fans prograded rapidly south and east (Fig. 4b). Both fans were dominated by sheetflood processes, with no evidence for fluvial continuity with the axially dominated Los Lobos System.

In the Mocatán System to the SW of the basin, small, coalescent fans fed a bajada, which contributed sediment to shallow, ephemeral channels in the more distal parts of the system (Fig. 4b).

In the central parts of the basin the drainage systems fed a large, west to east drainage, the Los Lobos System. This was characterised by a finer bedload than its feeder systems from the basin margins and a stronger development of lateral bar forms. Flow appears to have been more channelised in the western and middle parts of the basin, but much less constricted in the more distal (eastern) part of the system. This Los Lobos system drained out of the basin via an opening between the Sierras Alhamilla and Cabrera. Flow had also become established over the former palaeo-high of the Sorbas area during the middle Gochar (Fig. 4b).

Late Gochar
The existing fluvial systems reached their maximum development in the late Gochar (Fig. 4c). Even then, however, the systems were immature in that they were dominated by infrequent flood events, which typically produce simple unit bars as opposed to more complex braid bars, associated with mature systems.

3.3 Superimposition

Towards the end of deposition of the Gochar Formation deformation rates increased, particularly adjacent to the main areas of uplift at the margins of the basin (such as Cariatiz and the Alhamilla/Cabrera axis; Fig. 6). This deformation generated the pronounced unconformity apparent between the Gochar Formation sediments and the Quaternary river terraces which mark the episodic incision of the fluvial systems. The uplift was differential. As a result of this differential uplift the Sorbas basin was elevated in relation to the adjoining basins. Superimposition, stimulated by this uplift was first evident in the channelisation of parts of the Gochar System (Fig. 4c). In the Marchalico System the dominant feeder streams of the main fan were deflected down the steeper fan margins

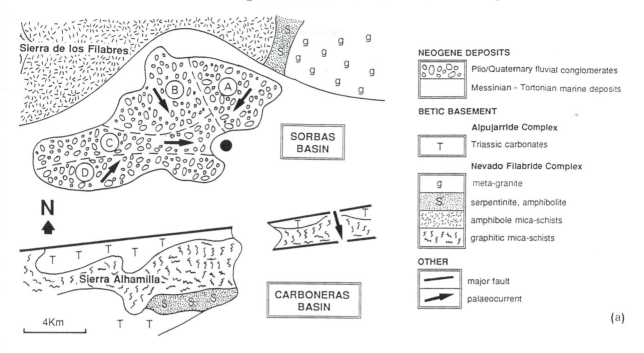

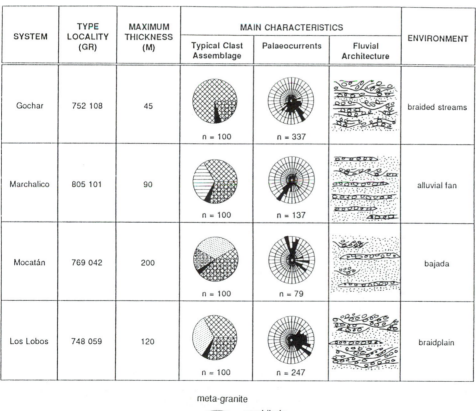

Figure 3. (a) Distribution of the main fluvial systems identified within the Gochar Formation of the Sorbas basin (modified from Mather 1993a) a: Marchalico System; b: Gochar System; c: Los Lobos System; d: Mocatan System. (b) Summary of main characteristics of the drainage systems within the Gochar Formation (modified from Mather 1993a).

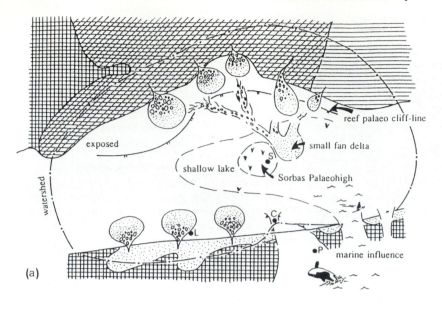

(a)

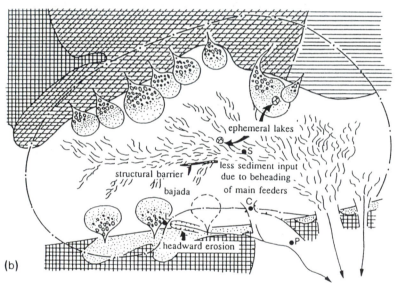

(b)

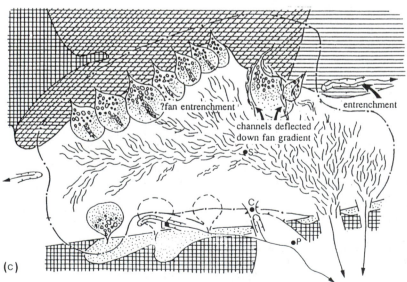

(c)

Figure 4. Schematic evolution of the Gochar Formation (drainage system expansion). S; Sorbas; C: Cantona; L: Lucainena; P: Po-lopos.

(a) Early Gochar Formation – development of marginal fans. The central parts of the basin were occupied by a shallow lake rapidly filled and subject to pedogenesis. Some marine influence from the Carboneras basin existed in the south.

(b) Middle Gochar Formation – main expansion of the fluvial system, with the development of a braidplain.

(c) Late Gochar Formation. Some superimposition. Rivers beginning to erode headwards into elevated areas and river capture becoming important.

to the east and the smaller fan was displaying evidence of channelisation of deposits.

Incision of the main Gochar Formation rivers continued, influenced by the end-Gochar landsurface, which comprised a series of moderately graded surfaces into the basin centre, and stimulated by post Gochar uplift. The main Quaternary rivers progressively incised, producing a series of fluvial terrace gravels and conglomerates which mark the episodic nature of the progressive incision. Over the Sierras to the south of the basin antecedence of the main transverse drainages such as the Feos and Gafares (a and b in Fig. 1) occurred with incision into the uplifted basement along the Sierra Alhamilla/Cabrera axis.

3.4 *Modification by capture*

The earliest recorded modification of the drainage system by capture was in the south of the area, in the upper part of the Mocatán System of the Gochar Formation (Early Quaternary), where aggressive headcutting was occurring along the line of the Rambla Lucainena/Alias, beheading former south to north drainage of the Rio Aguas (Fig. 4b and c). The capture events are recorded by clast assemblage changes (no longer containing material

derived from the former source area in the Sierra Alhamilla) and changes in sedimentary architecture (braided to meandering as a function of resulting discharge/sediment load changes) of the Mocatán sediments (Mather 1993a).

The episodic nature of the incision of the Quaternary river systems is documented by 4 main terrace levels (A-D; A being the oldest) which can be distinguished by maturity of soil development (Harvey et al. this volume). The terraces A to C are dominantly conglomeratic whilst the terrace D is locally a stage of aggradation and typically finer material (Mather et al. 1991; Harvey and Wells 1987; Miller 1991; Harvey et al. this volume). As the main drainage lines became incised the margins of the basin came under attack from aggressive subsequent streams from neighbouring basins. On the western margin of the basin tributaries of the Rambla de los Molinos (c in Fig. 1) captured streams of the southern and northern basin margins. The captured streams are known to have formerly contributed to the Sorbas drainage from both palaeocurrent and clast provenance of Gochar age sediments in this region (Mather 1991). The nature of the current drainage pattern, with pronounced elbows (Fig. 5a), similarly reflects river capture.

Along the southern margin of the basin the Rambla

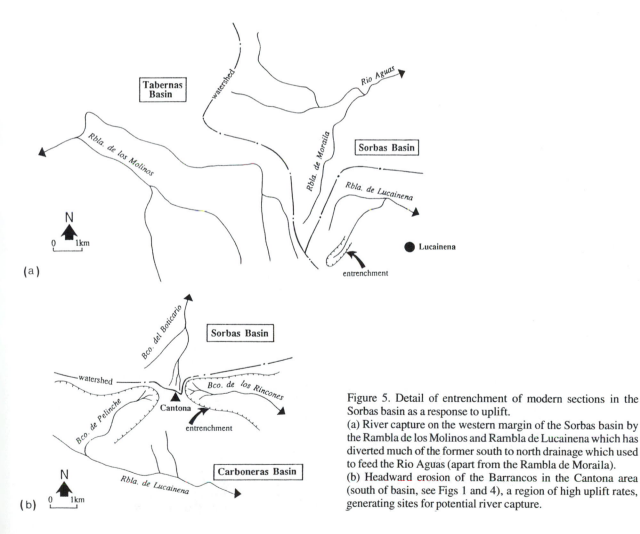

Figure 5. Detail of entrenchment of modern sections in the Sorbas basin as a response to uplift.

(a) River capture on the western margin of the Sorbas basin by the Rambla de los Molinos and Rambla de Lucainena which has diverted much of the former south to north drainage which used to feed the Rio Aguas (apart from the Rambla de Moraila).

(b) Headward erosion of the Barrancos in the Cantona area (south of basin, see Figs 1 and 4), a region of high uplift rates, generating sites for potential river capture.

Lucainena/Alias (d in Fig. 1) continued to headcut and captured former south to north drainage of the Rio Aguas system. The only remnant of the former south to north drainage today lies in the Rambla de Moraila (e in Fig. 1), whose course occupies an elevated position between the Rambla Lucainena and Molinos pirate streams (Fig. 5a).

Perhaps the most significant capture in terms of modification of the drainage pattern occurred towards the eastern margin of the basin, from the Vera basin (f in Fig. 1). This capture diverted the Sorbas basin drainage from its former course across the southern Sierras, to the east. The capture is documented by the river terrace sequence in this area (Harvey and Wells 1987; Harvey 1987a; Harvey et al. this volume), and a capture col at the head of the beheaded river, the Feos (a in Fig. 1). Rapid incision upstream of the capture site of the Rio Aguas System then ensued. A similar sequence of events may have also affected the Arroyo de Gafares further east.

The Rio Jauto of the Vera basin (g in Fig. 1) also aggressively headcut into the eastern margin of the Sorbas basin, following a fault line in this area (Mather and Westhead 1993) and capturing former north to south draining rivers.

Other captures within the area are imminent such as the Rambla de Moraila in the west of the basin (Fig. 5a), the solitary surviving uncaptured south to north draining Aguas tributary. Also in the vicinity of Cantona (h in Fig. 1, Fig. 5b), which forms a structural high, the Barrancos de Pelinche, del Boticario and de los Rincones are currently aggressively headcutting and in competition with each other.

4 CONTROLS ON THE DEVELOPMENT OF THE FLUVIAL SYSTEM

Controls on the evolution of the Sorbas basin fluvial system can be considered in terms of a) those factors extrinsic to the basin, and largely independent of the drainage basin characteristics (i.e. relative sea level, tectonics and climate) and b) factors dependent, in part, on the drainage basin characteristics, and intrinsic to the basin (i.e. topography, sediment supply).

4.1 *Extrinsic controls*

Sea level
Clearly the impact of sea level was most pronounced during the initial stage of drainage development when the relative proximity of the Mio/Pliocene Mediterranean Sea (in the Almeria/Carboneras basin) meant that any slight variations in relative sea level generated widespread incursions into the basin (such as the lacustrine and marine horizons). These incursions typically drowned the previous topography and fluvial sedimentation became focused at the periphery of the basin. In the upper parts of the alluvial fans sedimentation was little affected as a result of the lack of fluvial continuity between the untrenched fans and base level (Mather 1991; Harvey 1987b) and as a result of the relatively short lived

nature of the event. A relative fall in sea level in the Early Pliocene meant that any minor fluctuations in sea level or subsidence in the basin were not as effective in creating such dramatic palaeoenvironmental change in the later stages of basin evolution. This fall in sea level may in part have been eustatic, but tectonics also played a major role in this area, emphasising the sea level fall via regional uplift.

Re-emergence of the topography after these incursions led to a re-establishment of the previous drainage pattern, which developed in the wake of retreating marine conditions (in the case of the marine horizon), each time establishing a basinal drainage with outlets to the south.

Tectonics
The Sorbas basin is located in an area of active tectonics. The area has been undergoing compression throughout the period of fluvial system development highlighted in this paper (Mather and Westhead 1993). The overall pattern of uplift has elevated Lower Pliocene marine levels to some 500 m above sea level. Uplift rates were greatest at the basin margin, adjacent to the main axis of uplift (Alhamilla/Cabrera axis, Fig. 6). The overall regional uplift was also important in generating the overall relative elevation of the Sorbas basin compared to neighbouring basins.

During deposition of the Cariatiz Formation (Upper Messinian to Lower Pliocene), differential subsidence generated a series of topographic lows and highs. The resulting topography affected the initial drainage pattern of the drainage system. In the case of the Moras Member, tectonics near Cariatiz, along the northern margin of the basin generated a marginal angular unconformity adjacent to the basin margin, indicating active uplift of the Sierra de los Filabres at this time. Along this northern basin margin subsidence rates were such that they outpaced the short term rates of sedimentation, inhibiting fan progradation (Mather 1991). High rates of subsidence are indicated by rapid rates of deposition and the thick sediment pile. Although the fans would have formed relative topographic highs on the low gradient coastal plain, the lacustrine carbonates and associated fines are developed in relatively thick sequences within the fan environment, implying relatively high subsidence rates in order to transform the fans into areas able to accommodate these facies, similar to the situation envisaged in Death Valley by Blair (1987).

After this Early Pliocene phase of major uplift, tectonics became relatively quiescent during the deposition of the lower Gochar (early-mid Pliocene). The middle Gochar drainage development was considerably influenced by deformation, however. In the Los Lobos System at Pilarico west of Sorbas (Fig. 6), the sediments were affected by syn-sedimentary faulting and associated halokinesis. This deformation is recorded by prolific examples of synsedimentary subsidence and progressive unconformities (Mather and Westhead 1993) and led to local deflection and ponding as a result of drainage impediment. The Gochar System, in the northwest of the basin was largely unaffected by deformation at this stage.

The main phase of deformation in the basin occurred after deposition of the Gochar Formation, and prior to the incision of the fluvial systems. Again uplift was differential, and mainly focused around the basin margins (Cariatiz and the Alhamilla/Cabrera axis, Fig. 6). On a regional scale the Sorbas basin became elevated above its neighbouring basins, stimulating incision and headward erosion of the marginal drainage of these basins, capturing sections of the Sorbas basin drainage. The main deformation occurred in the sierras to the south (Fig. 6). Over the Alhamilla/Cabrera axis this further stimulated incision and antecedence of the main drainage lines such as the Rambla de los Feos (Fig. 6).

The youngest deformation recorded within the basin is Early to Late Würm. At Urra, southeast of Sorbas, an anomalous thickness of Quaternary fine sediments and more numerous terrace levels than seen elsewhere in the basin are developed (Mather et al. 1991). The deposits are confined to a series of closed synclines and show abundant evidence of syn-sedimentary deformation. Locally gypsum bosses are evident penetrating the underlying strata. The sediments are both temporally (Early to Late Würm) and spatially restricted. The deformation, which occurred after a major river capture, is thought to have rejuvenated the system and led to the geomorphic unroofing of the gypsum. This stimulated deformation by locally reducing effective stresses.

Climate

The effect of climate can be difficult to unravel from the general background of sedimentation affected by tectonics. Increased sediment yield, for example, may reflect source area rejuvenation via tectonics or climatic change. In this situation palaeosols can be more reliable indicators of local climate. The development of immature calcretes indicate a predominantly semi-arid climate in the Messinian. Within the Pliocene deposits the generally better drained pedogenic profiles in the lower parts of the sequence, lacking in carbonate (despite a prolific supply) indicate that leaching was operative at this time and that a relatively more humid climate may have developed. However in later Gochar times well developed calcretes are evident, suggesting a return to more arid conditions (Mather 1991). This is also reflected in Quaternary soils which are dominantly red and carbonate rich (Harvey et al. this volume). Cold, dry conditions prevailed in the Pleistocene (Sabelberg 1977) and semi-arid in the Holocene (Harvey 1990).

Even apparently minor climate changes can have significant implications in climatically marginal zones such as the semi-arid. It may be that the expansion and rapid sedimentation in Gochar times owes some of its existence to a more humid climate and more effective runoff generation. However, rejuvenation from uplift was also operative at this time and the increased sediment supply may thus indicate a combination of both factors.

Quaternary major sediment production periods in the Mediterranean have been related to glacials (Amor and Florschütz 1964) and incision to interglacials (Harvey 1990). Episodic incision and aggradation may thus in part reflect climatic fluctuations.

4.2 Intrinsic controls

Topography

Topographic controls were important in establishing the initial pattern of the drainage network. As this network is largely superimposed, the initial topography has thus also been influential in the final drainage pattern. Subsequently river capture became an important modifier of the local environment, changing the primary drainage patterns and locally affecting stream power and sedimentation rates (Mather 1993a; Harvey and Wells 1987; Harvey et al. this volume).

Sediment Supply

Sediment supply is important in governing the response of the fluvial system to tectonic influences. A long-term increase in the water to sediment ratio via a reduction in sediment supply would, for example, facilitate incision. Rates of sediment supply will be affected by lithological controls, amount of effective runoff generation and tectonic or climatic rejuvenation of the source area. Reduction in supply may reflect capture or exhaustion of the source area by reduced uplift rates or climatic change, reducing effective runoff. The nature of sediment supply (calibre, rate) will also be influential in governing sedimentary processes, fan morphology and resulting trenching thresholds (Harvey 1990).

5 THE PRESENT DRAINAGE SYSTEM

The main controls operative on the evolution of the fluvial system have been outlined above. Clearly these will vary in significance both spatially and temporally. Accordingly the current drainage system can be classified genetically, according to the nature of the evolution of the fluvial segments into 4 main types (Fig. 6). Examples of each are given below.

5.1 Original, consequent drainage (type 1)

Areas which have suffered minimum impact from external controls such as tectonics are least likely to deviate significantly from the original drainage pattern. Once established the network is more difficult to modify. Consequently the most tectonically stable portions of the basin, such as the northwest, reflect little evidence of modification of the drainage system, and the modern drainage pattern closely resembles the original drainage pattern (compare Figs 4 and 6).

5.2 Transverse antecedents (type 2)

There are many variants of transverse drainage development (Oberlander 1985). In the Sorbas basin type 2 streams are transverse to the axis of the mountain fronts

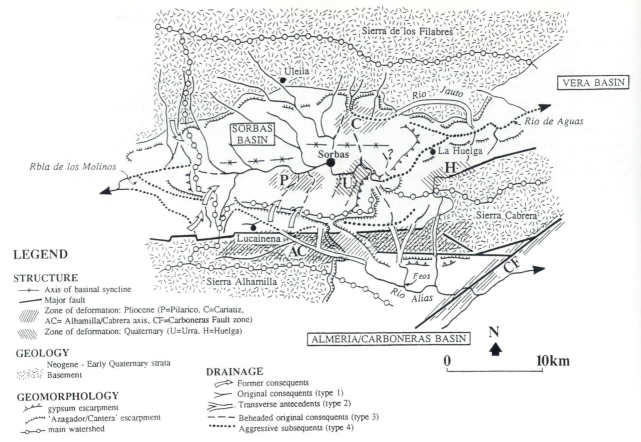

Figure 6. Main types of drainage developed in the Sorbas basin, principal areas of deformation and main geomorphological features.

and drain perpendicular to strike. They have maintained their courses across axes of uplift via incision, becoming superimposed/antecedent across these structures. The Feos for example maintained its course across the uplifting Alhamilla/Cabrera axis (Fig. 6).

5.3 *Beheaded original consequents (type 3)*

Type 3 streams reflect those streams which have been beheaded by aggressive subsequents (type 4). They suffer a reduced sediment and water input post-capture. This is recorded in early Quaternary sediments in the south of the basin, which in response to reduction in sediment and water input show a change in channel pattern (braided to meandering) and sedimentation rates (reflected in degree of pedogenesis).

5.4 *Aggressive subsequents (type 4)*

Post-superimposition capture modification characterises the evolution of the fluvial system in the marginal basin areas. This reflects the structural elevation of the Sorbas basin in relation to the neighbouring basins which has stimulated the aggressive behaviour in the headwaters of adjacent drainage systems. Perhaps the most significant of these capture events in terms of drainage pattern is that

of the Aguas/Feos during the Quaternary (Harvey and Wells 1987).

Similarly the Rambla Lucainena/Alias was originally a consequent from the withdrawal of the last, Early Pliocene marine incursion and incised into, and headcut across the Sierra Alhamilla. Once through the basement schists of the Sierra Alhamilla, the rambla was able to rapidly headcut along the strike of the lithologically weaker Tortonian marls, successively capturing former south to north flowing drainage of the Sorbas basin. The capture is first documented in the Plio/Quaternary deposits of the Gochar Formation in terms of a reduction in sediment supply and a change in the sedimentology (braided to meandering in response to input changes) and provenance (clast assemblages no longer contain material derived from the Sierra Alhamilla). Younger captures are evident in the western part of the basin. As the incision ensued, so anti-dip streams (streams flowing against the dip of the Neogene strata) developed. These are actively headcutting into the escarpment of Messinian strata on the north of the valley. Around Cantona, a centre of tectonic uplift, future river captures are imminent (Fig. 5b) as a result of the aggressive behaviour of these streams stimulated by the relatively high uplift rates (in excess of 0.16 mm a^{-1}; Mather 1991; Weijermars et al. 1985).

6 CONCLUSIONS

Even in an area subjected to tectonic activity, such as the Betics of southeast Spain, the original topography will have a significant influence on the final drainage pattern. As the amount of tectonic activity increases, so this pattern will become more significantly modified. In the Sorbas basin, with general uplift rates during the development of the fluvial system in the region of 0.16 mm a^{-1}, the most significant of the drainage pattern modifications has occurred via river capture. Regional uplift has been influential in affecting base levels and leading to overall drainage network modification (incision and river capture). More intensive structural controls appear to be only of localised significance to the drainage network development.

7 ACKNOWLEDGEMENTS

AEM would like to thank NERC grant GT4/87/GS/53 (University of Liverpool) and Worcester College Research and Development fund for the fieldwork support. We also acknowledge Bill and Lindy of Cortijo Urra Field Study Centre, Sorbas, for their hospitality.

REFERENCES

Amor, J.M. & F. Florschütz 1964. Results of the preliminary palynological investigation of samples from a 50 m boring in Southern Spain. *Bolétin de la Real Sociedad Española de Historía Natural (Geología)* 62:251-5.

Blair, T.C. 1987. Tectonic and hydrologic controls on cyclic alluvial fan, fluvial, and Lacustrine rift-basin sedimentation, Jurassic-lowermost Cretaceous Todos Formation, Chiapas, Mexico. *Journal of Sedimentary Petrology* 57:845-862.

Bourrouilh, R. & D.S. Gorsline 1979. Pre-Triassic fit and Alpine tectonics of continental blocks in the western Mediterranean. *Geological Society of America Bulletin* 90:1074-1083.

Bousquet, J.C. 1979. Quaternary strike-slip faults in southeastern Spain. *Tectonophysics* 52:277-286.

Coppier, G., P. Griveaud, F.D. de Laroziere, C. Montenat & Ph. Ott d'Estevou 1989. Example of Neogene tectonic indentation in the Eastern Betic Cordilleras: the Arc of Aguilas (Southeastern Spain). *Geodinamica Acta* 3:37-51.

Dronkert H. 1976. Late Miocene evaporites in the Sorbas basin and adjoining areas. *Mem. Soc. Geol. It.* 16:341-361.

Hall, S.H. 1983. Post Alpine tectonic evolution of southeast Spain and the structure of fault gouge. Ph.D thesis, University of London.

Haq, B.U., J. Hardenbol & P.R. Vail 1987. Chronology of fluctuating sea levels since the Triassic (250 million years ago to the present). *Science* 235:1156-1167.

Harvey, A.M. 1987a. Patterns of Quaternary aggradational and dissectional landform development in the Almeria region, southeast Spain: a dry region tectonically active landscape. *Die Erde* 118:193-215

Harvey, A.M. 1987b Alluvial fan dissection: relationships between morphology and sedimentation. In L. Frostick & I. Reid (eds.), *Desert Sediments: Ancient and Modern.* Geological Society Special Publication 35:87-103.

Harvey, A.M. 1990. Factors influencing Quaternary alluvial fan development in southeast Spain. In A.H. Rachocki & M. Church (eds.), *Alluvial Fans: A field approach.* John Wiley & Sons, 247-269.

Harvey, A.M., S.Y. Miller & S.G. Wells (this volume). Quaternary soil and river terrace evolution in the Aguas/Feos river systems, Sorbas basin, southeast Spain.

Harvey, A.M. & S.G. Wells 1987. Response of Quaternary fluvial systems to differential epeirogenic uplift: Aguas and Feos River systems, south-east Spain. *Geology* 15:689-693.

Mather, A.E. 1991. Caenozoic drainage evolution of the Sorbas basin, southeast Spain. Ph.D Thesis, University of Liverpool.

Mather, A.E. 1993a. Basin inversion : some consequences for drainage evolution and alluvial architecture. *Sedimentology* 40, 1069-1089.

Mather, A.E. 1993b. Evolution of a Pliocene fan delta: links between the Sorbas and Carboneras Basins, southeast Spain. In L. Frostick & R. Steel (eds),*Tectonic controls and signatures in sedimentary successions.* IAS Special Publication 20, 277-290.

Mather, A.E., A.M. Harvey & P.J. Brenchley 1991. Halokinetic deformation of Quaternary river terraces in the Sorbas basin, southeast Spain. *Zeitschrift Fur Geomorphologie,* Suppl. 82:97-97.

Mather, A.E. & R.K. Westhead 1993. Plio/Quaternary strain of the Sorbas basin, southeast Spain: evidence from sediment deformation structures. In L.A. Owen, I. Stewart & C. Vita-Finzi, *Neotectonics, Recent Advances.* Quaternary Proceedings 3:17-24.

Miller, S.Y. 1991. Soil chronosequences and fluvial landform development: studies in southeast Spain and NW England. Ph.D Thesis, University of Liverpool.

Montenat, C., Ph. Ott d'Estevou, F.D. de Larouziere & P. Bedu 1987. Originalité géodynamique des bassins néogène du domaine bétique oriental. *Total Compagnie Francaise des Pétroles, Paris.* Notes et Mémoirs 21:11-41.

Oberlander, T.M. 1985. Origin of drainage transverse to structures in orogens. In M. Morisawa & J.T. Hack (eds), *Tectonic Geomorphology.* Unwin, 155-82.

Ott d'Estevou, Ph. 1980. Evolution dynamique du basin Neogene de Sorbas (Cordilleres Betiques Orientales, Espagne). Ph.D. Thesis, University of Paris VII, Paris. 264p

Platt, J.P. 1982. Emplacement of a fold-nappe, Betic orogen, southern Spain. *Geology* 10:97-102.

Sabelberg, U. 1977. The stratigraphic record of late Quaternary accumulation series in southwest Morocco and its consequences concerning the pluvial hypothesis. *Catena* 4:204-215.

Sanz de Galdeano, C. 1990. Geologic evolution of the Betic Cordilleras in the Western Mediterranean, Miocene to the present. *Tectonophysics* 172:117-119

Schumm, S.A., M.P. Mosley & W.E. Weaver, 1987. *Experimental Fluvial Geomorphology.* Wiley.

Weijermars, R. 1985a. In search for a relationship between harmonic resolutions of the geoid, convective stress patterns and tectonics in the lithosphere: a possible explanation for the Betic-Rif orocline. *Phys. Earth Planet. Inter.* 37:135-148.

Weijermars, R. 1985b. Uplift and subsidence history of the Alboran basin and a profile of the Alboran Diapir (W. Mediterranean). *Geologie en Mijnbouw* 64:349-356.

Weijermars, R. 1991. Geology and tectonics of the Betic Zone, southeast Spain. *Earth Science Reviews* 31:153-236.

Weijermars, R., Th.B. Roep, B. Van den Eeckout, G. Postma &

K. Kleverlaan 1985. Uplift history of a Betic fold nappe inferred from Neogene-Quaternary sedimentation and tectonics (in the Sierra Alhamilla and Almería, Sorbas and Tabernas Basins of the Betic Cordilleras, southeast Spain. *Geologie en Mijnbouw* 64:397-411.

CHAPTER 7

Tectonics versus climate: An example from late Quaternary aggradational and dissectional sequences of the Mula basin, southeast Spain

A.E. MATHER
Department of Geographical Sciences, University of Plymouth, Drake Circus, Plymouth, Devon, UK

P.G. SILVA and J.L. GOY
Departamento de Geología, Facultad de Ciencias, Universidad de Salamanca, Salamanca, Spain

A.M. HARVEY
Department of Geography, Roxby Building, University of Liverpool, Liverpool, UK

C. ZAZO
Departamento de Geología, Facultad de Ciencias Natural, CSIC, Jose Guitierrez Abascal, Madrid, Spain

ABSTRACT: The affect and effect of climate and tectonics are often difficult to unravel in the sedimentary record. The Quaternary of the Mula basin, southeast Spain records the dissectional and aggradational behaviour of a fluvial system which is clearly independent of eustatic and anthropogenic controls. The sedimentary basin is located in a semi-arid area within the internal, left-lateral shear zone of the eastern Betic Cordilleras of southeast Spain. It is defined to the north and south by SW-NE orientated faults, the former of which delineates the boundary between the external and internal zones of the Betics. During the Late Pliocene the basin was subjected to north-south compression and uplift. Further intermittent uplift of the basin led to the initiation of the principal drainage network, the Rio Mula.

The installation and evolution of the Mula fluvial system was associated with the construction of 6 main fluvial terraces during 4 principal stages of drainage development: (1) Early aggradation (Late Pliocene/Early Pleistocene); (2) Dissection of up to 60 m, installing the Palaeo-Mula Valley (Early-Middle Pleistocene); (3) Large scale Late Pleistocene (pre Würm) aggradation infilling the palaeovalley of stage 2; (4) Holocene dissection of some 36 m down to the present day channel.

It is evident that the development of the aggradational and dissectional sequence reflects both climate and base level controls. These provide the overall framework on which was superimposed modification by local tectonic activity (during stage 3). The main part of stage 3 aggradation was triggered by the activity of a NNE-SSW strike-slip fault orientated across the path of the Rio Mula. Seismic activity along this fault stimulated mass-movement which impeded the fluvial system and created a palustrine palaeoenvironment upstream of the fault. The stage 4, Holocene dissection was stimulated by recoupling of the Rio Mula with the main drainage system, the Segura.

In particular, this study highlights the value of combining sedimentological and morphological perspectives and indicates the rapid response a semi-arid fluvial system can effect in response to external controls (36 m of incision over the Holocene).

KEYWORDS: Betics, aggradation, dissection, tectonics, base level, climate, southeast Spain.

1 INTRODUCTION

There is often much controversy over the relative impact of controls such as climate, sea level, tectonics and anthropogenic activity on the development of fluvial systems. Changes in the long-term record can often be effected by very different mechanisms which have very similar signatures (eg climate and tectonics; Frostick and Reid 1989; climate and sea level; Blum 1990). In order to better understand shifts in system behaviour stimulated by long-term controls it is necessary to combine information from the geological, morphological and structural aspects of sedimentary basin evolution in environments where the number of potential controlling variables are restricted.

The Mula basin of the Murcia Province, southeast Spain has good exposure through the sedimentological and morphological Quaternary fluvial sequence of the palaeo-Rio Mula. This enables the Quaternary to recent aggradational and dissectional behaviour of the fluvial system to be determined. Over this period the Rio Mula has been connected to the Rio Segura which has acted as the local base level for the Rio Mula. The effect of sea level fluctuations have thus been buffered by the lower reaches of the Rio Segura. In addition the development of the main dissectional and aggradational phase (Pleistocene) was before significant anthropogenic impact was felt within the Mediterranean region during the Holocene (Bell and Walker, 1992). The basin is however located within a tectonically active area of the Betic Cordillera and will have been subjected to the Quaternary climatic fluctuations which occurred over this time period (Harvey 1984; Amor and Florschütz 1964).

It is the aim of this paper to examine the evolution of the Rio Mula from installation in the early Quaternary to the present day, combining evidence from fluvial morphology and sedimentology in order to determine the relative impact of climatic and tectonic controls on the dissectional and aggradational behaviour of the system.

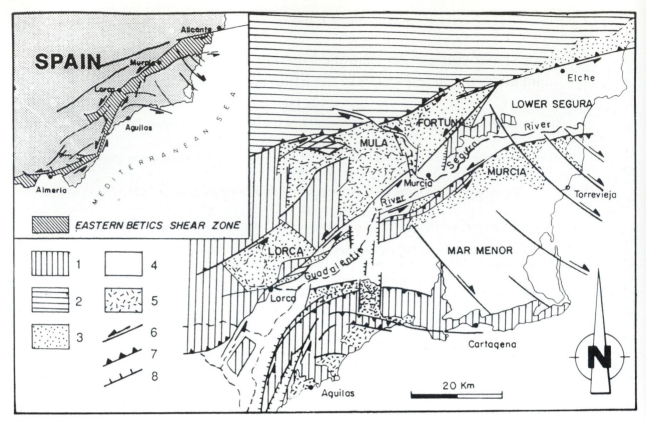

Figure 1. Location of the Mula basin in relation to regional tectonic framework, and other dissectional and aggradational Quaternary basins: 1. Internal Betic Zone; 2. External Betic Zone; 3. dissection dominated Quaternary basins; 4. aggradation dominated Quaternary basins; 5. Littoral basins; 6. strike-slip fault (left-lateral illustrated); 7. reverse fault; 8. normal fault.

2 REGIONAL SETTING

The Mula basin is located in the eastern part of the Betic Cordilleras (Fig. 1) and is one of a number of sedimentary basins which were generated in the eastern Betic Shear Zone during the Late Neogene. The eastern Betic Shear Zone is a large, left-lateral transcurrent corridor that crosses this part of the Betic Cordilleras from Almeria in the south to Alicante in the north, and has a SW-NE general orientation (Fig. 1).

The present morphological appearance of the shear zone is that of a sigmoidal, morphostructural corridor in which the main Quaternary basins are located (Goy et al. 1992). Late Neogene deformation in this transcurrent zone gave rise to intense magmatic activity and the formation of strike-slip marine basins (Montenat et al. 1987). These Neogene basins evolved along both sides of the present corridor, in which ancient Betic palaeomassifs were located separating the individual basins (Montenat 1973; Larouziere et al. 1988).

During the Pliocene important palaeogeographical changes occurred across the whole area. The stress field suffered a rotation from N170°E to N150°E (Montenat et al. 1987) inducing the generation of subsident troughs (aggradation-dominated Quaternary basins; Fig. 1) in those regions previously occupied by the ancient Betic

palaeomassifs (Silva et al. 1992) and triggering the uplift and dissection of Late Neogene basins located on both sides of the present corridor (dissection dominated Quaternary basins; Fig. 1).

3 BASIN STRUCTURE

The Mula basin is bounded to the north and south by two of the most important left-lateral strike-slip faults of the eastern Betics – the North Betic fault and the Lorca-Alhama faults respectively (Fig. 2). These faults have a northeast-southwest orientation and determine the boundary between the external and internal zones of the Betics in this area.

The eastern limit of the Mula basin is defined by a NW-SE orientated fault, the Villanueva fault (Fig. 2), a right-lateral strike-slip fault which controls the path of the Rio Segura (the dissectional base level of the Mula basin). The western margin of the Mula basin is dominated by the Sierra Espuña.

In the centre of the basin Neogene deposits were deformed along two locally important NNE-SSW trending left-lateral normal faults – the Limite fault and the Tollos Rodeos fault (Fig. 2). Both faults were active during the deposition of the Tortonian and Messinian

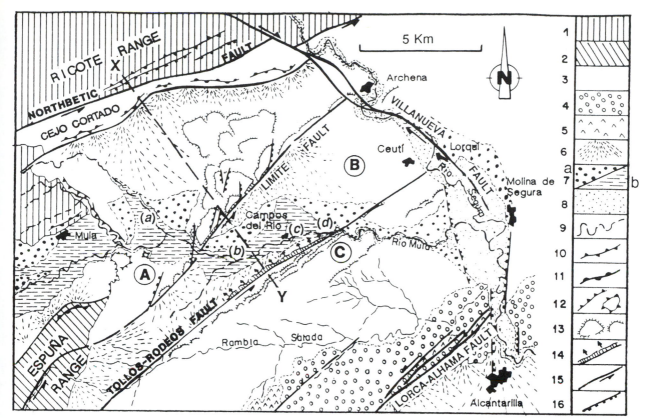

Figure 2. Tectonic and morpho-sedimentary sketch map of the Mula basin. Line XY indicates line of cross-section illustrated in Figure 3. Sedimentary domains are indicated by A (Mula area), B (Campo del Rio corridor) and C (Rodeos/Florida). Locations of sites discussed in the text are: (a) Baños de Mula; (b) Albudeite; (c) Campos del Rio; (d) Los Rodeos. Other key features: 1. Betic basement; 2. Subetic basement; 3. Tortonian and Messinian marine deposits; 4. Plio-Messinian fluvio-marine deposits (alluvial fan and fandelta); 5. Neogene volcanics; 6. Alluvial fans and lower Pleistocene colluvium; 7. (a) Early Pleistocene high fluvial terrace & (b) Late Pleistocene (pre-Würm) palaeovalley filling; 8. Upper Pleistocene/Holocene fluvial deposits; 9. river. 10. ridge; 11. marginal relief; 12. scarps/buttes; 13. cuestas/mesas; 14. Landslips; 15. transverse fault; 16. normal fault.

sediments, defining a structural corridor (the Campos del Rio corridor; Fig. 2) within the basin (Loiseau et al. 1990).

4 BASIN FILL

The Neogene sedimentary fill of the Mula basin is dominated by Tortonian and Messinian deposits. These are mainly interbedded marine marls and fine-grained sandstones with occasional beds of limestone and conglomerate (Jerez Mir et al. 1974). Older Tortonian sediments are present in the Cejo Cortado Range at the northern border of the basin (Figs 2 and 3).

Although the ancient sedimentary record contains important internal unconformities, the strata generally strike at N30°E – N50°E with a general dip towards the south-east (Loiseau et al. 1990). Influenced by the activity of SW-NE trending strike-slip faults present within the basin, the depocentre of the basin fill shows a progressive migration towards the southeast from the Tortonian until the Pliocene.

As a result of this progressive migration of the focus of sedimentation the Plio-Messinian deposits are largely restricted to the southeast of the basin. They are mainly dominated by fan delta to alluvial fan systems (Loiseau et al. 1990) fed by the ancient Betic palaeo-massifs which existed to the south. Fossil mammals found in these deposits (Agusti et al. 1985) indicate an Early Pliocene age for the uppermost parts of the sequence.

Further uplift of the basin in relation to the adjacent subsident corridor (Segura and Guadelentin valleys, Fig. 1) facilitated the installation and incision of the drainage networks in the Mula basin. The subsident corridors thus acted as dissectional base-levels.

Within the Mula basin the two structural alignments of the Mula Range and Los Tollos-Rodeos are important intra-basinal reliefs which control the distribution of the Quaternary deposits. Where they cross the path of the Rio Mula they geographically divide the Quaternary deposits into three morphological and sedimentological domains (Figs 2 and 4):

1. The Mula area (A in Fig. 2), west of the Mula Range.

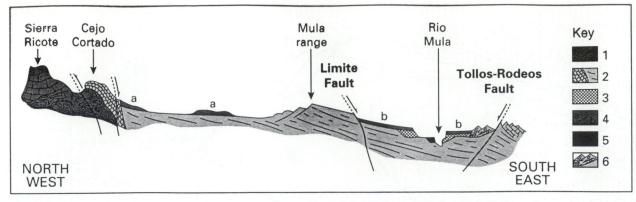

Figure 3. Schematic NW-SE cross-section of the Mula basin (see Fig. 2 for location). 1. Betic Basement; 2. Tortonian marine deposits; 3. Late Pleistocene palaeovalley filling; 4. colluvium: (a) lower Pleistocene and (b) upper Pleistocene; 5. Holocene Fluvial terraces; 6. Landslips. Note sketch is not to scale.

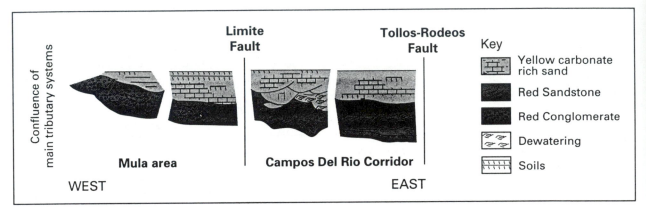

Figure 4. Schematic representation of the spatial arrangement of the facies associated with the +32-36 m terrace in the Mula area and Campo del Rio corridor (for explanation see text). Not drawn to scale. Basal erosion surface is set in Tortonian sediments.

2. The Campos del Rio corridor (B in Fig. 2), defined by the Limite fault to the west and Tollos-Rodeos fault to the east.

3. Los Rodeos/Florida (C in Fig. 2), the region east of the Tollos-Rodeos lineament.

5 CLIMATE

The fluvial system developed under a semi-arid Pleistocene climate. Although this climate was dry, especially during northern European glacials, it was more humid than the present (Amor and Florschüzt 1964). In general, alluvial fan sequences in southeastern Spain appear to indicate a progressive aridification over the Quaternary period (Harvey 1984). Today the climate of the Mula area is still semi-arid with precipitation totals ranging from 300 mm in the headwaters (Espuña Range) to 270 mm in the lower parts of the basin (Romero Diaz and Lopez Bermudez 1985).

6 QUATERNARY TERRACE SEQUENCE

The drainage system of the Mula basin was initiated in the Pliocene. The present drainage network is deeply incised into both the weak Neogene sediments and the Pleistocene river terraces and feeds the axially draining Rio Mula, a tributary of the Rio Segura (Fig. 2).

The aggradational and dissectional history of the Rio Mula is reflected by the development of 6 main fluvial terraces located at c. +65 m, +40 m, +32-36 m, +15 m, +5 m and +2 m above the present river thalweg. These terraces generally show a simple staircase arrangement. The +2 m level can be considered the present day floodplain of the River Mula, and comprises fine-grained sands.

Deposits related to the +65 m, +40 m, +15 m and +2 m terraces are typically fluvial with thicknesses of up to 3 m and sediments reflecting gravel bar, channel and floodplain environments. The +65 m terrace can be correlated with the early Quaternary large alluvial platforms (7a and 6 respectively in Fig. 2) fed by the most important marginal reliefs of the basin (the western, Espuña and northern, Ricote ranges). This terrace thus reflects the first principal drainage base-level associated with the Mula basin. Near Baños de Mula the sediments of this terrace level differ from those elsewhere. In this area, located upstream of the Mula Range, the Limite fault crosses the path of the Rio Mula (Fig. 2). This fault has been dominated by strong hydrothermal activity during most of the Quaternary (Rodriguez Estrella 1979). Tufa and travertine deposits dominate the fluvial deposits in this area as a consequence of carbonate-rich groundwater associated with this fault zone (Romero Diaz et al. 1992).

It is the +32-36 m levels which appear anomalous within the general terrace sequence, in terms of thickness (up to 18 m), sedimentary characteristics (containing more palustrine than fluvial style sedimentation) and spatial distribution (only located upstream of the Los Tollos Rodeos fault zone in the Mula area and Campos del Rio corridor).

These fluvio-palustrine deposits are located within well defined palaeovalleys delineated by the +65 m and +40 m terrace levels. The top of the fluvio-palustrine sequence has been dated to Late Pleistocene, pre-Würm (Agusti et al. 1990; see below).

The top of this fill is buried by a thin veneer of colluvial deposits (4b in Fig. 3) sourced from the intrabasinal reliefs mentioned above. These deposits post-date the major phase of landslides along the western side of the Los Tollos Rodeos structural alignment, which were probably contemporaneous with, or older than the fluvio-palustrine deposits of the +32-36 m terrace level.

7 SEDIMENTOLOGY AND PALAEOENVIRONMENT OF THE +32-36 m TERRACE DEPOSITS

The facies of the +32-36 m terrace can be grouped into two areas according to their sedimentological and morphological domain (the +32-36 m terrace level is absent from the Los Rodeos/Florida area):

1. The Mula area (A in Fig. 2) where it is dominated by fluvial sediments with subordinate palustrine deposits at the top.

2. The Campos del Rio corridor (B in Fig. 2), dominated by palustrine deposits with only subordinate fluvial deposits, typically in the lower parts of the sequence.

7.1 *Mula area*

Sedimentology

The sediments in the west of the Mula area are dominated by red fluvial conglomerates and sandstones (Fig. 4). The conglomerates are mainly derived from the Sierra de Espuña and Cejo Cortado and comprise a series of lenticular, channel deposits of Gm (see Table 1 for lithofacies codes) with subordinate Gp. These indicate transportation of the conglomerates in gravel sheets and small, transverse bars (Smith 1974). The channels may be draped with sandstone which was probably the product of deposition from the high suspended sediment concentrations typical of flash floods (Reid and Frostick 1987). The coarser material is generally concentrated in palaeoval-

Table 1. Facies codes used in the text and diagrams (after Miall 1985).

Facies code	Lithofacies	Sedimentary structures
Gm	Massive or crudely bedded gravel	None
Gt	Gravel, stratified	Trough cross-stratification
Gp	Gravel, stratified	Planar cross-stratification
St	Sand, medium to very coarse	Trough cross-stratification
Sp	Sand, medium to very coarse	Planar cross-stratification
Sr	Sand, medium to very coarse	Ripple cross-lamination
Sh	Sand, medium to very coarse	Horizontal lamination
Sl	Sand, fine	Low angle (< 10°) cross-stratification

leys marginal to the main +32-36 m terrace level and in more western (upstream) localities. In central and eastern parts of the Mula area, further east, the red sandstones are overlain by laminated white/yellow sediments (carbonate rich sand in Fig. 4), both probably sourced largely from the local Tortonian and Messinian sands and marls. These contain facies of Sh and minor Sp. These are likely to reflect deposition away from the active channel.

Pedogenic modification

Sandstone sequences are typically of Sh, Sm (Table 1) and show evidence of pedogenesis: mottling, weak carbonate and iron accumulation (gradational colour changes) and root casts. These palaeosols have a lack of gley (red mottles in a grey matrix) and pseudogley features (grey mottles with red haloes, Retallack 1990) implying that they were dominantly well drained. In the east of the area (Baños de Mula, Fig. 5a) mottling within the sediments indicates a change from better-drained soils in the lower profile (weak pseudogleys) to more poorly drained soils (gleys) in the upper part (Fig. 5a).

Fauna

In the uppermost part of the eastern sections grey silts and clays form two dominant horizons in which mammalian remains of Late Pleistocene but pre-Würm age (Agusti et al. 1990) have been found.

Palaeoenvironment

The deposits of the +32-36 m terrace indicate that the river system was braided and occupied a well defined palaeovalley. Sediment supply was dominated by an abundance of sands from the Tortonian and Messinian strata. Conditions were typically well drained but in the lower reaches there is the suggestion that local waterlogging of the sediments occurred (gley palaeosols).

7.2 *Campos del Rio corridor*

Sedimentology

Conglomerates are relatively rare and typically concentrated in marginal and more western (upstream) parts of the +32-36 m terrace deposits. In the west of the Campos del Rio corridor, near Albudeite, the basal parts of the

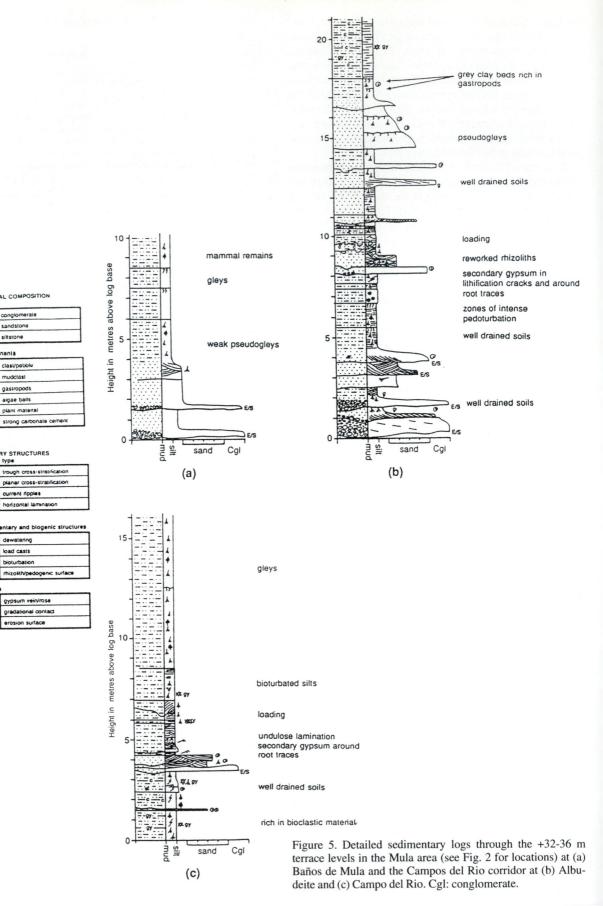

Figure 5. Detailed sedimentary logs through the +32-36 m terrace levels in the Mula area (see Fig. 2 for locations) at (a) Baños de Mula and the Campos del Rio corridor at (b) Albudeite and (c) Campo del Rio. Cgl: conglomerate.

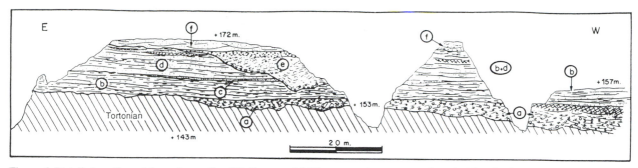

Figure 6. Detailed section of the palaeovalley fill, Campos del Rio: a. conglomerates and sands (Gp, Gt, Sp, St) of fluvial bars; b. laminated and bioturbated sheetflood deposits (Sl); c. sands (Sp, Sr, and Sl) with abundant freshwater gastropods; d. sands (Sp, St) with freshwater gastropods and containing dewatered sediments (dish and pillar structures); e. sand channel fill (Sh, Sp, St); f. colluvium (St, Sp).

sections are dominantly conglomerates and red sandstones erosive into the underlying Tortonian sediments (Fig. 4). These red sediments are overlain by, and intercalated with white sands in small (2 m deep, 8 m wide) channels belonging to the same system. Within these sections facies of Sp, Sh, and St are evident. Ripples and megaripples are present in some channels (Albudeite, Fig. 5b) which contain reworked rhizoliths. The facies suggest the presence of a sand-dominated braid system with small slip-face bars developed in the channels. The dominance of horizontal lamination and sandstone drapes suggests ephemeral streams (Reid and Frostick 1989). More significant, larger erosion surfaces are also present (Fig. 6). Further east the sequence is dominated by yellow silts and sands. Two prominent grey, illite-rich beds (Fig. 5b) can be detected towards the top of the sections. These clay-rich beds must have been deposited in standing bodies of water.

Pedogenic modification
Pedoturbation is common throughout the sequence and indicates a change from moderately well drained soils (ferruginous mottles) to pseudogleys (mottles with reduced, grey cores and ferruginous haloes) in the more upstream localities (Albudeite, Fig. 5b).

The uppermost parts of the sequence at Campos del Rio are dominated by poorly exposed gley palaeosols. These overlie a major erosion surface inset within the Quaternary sediments (Fig. 6). Root traces are commonly infilled with secondary gypsum or ferruginous sand. Secondary gypsum associated with root traces and gypsum roses are also common (Fig. 5b and c).

Biogenic reworking
At Los Rodeos large (dm-wide, metre-scale in vertical extent) burrows, some containing weak spreite (Ekdale et al. 1984; internal laminations resulting from the excavation/feeding behaviour of the animal) are evident within the dewatered red sandstone, typically developed from the planar erosion surface which separates the lower, dewatered red sandstones and overlying laminated buff coloured silts. Features such as this typically reflect the burrowing activity of arthropods (Ekdale et al. 1984).

Fauna
In the lower parts of the sequence, at Campos del Rio a number of Quaternary freshwater gastropods (*Lymnaea*, *Bythinia* and *Hydrobia* being the most common; C. Paul, pers. comm. 1993) are present.

At Albudeite, balls of freshwater algae are present in the lower parts of the section (Fig. 5b). Towards the top of the section, two prominent dark grey illite rich levels are developed, rich in gastropods (*Melanopsis*, *Theodoxos* and *Hydrobia* being the most common).

Both faunal assemblages suggest the presence of well oxygenated, fairly permanent bodies of water. The Albudeite assemblage generally prefers more agitated, flowing water but is more tolerant of saline conditions. *Theodoxos* in particular usually prefers a habitat containing conglomeratic material.

Soft sediment deformation
The Campos del Rio corridor is characterised by a bed of soft-sediment deformation (Fig. 4) which forms a discrete, well defined unit in the east (Los Rodeos; (d) on Fig. 2) and becomes less clear further west (Albudeite and Campos del Rio; (b) and (c) on Fig. 2).

In the west the deformation affects a unit 1-2 m thick and c. 10 m below the surface of the +32-36 m level. The unit contains undulose laminations and cross-stratification which locally lose definition. Pseudonodules (Allen 1982) are developed at Albudeite and weakly defined pillar structures (Lowe 1975) can be observed at Campos del Rio.

At the extreme east of the Campos del Rio corridor (Los Rodeos) the zone is some 3-4 m thick and clearly defined in red sandstones overlain by horizontally laminated cream coloured silts. The red sandstone contains broad, draped, shallow troughs. The drapes thicken into the centre of the troughs and lose definition in places. Laminations traced laterally show similar characteristics, indicating that liquefaction and localised fluidisation (Lowe 1975) have modified the original sedimentary structures.

Palaeoenvironment
The Campos del Rio corridor comprises much finer sed-

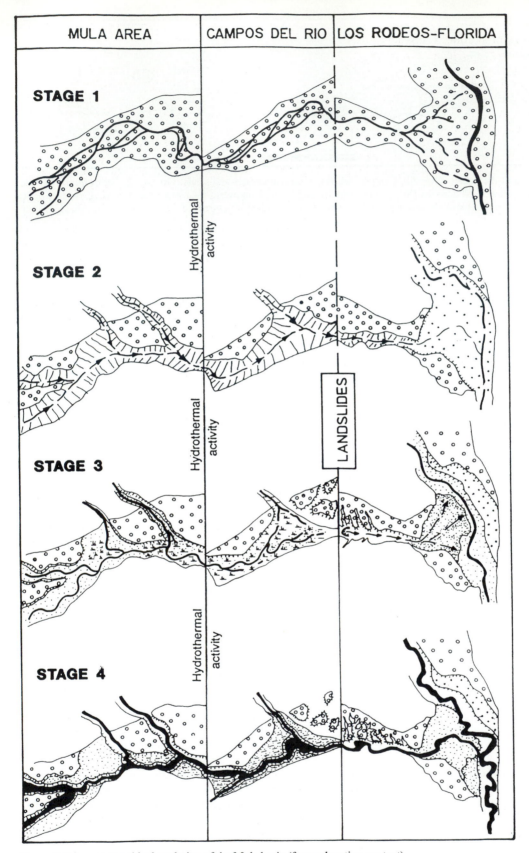

Figure 7. Palaeogeographical evolution of the Mula basin (for explanation see text).

iments than the Mula area, which may be expected as it lies in a more downstream location. The sedimentology of the Campo del Rio corridor suggests a sandy braid system which occupied a well defined palaeovalley. The dominance of horizontal lamination suggests mainly ephemeral flows. The fauna, however, implies the presence of a fairly permanent running water source, probably in a minor channel. Locally, standing bodies of water existed (Albudeite and Campo del Rio grey clay beds). The gastropod assemblage found in the grey clays at Albudeite usually prefer fast flowing water, and *Theodoxos* in particular prefers a mud free environment. It seems likely, therefore, that the gastropods were flushed into the standing body of water from an adjacent flowing water source. The fact that the Albudeite assemblage is also more tolerant of brackish conditions than those gastropods found lower in the sequence at Campos del Rio, suggests that this water may have been rich in salts. The carbonate rhizoliths indicate an abundance of carbonate.

Temporally the palaeosols indicate a change from well drained soils to hydromorphic soils. Pseudogleys develop in more upstream locations (Albudeite) and gleys further downstream (Campo del Rio).

The overall setting, then, appears to be of a sandy braided system, with some channels occupied by permanent, if low discharge, low quality water, feeding into local pools of standing water which were periodically evaporated and modified by pedogenesis (i.e. a palustrine environment).

The presence of a well defined package of sediments exhibiting soft-sediment deformation across the area (Fig. 4) suggests a period of seismic shocking of water rich sediments during the last major aggradation phase.

8 PALAEOGEOGRAPHICAL EVOLUTION

Using the palaeoenvironmental information provided by the geomorphology and sedimentology of the +32-36 m terrace level it is possible to reconstruct the palaeogeographical evolution of the Mula fluvial system from the early Quaternary to the present day. This evolution can be characterised by four principal stages (Fig. 7).

Stage 1. Early aggradation
This stage was dominated by the creation of the first local base level in the centre of the basin (+65 m terrace). This was an extensive braidplain connected to the alluvial aprons of the main marginal reliefs, located at the northern (Cejo Cortado Range) and at the western (Espuña Range) boundaries of the basin (Fig. 2). The ancient palaeo-Mula channel was a poorly incised axial drainage in the centre of the basin. The early aggradation appears to date from the Late Pliocene/Early Pleistocene.

Stage 2. Dissection
Stage 2 was dominated by an incision of up to 60 m. During this dissectional stage smaller aggradational phases also occurred creating the +40 m terrace. It is during this period that the main, well defined palaeo-

Mula valley developed. Dissection appears to have dominated the early-mid Pleistocene.

Stage 3. Late Pleistocene aggradation
This corresponds to the filling of the palaeovalley generated during stage 2, creating up to 18m thick terraces associated with the +32-36 m level. The distribution of sedimentary facies related to this palaeovalley infill indicates the presence of three main sedimentary domains, defined by the two main structural lineaments that cross the Mula Valley. Floodplain deposits dominate in the Mula zone with palustrine deposits only becoming significant adjacent to where the palaeo-Mula crosses the Limite Fault Zone. The Campos del Rio corridor was dominated by significant fluvio-palustrine sedimentation, whilst the Los Rodeos/Florida area was only intermittently connected to the upstream zone which acted as a sediment trap, giving rise to fluvial deposits of finer calibre in this zone. Filling of this palaeo-Mula valley terminated in the Late Pleistocene, as is indicated from mammal (Agusti et al. 1990) and gastropod fauna found at the top of this sequence.

Stage 4. Holocene dissection
During stage 4 dissection occurred down to the present day channel, with incision of at least 36 m. During this stage the +15 m and +5 m terraces were generated, reflecting minor aggradational stages.

9 CONTROLS ON DRAINAGE EVOLUTION

The factors which influence the drainage development within the Mula basin can be identified as (a) those which influence the location of the axial fluvial system, (b) those which influence sediment production from the source area and thus the aggradational and dissectional behaviour of the fluvial system and (c) those which influence the local and external base levels. These factors can be grouped into tectonic, climatic and base-level related controls.

9.1 *Tectonics*

The nature of the tectonic controls can be sub-divided into those which were inactive during the development of the Mula drainage system but which indirectly influenced fluvial development in some way (passive controls), and those which were active contemporaneous with the development of the system (active controls).

Passive controls
The NE-SW fault systems of the Mula basin were important in topographically subdividing the Quaternary basin into the three sedimentary domains outlined above. The structurally-defined topography enabled only restricted fluvial communication between these domains, facilitating the development of the different palaeoenvironments.

In addition the pre-existing Limite fault was significant

in accounting for the distribution of facies within individual sedimentary domains, in particular in the Campo del Rio corridor. In the later Pleistocene travertines were deposited from groundwater associated with the fault splays of the Limite fault Zone. Today the area is associated with hot springs (36°C) at Baños de Mula (Romero Diaz et al. 1992). The location of hot springs in this area during most of the evolution of the Mula drainage system may well account for the concentration of carbonate rich palustrine sediments and apparently poorer water quality (indicated by the gastropods) in this area during the sedimentation of the stage 3 deposits.

Active controls

In the case of the Tollos-Rodeos lineament, activity of this structure approximately contemporaneous with stage 3 aggradation is evident, stimulating the generation of large landslides. These pre-date the terminal phase of stage 3 aggradation but their exact age is unclear. It is likely that the slides are of similar age to the main aggradation. This fault is also associated with more significant surface expression than the Limite fault, despite being developed in easily erodible marls and sandstones, suggesting its more recent activity.

The sedimentology of the stage 3 aggradation hints at some form of cyclic shocking focused around the Tollos-Rodeos fault. Dewatered sediments form a defined sedimentary unit which can be traced in the finer sediments from Los Rodeos to the lower parts of the Mula area (Fig. 4). The intensity of this deformation decreases with distance from the Tollos Rodeos fault. There is nothing to imply a sedimentary origin for the dewatering of the sediments (e.g. rapid sedimentation, Lowe 1975). The temporal and geographical restriction of this unit implies that this feature was stimulated by seismic activity focused in the vicinity of the Tollos-Rodeos fault, indicating seismic activity contemporaneous with development of the fluvial system.

9.2 Climate

Climatic change can have significant implications for semi-arid areas (Frostick and Reid 1989). In SE Spain major episodes of sediment production and therefore aggradation within fluvial and alluvial systems is typically associated with arid Pleistocene glacials (Amor and Florschütz 1964) whilst episodes dominated by dissection seem to be related to interglacials (Harvey, 1990).

It is thus possible to relate some of the last major phase of sediment production (stage 3 aggradation in the Palaeo-Mula valley) to such late Quaternary fluctuations. As a corollary to this, most of the stage 4 dissection of the Rio Mula could be related to the Würm and Holocene dissectional episode which has been recorded in alluvial fan sequences (Harvey, 1984).

It is, however, difficult to envisage how climate alone could account for such features as the extensive bed of dewatering and segmentation of the sedimentary environments during the same time interval.

9.3 Base level

Relative base level is important in governing aggradational and dissectional behaviour within fluvial systems. No doubt many of the most recent terrace stages (+15 m, +5 m and +2 m) reflect changes in base level associated with the Rio Segura. However, the stage 3 deposits of the palaeo-Mula (+32-36 m) appear to reflect discontinuities within the system, particularly around the Tollos-Rodeos lineament. This seems to indicate the importance of local base levels and temporal disconnection of the Mula and Segura fluvial systems. Re-establishment of this connection with the Segura base level would give rise to base-level induced dissection of the palaeovalley fill.

10 CONCLUSIONS

The development of the Mula basin drainage is characterised by 4 main stages of evolution. Of these, the Late Pleistocene stage 3 aggradation is associated with the development of an extensive depositional surface, 32-36 m above the current river bed. This level is only present upstream of the Tollos-Rodeos fault zone which intersects the path of the Rio Mula. It is worth at this point reiterating two of the key features associated with this stage 3 aggradation:

1. The development of a temporally and geographically restricted bed of dewatering (approximately coeval with landslides generated on the Tollos-Rodeos fault and most prolific in the vicinity of the fault);

2. A change in palaeoenvironmental conditions (from fluvial to palustrine dominated) upstream of the Tollos-Rodeos fault.

Fault activity along the Tollos-Rodeos fault could account for these features. Mass movement along the fault, stimulated by seismic activity would be sufficient to impede drainage within the palaeo-Mula system, generating local base levels and dewatering existing sediments. The stage 4 Holocene dissection down to the present stream bed was probably stimulated by recoupling of the Rio Mula and Rio Segura fluvial systems.

This example serves to illustrate the value of combining morphological, sedimentological and structural data to examine long-term controls on a developing fluvial system. It appears that whilst some aspects of the aggradation and dissection of the Mula basin fluvial system may be explained by climatic and base level influences, these only provided the general framework which was subject to modification by local tectonic activity. Of particular significance to other studies is the rapidity with which a semi-arid river such as the Rio Mula has responded to changes in its environment (36 m of incision over the Holocene, facilitated by the re-excavation of the palaeo-valley of the Quaternary Rio Mula).

11 ACKNOWLEDGEMENTS

This research has been supported by Spanish DGICYT

projects PB88-0125 and PB89-0049. AEM would like to thank Worcester College Research and Development fund for the fieldwork support, Professor C. Paul (University of Liverpool) for specimen identification and palaeoecological advice and Tim Absalom (University of Plymouth Cartography Unit) for assistance with the diagrams.

REFERENCES

Agusti, J., S. Moya-Sola, J. Gibert, J. Guillen & M. Labrador 1985. Nuevos datos sobre la biostratigrafía del Néogeno continental de Mucria. *Paleont. i. Evol.* 18:83-94.

Agusti, J., M. Freudenthal, J.L. Lacombat, E. Martin & C. Nageli 1990. Primeros micromamiferos del Pleistoceno superior de la Cuenca de Mula (Murcia, España). *Rev. Soc. Gel. España* 3:289-293.

Allen, J.R.L. 1982. *Sedimentary Structures.* Vol. II. Elsevier.

Amor, J.M. & F. Florschütz 1964. Results of the preliminary palynological investigation of samples from a 50 m boring in Southern Spain. *Bolétin de la Real Sociedad Española de Historía Natural (Geología)* 62:251-5.

Bell, M. & M.J.C. Walker 1992. *Late Quaternary Environmental Change: Physical and human perspectives.* Longman.

Blum, M.D. 1990. Climatic and eustatic controls on Gulf Coastal Plain fluvial sedimentation: An example from the Late Quaternary of the Colorado River Texas. *GCSSEPM Foundation Eleventh Annual Research Conference Program of Abstracts*, 71-83.

Ekdale, A.A., R.G. Bromley & S.G. Pemberton 1984. *Ichnology: Trace fossils in sedimentary and stratigraphy.* SEPM Short Course 15.

Frostick, L.E. & I. Reid 1989. Climatic versus tectonic controls of alluvial fan sequences: Lessons from the Dead Sea, Israel. *Journal of the Geological Society* 146:527-538.

Goy, J.L., P.G. Silva, G. Somoza, C. Zazo & T. Bardaji 1992. Morphological response to an intraplate transcurrent zone (Eastern Betics, SE Spain). Quaternary basins types and neotectonics. *Quaternary Proceedings.*

Harvey, A.M. 1984. Aggradation and dissection sequences on Spanish alluvial fans: Influence on morphological development. *Catena* 11:289-304.

Harvey, A.M. 1990. Factors influencing Quaternary alluvial fan development in southeast Spain. In A.H. Rachocki & M. Church (eds), *Alluvial fans: A field approach.* John Wiley & Sons, 247-269.

Jerez Mir, L., F. Jerez Mir & G. Garcia Monzon 1974. Mapa Geológico de España escala 1:50 000 (MAGNA). Hoja 912. Mula IGME.

Larouziere, De F., J. Bolze, P. Bordet, J. Hernandez, Ch. Montenat & Ph. Ott d'Estevou 1988. The Betic segment of the lithospheric Trans-Alboran shear zone during the Late Miocene. *Tectonophysics* 152:41-52.

Loiseau, J., Ph. Ott d'Estevou & Ch. Montenat 1990. Le secteur d'Archena-Mula. *Doc. et trav. IGAL.* 12-13:287-301.

Lowe, D.R. 1975. Water Escape structures in coarse grained sediments. *Sedimentology* 22:157-204.

Miall, A.D. 1985. Architectural-element analysis: A new method of facies analysis applied to fluvial deposits. *Recognition of fluvial depositional systems and their resource potential.* SEPM short course No. 19, 33-81.

Monenat, Ch. 1973. Les formations Néogenes et Quaternaires du Levant Espagnol. Thèse d'Etat Université Paris-Orsay, 1170 pp.

Montenat, Ch., Ph. Ott d'Estevou & P. Masse 1987. Tectonic sedimentary characters of the betics Neogene basins evolving in a crustal transcurrent shear zone (SE Spain). *Bull. Centre Rech. Explor. Prod. Elf-Aquitaine* 11:1-22.

Retallack, G.J. 1990. *Soils of the past: An introduction to palaeopedology.* Unwin.

Reid, I. & L.E. Frostick 1987. Flow dynamics and suspended sediment properties in arid zone flash flood. Hydrological Processes 1:239-253.

Reid, I. & L.E. Frostick 1989. Channel form, flows and sediments in deserts. In D.S.G. Thomas (ed), *Arid Zone Geomorphology.* Belhaven & Halsted, 117-135.

Rodriguez Estrella, T. 1979. Geología e hidrogeología del sector Alcaraz-Lietor-Yeste (probacete). Sintetis geologica de la zona Prebetica. *Mem. Inst. Geol. Min. España* 97:560 pp.

Romero Diaz, M.A. & F. Lopez Bermudez 1985. Los procesos de erosion de las Cuencas néogeno-cauternarias. *Guía de itinerarios geográficos de la Región de Murcia*, 85-97. AGE-Universidad de Murica, Murcia.

Romero Diaz, M.A., F. Lopez Bermudez, P.G. Silva, T. Rodriguez Estrella, F. Navarro Hervas, F. Diaz Del Olmo, J.L. Goy, C. Zazo, R. Baena Escudero, L. Somoza, A.E. Mather & F. Borja Barrera 1992. Geomorfología de las cuencas néogeno-cauternarias de Mula y Guadalentín. Cordilleras Béticas, Sureste de España. In *Estudios de Geomorfologia en España. Murcia 1992*, 749-786.

Silva, G., A.M. Harvey, C. Zazo & J.L. Goy 1992. Geomorphology, depositional style and morphometric relationships of Quaternary alluvial fans in the Guadelentín depression (Murcia, southeast Spain). *Z. F. Geomorph.* 36:325-341.

Smith, N.D. 1974. Sedimentology and bar formation in the Upper Kicking Horse River, a braided outwash stream. *Journal of Geology* 82:205-224.

CHAPTER 8

Late Quaternary controls on river behaviour in the eastern part of the Argive plain, eastern Peloponnese, Greece

K. GAKI-PAPANASTASSIOU and H. MAROUKIAN
Department of Geography and Climatology, University of Athens, Athens, Greece

ABSTRACT: In this study examples of river response to tectonic activity, differential erosion and climatic change are examined to the east of the Argive plain, in the eastern Peloponnese. The human influence on the fluvial environment is also considered and the fluvial deposits of the area are dated by archaeological means. The rivers Dafnorema, Megalo Rema and lower Ksirias are located in a tectonic depression, and are in an advanced stage of development as they have evolved on easily erodible lithological units generating low relief. The upper sections of the rivers Ksirias, Koulouvinas and Berbadiotis are developed on the limestone planation surface located north of the tectonic depression. In their middle and/or lower parts they are rejuvenated and form deep gorges. The reactivation of a fault along Havos stream, east of the Acropolis of Mycenae, blocked the stream forcing aggradation and the associated sediments have been dated to Mycenaean to Roman times. The presence of a Mycenaean dam in Megalo Rema demonstrates that most of the downcutting had already occurred before its construction and probably extends into the Late Pleistocene. A terraced channel fill unit of 2-3 m thickness is found in most of these rivers and has a post-Roman age.

KEYWORDS: River terraces, tectonism, erosion, climate, archaeogeology, late Quaternary, Argive plain, Greece.

1 INTRODUCTION

A significant percentage of the drainage systems of Greece are in a dynamic stage of development. In most cases this is the result of recent tectonic activity (uplift) which forces the rivers to actively dissect the newly formed highlands and create considerable relief. However, some parts of Greece have evolved differently from others, primarily because of differences in lithology. For example, marked contrasts in the rates of denudation between resistant limestones and easily erodible flysch formations result in completely different relief (e.g. Woodward et al. 1992). The climatic changes of the Quaternary have also contributed to the differentiation of the Greek landscape. Furthermore, Greece is an area that has been inhabited for thousands of years and human influence on the environment has been very prominent, with ancient impacts still in evidence today. All the above factors have contributed in some degree in the present study area along the eastern edges of the Argive plain in eastern Peloponnese. In this part of Greece the great Mycenaean civilization flourished from 1600 to 1100 BC and some of the best known classical sites are located here, namely Mycenae, Midea and Tiryns.

This paper documents the results of a recent study of the fluvial systems located in the east of the Argive plain. This investigation was concerned with examining the roles of tectonic activity, differential erosion and climatic change as well as human interference, in the development of the fluvial landscape. In order to interpret the palaeogeographic evolution of this area data were obtained mainly from topographic maps, air-photos and extensive geomorphological field surveys, which also included the study of several archaeological sites.

2 REGIONAL GEOLOGICAL SETTING

The Argive plain is a tectonic depression of Plio-Pleistocene age (Fig. 1). This study focuses upon the major river systems which are filling the eastern part of the depression with sediments. These are, from north to south; the Havos, Berbadiotis, Koulouvinas, Ksirias, Megalo Rema and Dafnorema rivers. The highlands located to the east of the Argive plain are composed mainly of massive limestones of Triassic to Cretaceous age and tectonically overlie, as they are thrust over, the younger flysch formations of Late Cretaceous age. At lower elevations there are Plio-Pleistocene marls and conglomerates of lacustrine, fluvial or flash flood origin. Finally, the lower parts of the study area are covered by various types of alluvial material of Quaternary age including alluvial cones and fans and recent floodplain sediments.

The major faults within the study area have a NW-SE direction bounding the eastern part of the tectonic depression of the Argive plain. In addition, secondary faults having roughly E-W and NE-SW directions are observed in the mountainous area. In many cases the most recent faults control the evolution of the main channels of the drainage networks of the area.

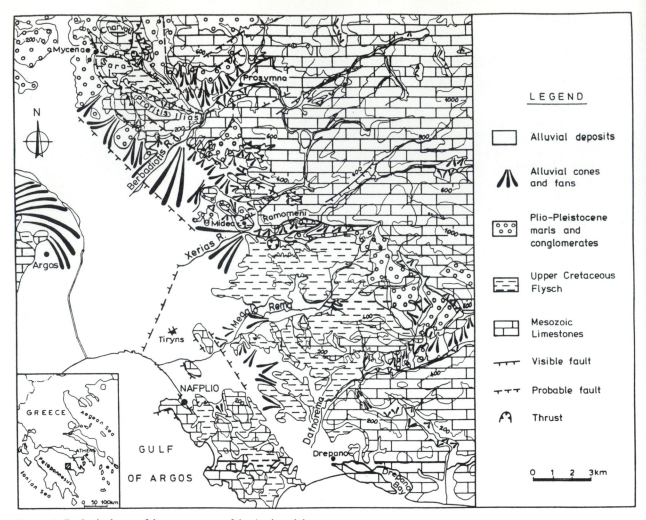

Figure 1. Geological map of the eastern part of the Argive plain.

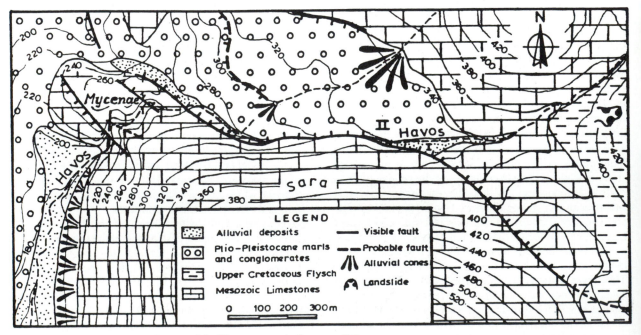

Figure 2. Large scale geotectonic map of the Havos stream in the area of Mycenae.

3 THE HAVOS STREAM

The Havos stream separates the hill of the Acropolis of Mycenae from the main mountain mass of Sara. A large part of this drainage basin is composed of Mesozoic limestones although there are also easily erodible flysch formations in the upper parts of the basin. In the north a small portion of the catchment is covered by Plio-Pleistocene marls and conglomerates. Alluvial deposits are found only in the lower reaches of the Havos stream.

The area of Mycenae was affected by the reactivation of a normal fault which is believed to have contributed to the decline of the Mycenaean civilization (Maroukian et al. 1991). This fault cuts primarily through limestones, however at one point, near elevation 327 m, it has cut across the Havos stream (Fig. 2, site I) in such a way that it has blocked its normal course. The old course of the Havos ran along the contact of the limestones and the Plio-Pleistocene marls and conglomerates. When the fault was reactivated, the blocked main channel was forced to aggrade (site I). At the same time the active channel moved northeastwards until it found a place to overtop the fault scarp and join its old channel (site II). At present, the inactive channel from site I to site II is clearly visible running behind the fault scarp (Fig. 2).

In the area that was filled by the flood sediments of the Havos stream a number of pot sherds were found ranging in age from Mycenaean to Roman. The reactivation of the fault could have caused a landslide observed in the flysch formations of the upper course of the river thus accelerating sedimentation rates in the area where the stream was blocked by the fault.

4 THE BERBADIOTIS REMA

East of Mycenae lies the drainage network of the Berbadiotis which extends primarily over limestones of Triassic-Jurassic age (Fig. 3). In its upper reaches this fluvial system is developed on a karst planation surface of Neogene age whose general slope is to the southeast. The highest elevations are found at Malia Leni (1046 m) and Trapezona (1137 m) and the lowest at Mavrovouni (655 m) and Gilgioni (583 m). The flow direction of the drainage network is in agreement with the general slope of the planation surface.

In the area of Prosymna-Limnes, along the western edge of the limestone mass, there is intensive downcutting which has formed a deep gorge of at least 200 m, with headward erosion towards Limnes. The rejuvenation of this part of the Berbadiotis system results from a lowering of the local base level of the interior Plio-Pleistocene lake at Prosymna. This lake found an outlet towards the Argive plain through the newly-opened gorge of Berbadiotis at Klissoura, aided by two parallel faults with a NNE-SSW orientation. On the sides of the gorge, some residual

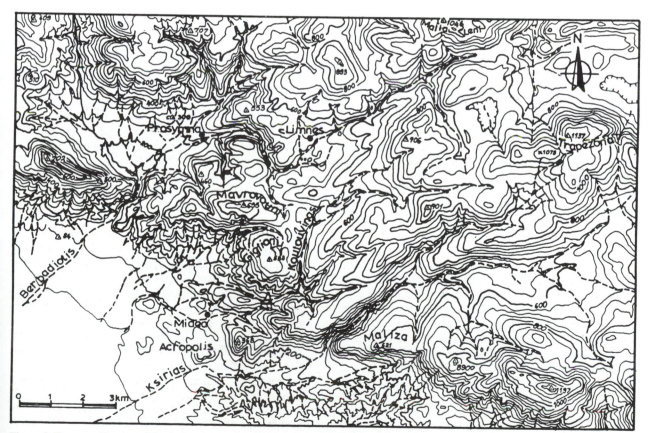

Figure 3. Topographic map of the drainage systems of the Berbadiotis, Koulouvinas and Ksirias rivers.

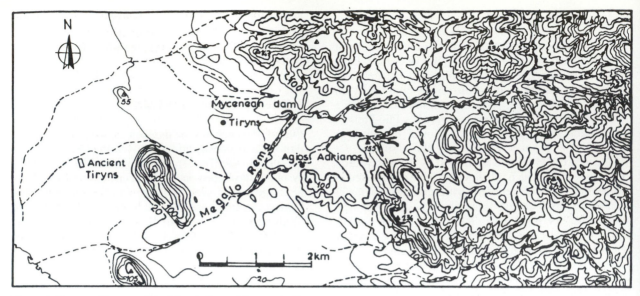

Figure 4. Topographic map of the Megalo Rema River and adjacent areas.

alluvial cones were observed which terminate approximately 10 m above the present main channel and 7 m above the terrace level. These sediments are flash flood deposits attaining a thickness of more than 5 m with many angular particles of up to 30 cm in size.

Along the main channel of the Berbadiotis there is no channel fill terrace, as seen in most other river systems (Dafnorema, Megalo Rema, Ksirias), due to the limited surface runoff over the limestone strata.

5 THE KOULOUVINAS REMA

East of the village of Midea (Fig. 3) extends the drainage network of the Koulouvinas Rema that starts from the palaeosurface of the Triassic limestones and flows to the west through the Cretaceous limestones of Midea down to the alluvial plain. In the upper, mountainous part of the Koulouvinas, the drainage system is rejuvenated with intense downcutting. At site A (Fig. 3) the main channel comes very close to the adjacent stream of Midea. On the geological map (1:50.000) of the Geological Institute of Greece (1970) the active channel of the Koulouvinas is shown to be flowing towards Midea. At site A, on the northern slope of the Koulouvinas Valley we observed talus deposits which are 10-15 m above the present channel and correspond in elevation to the col that separates the two streams. This means that the flow of the Koulouvinas River, as it is shown in the geological map, corresponds to an older course of the river. This was probably the case in the Middle to Late Pleistocene, when the talus cones were active and prevented the flow of the Koulouvinas along the present northern channel.

In the Late Pleistocene the stream returned to its present channel eroding the talus cones to a depth of 15 m. This channel shift could have been assisted by the presence of an active NE-SW trending fault which starts from area A (Fig. 3) and extends as far as the Mycenaean

Acropolis of Midea. The recent activation of the fault is proven by the presence of two knickpoints whose youth is demonstrated by the absence of headward movement of a scarp (knickpoint) upstream from the fault line.

6 THE KSIRIAS REMA

The drainage network of the Ksirias Rema extends east of the Mycenaean Acropolis of Midea (Fig. 3). It can be divided into two sections; the upper and older part found on Triassic limestones which form a palaeosurface and evolved in an older erosion cycle, and the lower and younger part located in the graben area. The main channel of Ksirias Rema has deeply dissected the upper palaeosurface creating a deep gorge which, at the point of its exit from the limestones, has a depth of more than 300 m. In this area an E-W trending fault forms the northern boundary of the tectonic depression. The lower section of the Ksirias traverses Cretaceous limestones, flysch formations, Plio-Pleistocene marls and conglomerates, and Pleistocene alluvial deposits.

The presence of extensive flood deposits along the eastern edge of the Ksirias drainage basin, found within the graben and reaching an elevation of 600 m in the limestones (site Maliza), suggest that an older outlet of Ksirias existed towards the tectonic depression during the Plio-Pleistocene. The location of the contemporary channel in the mountainous section of Ksirias (gorge) could have been favoured by the presence of a fault of the same NE-SW orientation. The absence of knickpoints along the main channel in the gorge shows that the fault has not been reactivated in recent times. Upstream of the gorge the cycle of erosion seems to have been continuous during the Quaternary.

In the lower section of the Ksirias River, valley floor evolution is more complex and fluvial terraces are present. The oldest terrace reaches 20 m above the present channel and the youngest is of the channel fill type of post-Roman age with an elevation of 2 m above the present channel. The age of this latter alluvial fill seems to be concurrent to those reported from other parts of Greece (Vita-Finzi 1969; Dufaure 1976; Dufaure and Fouache 1988; Genre 1988; Zamani et al. 1991).

7 THE MEGALO REMA

The drainage network of the Megalo Rema is developed mainly upon flysch formations and is characterized in its greater part by intense downcutting (Fig. 4). The flysch is located in an E-W running tectonic depression between two higher limestone horsts. In the Plio-Pleistocene most of the eastern part of the flysch was covered by flood deposits. During the Pleistocene, the drainage network of the Megalo Rema was formed and eroded in part into the Plio-Pleistocene deposits, and through continued downcutting into the flysch formations. All this eroded material was ultimately deposited in the tectonic depression of the Argive plain.

In the Late Pleistocene the network entered a depositional phase resulting in the accumulation of a 10-15 m thickness of alluvial sediments. These have a colour of 10YR 5/5 or 4/5 and are typically of a sandy/silt texture. At least four palaeosol horizons have been observed having a colour of 5YR 4/8, mostly of sandy/clay texture with a $CaCO_3$ content of about 53% and a pH of about 6.9. Near the end of this depositional phase a brief erosional phase is observed in the form of coarse grained channel fills. The $CaCO_3$ content of the fine fraction of these materials is 14%. A period of intense downcutting followed which reaches down to 15 m at some points, and into the flysch formations. This period of downcutting continued into the Holocene. The presence of a Mycenaean dam located in the main channel of Megalo Rema indicates that the gorge was present at that time. A depositional period in post-Roman times left a channel-fill type terrace of 3-4 m. These deposits are brown in colour and the grain size is rather fine reflecting the lithological composition of the flysch source rocks. The gravels are derived from limestone intercalations found within the flysch formations.

8 THE MYCENAEAN DAM OF TIRYNS

The presence of a Mycenaean dam 4 km east of ancient Tiryns in the middle reaches of Megalo Rema provides valuable clues to aid in the interpretation of the evolution of this area during historical times (Fig. 5). The dam is located immediately downstream of the confluence of the main channel with two tributary streams. The outcropping of the underlying flysch rocks at this point was favourable for the stability and impermeability of the dam. At the time of construction, the main channel of the Megalo Rema had already cut down to a depth of about 13 m which favoured dam building. In this way, the downstream main channel of Megalo Rema became

inactive but it is still clearly visible today (Fig. 5). Southeast of the dam is the artificially opened channel which discharges the flow of Megalo Rema to the adjacent channel of Aghios Adrianos Rema in the southeast.

The purpose of the dam is thought to be the protection of the Tiryns area from catastrophic flooding (Balcer 1974). The discovery of an accumulation of up to 5 m of alluvium east of Tiryns is reported by Zangger (1991) and attributed to an extraordinary, high magnitude ephemeral flood which occurred around the 12th century BC. Zangger (1991) also suggests that the deposition was caused by an earthquake that affected the upper part of the drainage basin of Megalo Rema and was followed by an extensive landslide. However, this part of the basin rarely exceeds elevations of 500 m and is characterized by relatively low relief. Moreover, extensive field survey did not reveal any evidence of a large landslide event. We believe that the dam was constructed to create a small pond for the purpose of irrigation as well as for flood control. The fact that the dam is resting upon the impermeable flysch supports this view. During the Late Helladic period, the dam structure failed causing extensive destruction in the area of ancient Tiryns and depositing a thick layer of alluvium. The cause for this could be attributed either to an exceptional flood after heavy rainfall or to an earthquake which seriously damaged the dam. The possibility of an earthquake is more plausible as numerous earthquakes have been reported around the time of the dam failure (Mylonas 1966; Kilian 1980; 1985; Shear-Mylonas 1987; Maroukian et al. 1991).

After its destruction an artificial diversion channel was

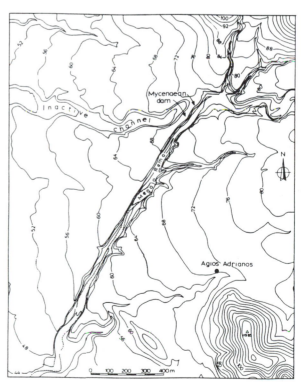

Figure 5. Large scale topographic map of the Mycenaean dam at Megalo Rema.

opened which lead to the nearby channel of the Aghios
Adrianos in the southeast, thus avoiding flooding. The
depth of this artificial channel was probably no greater
than 6-7 m. There is a small stream 400 m downstream
from the dam which traverses the artificial channel and
has a depth of about 5 m. Within the artificial channel, a 3
to 4 m historical terrace is present which corresponds
well to those found in other valleys in the area. Although
no dating was possible, its age, in common with a similar
feature in the nearby river of Dafnorema, is likely to be
post-Roman.

9 DAFNOREMA

The Dafnorema River occupies the southern part of the
same tectonic depression where the Megalo Rema is
situated. The main channel of the Dafnorema flows along
the tectonic contact of Triassic limestones (horst) and
flysch formations with Plio-Pleistocene and more recent
alluvial deposits (graben) (Fig. 6). The development of
the drainage network of the Dafnorema is unlike those
previously described as its tributaries extend north of its
main channel almost entirely within the depression, and
the catchment is restricted in its southern part by the
precipitous limestone slopes.

Within the limestones a well developed channel exists
whose upstream part comes almost in perpendicular contact with the main channel of the Dafnorema with a

vertical difference of 60 m. This palaeochannel formed
part of a lower section of an ancestral Dafnorema network which traversed the limestones down to the area of
Candia. The modern stream, the Giannakaki Rema, is
characteristically underfit. This older flow direction of
the Dafnorema was probably active from Pliocene times
given that only 1.5 km north of the upper edge of the
palaeochannel there is a remnant of Plio-Pleistocene conglomerates overlying flysch deposits and having a contact elevation of 200 m that is about 30 m above the
palaeochannel edge. The channel would have been cut
off in the Pleistocene, coinciding with the cessation of
conglomerate deposition and the commencement of the
erosive phase of the Dafnorema drainage network. This
cut off would have been favoured by the easily erodible
flysch and conglomerates which were incised about 60 m
during the second half of the Pleistocene. In addition, the
fact that the Dafnorema has developed along the southern
tectonic contact of the depression favours accelerated
erosion. It is also highly likely that the fault running along
the southern boundary of the depression was reactivated
in the second half of the Pleistocene thus contributing to
the 60 m height difference between the modern channel
and the ancient course of the Dafnorema.

Finally, the possibility of piracy by a stream located on
the present Dafnorema alignment, but downstream from
the palaeochannel (Fig. 6, sites A and B), should not be
ruled out. Headward erosion could have assisted in the
final cut off of the Dafnorema from its ancestral flow

Figure 6. Topographic map of the Dafnorema
River and surrounding area.

direction through the limestones. Area A-B shows evidence of recent erosive activity as the stream is still downcutting. The upper part of the Dafnorema, upstream of A, is characterized by the presence of extensive talus cones. In the lower and younger reaches of the Dafnorema there is a channel fill and a 2-4 m high terrace which is dated to the post-Roman period on the basis of archaeological sherds found within the deposits. In the upper section of the river this terrace does not exist.

10 CONCLUSIONS

The drainage networks of the eastern part of the Argive plain are still evolving because they are still affected by the active tectonic regime of the area. In the tectonic depression of the Ksirias, Megalo Rema, and Dafnorema river systems, erosion has proceeded faster than in the adjacent areas due to the presence of readily erodible lithological units (flysch, Plio-Pleistocene marls and conglomerates, and alluvial deposits) producing undulating and comparatively low-relief topography.

In contrast, to the north of this tectonic depression a planation surface composed of limestones is evolving at much slower rates and includes the upper parts of the Ksirias, the Koulouvinas and the Berbadiotis rivers. Around the margins of the planation surface these rivers are rejuvenated, forming deep gorges, due to base level lowering during the Early Pleistocene. On the northwestern edge of this limestone mass along the Havos stream, the reactivation of a fault has been confirmed.

The deep Late Pleistocene/Holocene downcutting observed in the lower sections of all the studied rivers is demonstrated by the presence of a Mycenaean dam in the Megalo Rema whose downstream inactive channel has a depth of 12 m. In post-Roman times a short depositional phase occurred in most of these rivers and this resulted in the formation of a channel fill terrace of 2-4 m. This terrace was the product of drier conditions in post-Roman times.

REFFERENCES

Balcer, J. 1974. The Mycenean dam at Tiryns. *American Journal of Archaeology.* 78: 141-149.

Dufaure, J.J. 1976. La terrasse holocene d'Olympie et ses equivalents mediterraneens. *Bull. Assos. Geogr. Fr.* 433: 85-94.

Dufaure, J.J. & E. Fouache 1988. Variabilite des crises d'age historique le long des vallees d'Elide (Ouest du Peloponnese). *Etudes mediterraneennes* 12, CIEM, Poitiers.

Finke, E. 1988. Landscape evolution of the Argive plain, Greece: Paleoecology, Holocene deposition history and coastline changes. Ph. D. Thesis. Stanford University.

Gaki-Papanastassiou, K. 1991. The geomorphological evolution in and around the Argive plain since the Quaternary. Ph.D. Thesis. University of Athens, (In Greek).

Genre, Ch. 1988. Les alluvionnements historiques en Eubee, Grece. *Etudes Mediterraneennes* 12, CIEM, Poitiers.

Institute of Geology and Mining Research (IGMR), 1970. *Geological sheet of Nafplio.* 1:50000, Athens.

Kilian, K. 1980. Zum Ende der Mykenischen Epoche in der Argolis. *Jahrbuch des Romisch-Germanischen Zentralmuseums, Mainz* 27: 166-196.

Kilian, K. 1985. La caduta dei Palazzi Micenei continentali: aspetti archaeologici. D. Musti (ed) *Le origini dei Greci. Dori e mondo Egeo.* Roma: Laterza. 73-115.

Maroukian H., K. Gaki-Papanastassiou & D. Papanastassiou 1991. The relation of geomorphologic-seismotectonic observations to the catastrophes at Mycenae. International Interdisciplinary meeting *'Earthquakes in the archaeological record: Palaeoseismological and archaeological aspects'*. Athens, 13-15 June 1991, (Abstract).

Mylonas, G. 1966. *Mycenae and the Mycenaean Age.* Princeton, New Jersey: Princeton University Press.

Shear-Mylonas, I. 1987. *The Panagia Houses at Mycenae.* The University Museum, Philadelphia, Monograph 68, University of Pennsylvania, Philadelphia.

Vita-Finzi, C. 1969. *The Mediterranean Valleys.* Cambridge: Cambridge University Press.

Woodward, J.C., J. Lewin & M.G. Macklin 1992. Alluvial sediment sources in a glaciated catchment: The Voidomatis basin, northwest Greece. *Earth Surface Processes and Landforms* 17, 205-216.

Zamani, A., H. Maroukian & K. Gaki-Papanastassiou 1991. Rythmes de depot et de creusement pendant les temps historiques dans le cadre des sites archeologiques de la region d'Argos (Grece). *Physio-Geo.* Paris, 22/23: 81-88.

Zangger, E. 1991. Prehistoric earthquakes and their consequences as preserved in Holocene sediments from Volos and Argos. International Interdisciplinary meeting *'Earthquakes in the archaeological record: Palaeoseismological and archaeological aspects'*. Athens, 13-15 June 1991, (Abstract).

PART 2

Archaeology and human-river environment interactions

CHAPTER 9

Archaeology and human-river environment interactions

JAMIE C. WOODWARD

Department of Environmental and Geographical Sciences, Manchester Metropolitan University, Manchester, UK

Reconstructions of Quaternary river behaviour and landscape evolution are not solely of interest to the geomorphologist and the geologist because the alluvial sequences and landforms of many environments are also of considerable archaeological value (see Butzer 1982; Gladfelter 1985; Needham and Macklin 1992). Information on the nature of ancient river environments is of particular relevance to archaeology because alluvial settings were frequently the focus of major early human settlements and, in certain cases, such as the Nile, Tigris-Euphrates, and Indus, the effective management of floodplain resources on a large scale proved crucial to the rise of irrigation agriculture and urban civilization (see Renfrew and Bahn 1991). The archaeological significance of river environments derives partly from the fact that the water- and soil-based resources of alluvial floodplains are of central importance to many human societies (Gilbertson et al. 1992). This is of particular importance in the Mediterranean basin because parts of the region were host to some of the earliest known examples of floodwater farming (Evenari et al. 1982), direct river channel management and elementary flood control (Biswas 1970), and inter-basin water transfers (Vita-Finzi 1978). More generally, river behaviour is of direct relevance to much archaeological research because patterns of river erosion and sediment deposition may not only have influenced the ways in which the riverine environment and the surrounding area were used by prehistoric and more recent populations, but fluvial processes will often control the actual preservation and subsequent exposure of the archaeological record (cf. Waters 1988).

As many of the papers in this volume demonstrate, the river systems of the Mediterranean region have provided diverse and dynamic settings for both hunter-gatherer and sedentary peoples throughout the human occupation of the region. For much of the Palaeolithic, for example, the waxing and waning of Pleistocene climate forced repeated changes in catchment vegetation and river flood regimes offering a changing environment and resource base for hunter-gather communities. During the Holocene, particularly since the advent of Neolithic cultures and the spread of agricultural activities, basin-wide modifications to vegetation cover and catchment water balances have tended to increase runoff rates, soil erosion and suspended sediment yields – perhaps even rendering many of the Holocene river systems more susceptible to

change during climatic perturbations such as the Little Ice Age. Recent research has suggested that the cultivation of catchment slopes was not always associated with accelerated soil loss because another important human modification of Mediterranean river basins – the introduction of terracing on slopes – served to check soil movement downslope to valley floors (see Butzer 1982; Pope and van Andel 1984). Stone-walled terraces have been a conspicuous feature of Mediterranean valley sides for many millennia and a number of studies have reported a close correspondence between the degree of maintenance of terrace structures and the rate of transfer of sediment from the slope to the channel system (e.g. Wise et al. 1982; van Andel et al. 1986).

It has often been stated that much of our knowledge of the history of human relationships with the environment has resulted from close, long-standing links between the disciplines of archaeology and geography (cf. Renfrew 1983; Goudie 1989). Much of this joint research activity has taken place outside of the Mediterranean region in both the Old and New Worlds, but it is true that some of the earliest and best known examples of archaeological research in alluvial settings come from the Mediterranean basin (cf. Vita-Finzi 1969; 1978). The continuation of this close relationship in the context of alluvial settings is evident in all six of the papers in this part of the volume and such collaboration may come about in several ways. In the following sample of papers some of the studies of alluvial history were commissioned by archaeologists seeking information on the history of a nearby stream which seemed to be important for site or survey interpretation. Often approaches are made in the other direction, with geomorphologists – in search of both dating control and evidence for episodes of landscape stability – seeking guidance on the age of artefacts such as flints, potsherds or building structures which may be buried within or scattered upon alluvial materials.

It has been estimated that approximately one third of the world's population presently live within floodplain environments. It is therefore of great importance that both contemporary floodplain processes and the history of human settlement and use of river depositional environments are fully researched and understood. The advantages of riverine settings for human occupation are numerous and include such basic requirements as access to water and food resources as well as the opportunities presented by fording points, level ground and the presence of fertile alluvial soils. River valleys also provide natural routeways through the steepland terrain of the Mediterranean region and the navigable middle and lower reaches of many rivers offered attractive possibilities for contact and trade with the coastal zone and beyond. Conversely, it is also important to appreciate the potential hazards of such locations including, for example, high magnitude floods and the dangers of salinization and overdependence on seasonal irrigation waters in semi-arid environments. Butzer (1982) outlines the factors behind the formation of the malarial coastal lowlands of the Mediterranean region which began to form around 2000 years ago. Increases in the frequency of overbank flows following anthropogenic catchment disturbance created extensive tracts of less productive, waterlogged land in areas of formerly high agricultural productivity. It has been suggested that this deterioration of lowland alluvial terrain and the coeval increase in epidemic disease contributed to the decline of the great classical civilizations.

From the examples discussed above and in the papers that follow, it is clear that the nature of the river environment – including both water quality as well as quantity factors and their variation over time – is of fundamental importance to the well-being of many economies. Indeed, the realisation of the potential of Mediterranean floodplains can justifiably be regarded as a major achievement of past and present communities. The relationship between past river behaviour and the archaeological record of human activity may be observed across a range of spatial and temporal scales – the following papers explore parts of the archaeological timescale from Middle Palaeolithic to Medieval times and, depending on the objectives of the particular research, the spatial dimension varies from the catchment or river basin scale down to the reach scale.

In the next chapter Macklin and Passmore describe the alluvial sedimentary sequence in the Guadalope River basin of northern Spain where they have identified eleven alluvial units and river terraces which represent a complex sequence of cut and fill events during the middle and late Quaternary. The Guadalope is a major tributary of the Ebro (the largest river in Spain to flow into the Mediterranean Sea) and the alluvial sequence relates to changes in climate (mainly precipitation regime) and progressive Pleistocene incision forced by long-term regional tectonic uplift. The authors also discuss some of the wider problems involved in achieving age control for alluvial sequences and argue that direct sediment-based dating techniques such as luminescence offer many advantages over methods such as radiocarbon which have been used to age 'bracket' episodes of alluviation. Provisional dating of a major aggradational episode to a cold but wet phase of the last interglacial complex points to the importance of changes in the annual total and seasonality of precipitation as the major control on basin sediment and water yields.

Middle and Upper Palaeolithic artefacts have been recovered from two of the alluvial units (T7 and T9) in

the Guadalope sequence and lithic materials (mainly flint and chert) are also present on the surface of some of the river terraces demonstrating human presence in the valley during various stages of the Palaeolithic. Two sites are described where lithic artefacts have been found stratified within alluvial materials. A Levallois core within the T7 alluvial unit indicates a last interglacial age for this alluviation and this find is in good agreement with the IRSL date of 115,000 (± 17,000) years BP obtained on fine sediments from the morphologically older T6 unit. A much larger assemblage of lithic fragments (debitage debris) stratified within the younger T9 alluvial unit provides little chronological control (as these artefacts alone are not diagnostic of a particular industry or tool technology) and age control for this phase of alluviation is provided by a date of 28,000 (± 4000) years BP indicating an Upper Palaeolithic age for the human activity which produced the worked flint flakes. Whilst all occurrences of worked flint and other artefacts discarded upon alluvial surfaces (which may be buried by later alluvial materials) provide evidence of human presence, not all of these are useful for dating purposes.

The impact of glacial activity on Quaternary river environments in the Mediterranean has not received a great deal of attention. However, in the Voidomatis River basin of northwest Greece the ice age legacy is particularly prominent. An extensive area of glaciated terrain is present in the catchment headwaters and terraced glaciofluvial sediments form an important Upper Palaeolithic landsurface. Woodward, Lewin and Macklin (Chapter 11) argue that glacial activity was an important process in landscape modification in parts of the Mediterranean region. This glacial influence extended well beyond ice margins by effecting major changes in river discharge regimes and fluvial depositional environments. Moreover, from an archaeological viewpoint, there is also a growing body of evidence to suggest that, in marginal upland (Alpine Mediterranean) environments, such as the Lower Vikos Gorge of the Voidomatis River basin, glacial activity may also have influenced the timing of Palaeolithic settlement. Evidence for Palaeolithic settlement before c. 20,000 BP is widespread in Epirus and is also present in the lower reaches of the Voidomatis River on the Konitsa Plain. However, the rockshelters of the Lower Vikos Gorge show evidence of occupation only *after* the Last Glacial Maximum even though the rockshelters were present and available for use (cf. Bailey 1992). The oldest radiocarbon dates from the Klithi rockshelter suggest that the earliest occupation began around 16,000 years BP. It seems clear that, within the gorge (upstream of the Konitsa Plain), the cold stage braided river environment did not present a favourable habitation area until after the beginning of incision following the Last Glacial Maximum. The authors suggest that, as far as the archaeological record of the close of the last cold stage is concerned, more information is needed from sites at intermediate altitude in the upland interior of the Mediterranean zone.

Remaining in continental Greece, this time to the east of the Pindus Mountains, the chapter by van Andel, Gallis and Toufexis (Chapter 12) describes the results of recent research into the alluvial palaeoenvironments associated with Neolithic settlement on the plain of Thessaly. This work has focused on the Neolithic mounds and associated Holocene river deposits and soils of the Peneios River floodplain in the northeast Trikala basin and, largely on the basis of stratigraphic observations from a detailed programme of sediment coring, they conclude that floodplain (rather than dryland) farming was the norm in this region in early Neolithic times. It seems likely that particular floodplain sites were selected and farmed as they allowed post-flood cultivation of recently deposited fine-grained alluvium. Van Andel and his co-workers make use of evidence from floodplain soils and are able to demonstrate a close relationship between sediment-charged flood events and riparian farming activities during the Neolithic. The collective reliance of such settlements upon the annual flood regime of the Peneios River provides a good example of the typically close relationship between human well-being and river behaviour.

Chapter 13 by Barker and Hunt outlines some of the results from one of the largest archaeological surveys yet conducted in the Mediterranean basin which was linked to an investigation of the Holocene alluvial record in the Biferno Valley in eastern Italy. Attention is focused on Neolithic to Roman times and the Holocene alluvial sequence embraces at least seven sedimentary units which range in age from Neolithic/Bronze Age to after the middle of the 20th century. Dating control is largely based on the archaeological evidence (mostly pottery found within the alluvial units) supplemented by pollen and plant macrofossil data. Much of the palaeobiological evidence points to phases of woodland clearance and more intensive cultivation as the mechanisms responsible for initiating increases in runoff and sediment delivery and episodes of channel aggradation. More generally, Barker and Hunt stress the need to design appropriate integrated methodologies which demand close collaboration between archaeologists and geomorphologists at all stages of project planning and implementation. This study records a close correspondence between major phases of Holocene stream aggradation in the Biferno Valley and the disturbance of catchment slopes through the expansion and intensification of agricultural activities. Indeed, by way of a modern analogue for earlier episodes of catchment disturbance, they note the dramatic increase in rates of soil erosion over the last two decades following the introduction of damaging techniques such as deep ploughing.

In Chapter 14 Provansal also makes use of archaeological survey data and relates this information to geological evidence for marked changes in sediment yield over the Holocene Period at the river basin scale. In contrast to the other papers in this part of the volume, rather than using the alluvial deposits of the valley floor to reconstruct changes in river behaviour, Provansal utilises the Holocene sedimentary record of a Mediterranean lagoon and delta system at the mouth of the Arc River in Provence, southern France. By combining geomorphological fieldwork in the river basin and geological investigations in

the delta with the archaeological record, it has proved possible to relate patterns of land use change and varying settlement densities with the nature and rate of sedimentation at the basin outlet. The rate of delivery of fine sediment to the catchment outlet is determined by radiocarbon dating of core sequences and sediment properties are used to infer particular catchment sources (cf. Walling and Woodward 1992). Provansal concludes that human-induced topsoil disturbance provided the main source of fluvial suspended sediment which promoted delta build up. The delivery of fine sediment to the basin outlet was enhanced during phases of increased precipitation and runoff and it was only when such episodes coincided with periods of soil disturbance that net delta growth took place. Two main periods of deltaic sedimentation took place during the Holocene. The first of these commenced during prehistory up to the Versilian transgression when infilling of marshy littoral plains began. The second major phase began at the end of the Middle Ages when a combination of intensive agricultural densities with the net climatic deterioration of the Little Ice Age produced highly favourable conditions for increased sediment supply and delta growth. Sedimentary horizons showing high concentrations of weathered (degraded) clay minerals are interpreted as indicating phases of topsoil erosion from the catchment slopes as these clay minerals are found in the upper part of soil profiles.

In the final chapter of this section, Hunt and Gilbertson present evidence for rapid changes in late Holocene river behaviour following episodes of forest clearance in the Feccia Valley in Tuscany, Italy. Three phases of coarse sediment deposition have been related to increases in hillslope runoff and sediment supply following land use changes. Hunt and Gilbertson make use of pollen, mollusc and plant macrofossil evidence and suggest that deliberate land management strategies within the Feccia Valley, particularly woodland clearance, were important for triggering episodes of slope erosion and river aggradation. The alluvial response was rapid, and it has been demonstrated that these changes took place within a relatively brief interval after the fifteenth century AD. The authors observe that late Holocene episodes of stream channel aggradation in the Feccia Valley are out of phase with known climatic episodes recorded in southern Europe over the last millennium and conclude that climate change is not likely to have been responsible for major changes in river behaviour. Furthermore, the palaeobotanical evidence for deforestation and accelerated soil erosion is in good agreement with the archaeological and historical records of land use intensification. In common with research conducted further south in the Biferno Valley in the Molise region, a positive relationship has been identified between phases of woodland clearance and alluviation.

Chapters 13 and 15 underscore the important role of palaeoecological analyses in the reconstruction of Quaternary river environments in contexts which favour the preservation of organic remains. Fine-grained channel fills often contain rich assemblages of pollen and plant macrofossil remains which provide valuable evidence to supplement and refine lithological and stratigraphical observations. Interestingly, the papers by Hunt and Gilbertson and Provansal both point to the significance of riparian vegetation for trapping sediment eroded from upslope and thus checking sediment delivery to the fluvial system.

It is beyond the scope of this brief introduction and the following series of papers to provide an exhaustive review of all aspects of human-river environment interactions in the Mediterranean region. The main objectives of this part of the volume are to present a sample of recent research activity in Mediterranean alluvial archaeology, to illustrate the use of a range of field approaches in contrasting Mediterranean river depositional environments over a variety of timescales and, most importantly, to highlight some of the potential impacts of human activity on river behaviour and *vice versa*. It is hoped that continued close cooperation between geomorphologists and archaeologists will further enrich our understanding of the history of the Mediterranean landscape and promote greater awareness of appropriate and sustainable land management strategies to minimise the undesirable effects of human activity.

REFERENCES

Bailey, G.N. 1992. The Palaeolithic of Klithi in its wider context. *The Annual of the British School of Archaeology at Athens* 87, 1-28.

Biswas, A.K. 1970. *History of Hydrology.* Amsterdam, North Holland Publishing Company.

Butzer, K. 1982. *Archaeology as Human Ecology.* Cambridge, Cambridge University Press.

Evenari, M., L. Shanan & N. Tadmor 1982. *The Negev: The Challenge of a Desert*. Cambridge, Mass, Harvard University Press.

Gilbertson, D.D., C.O. Hunt, R.E. Donahue, D.D. Harkness & C.M. Mills, 1992. Late Holocene geomorphology, palaeoecology and land use on the floodplain of the Pian di Feccia,

Tuscany, *Archeologia del Paesaggio*, 231-248.

Gladfelter, B. 1985. On the interpretation of archaeological sites in alluvial settings. In J.K. Stein & W.R. Farrand (eds), *Archaeological Sediments in Context* (Peopling Americas Edited Volume Series, 1). Centre for the study of early man, Institute of Quaternary Studies: University of Maine Orono, 41-52.

Goudie, A.S. 1989. The changing human impact. In L. Friday & R. Laskey (eds), *The Fragile Environment* (The Darwin College Lectures), Cambridge, Cambridge University Press, 1-21.

Needham, S. & M.G. Macklin (eds) 1992. *Alluvial Archaeology in Britain*, Oxbow Monograph 27, Oxford.

Pope, K.O. & T.H. van Andel 1984. Late Quaternary Alluviation and Soil Formation in the Southern Argolid: Its History, Causes and Archaeological Implications, *Journal of Archaeological Science* 11, 281-306.

Renfrew, C. 1983. Geography, Archaeology and Environment, *The Geographical Journal*, 149, 316-323.

Renfrew, C. & P. Bahn 1991. *Archaeology: Theories, Methods and Practice.* London, Thames and Hudson.

van Andel T.H., C.N. Runnels & K.O. Pope 1986. Five thousand years of land use and abuse in the southern Argolid, Greece, *Hesperia*, 55, 103-128.

Vita-Finzi, C. 1978. *Archaeological Sites in their Setting.* London, Thames and Hudson.

Vita-Finzi, C. 1969. *The Mediterranean Valleys: Geological Changes in Historical Times.* Cambridge, Cambridge University Press.

Walling, D.E. & J.C. Woodward 1992. Use of radiometric fingerprints to derive information on suspended sediment sources. In *Erosion and sediment transport monitoring programmes in river basins* (Proceedings of the Oslo Symposium – August 1992). IAHS Publication No. 210, 153-164.

Waters, M.R. 1988. Holocene alluvial geology and geoarchaeology of the San Xavier reach of the Santa Cruz River, Arizona, *Geological Society of America Bulletin*, 100, 479-491.

Wise, S.M., J.B. Thornes & A. Gilman 1982. How old are the badlands? A case study from southwest Spain. In R.B. Bryan & A. Yair (eds), *Badland geomorphology and piping.* GeoBooks, Norwich: 259-277.

CHAPTER 10

Pleistocene environmental change in the Guadalope basin, northeast Spain: fluvial and archaeological records

MARK G. MACKLIN
School of Geography, University of Leeds, Leeds, UK
DAVID G. PASSMORE
Department of Geography, University of Newcastle upon Tyne, Newcastle upon Tyne, UK

ABSTRACT: Geomorphological and archaeological investigations of Quaternary fluvial deposits in the Guadalope Valley, northeast Spain, have revealed an extended record of river and drainage basin response to long-term environmental change. Eleven Quaternary alluvial terraces ranging from 4 to 81 m above present river level have been identified along a 13 km reach of the Rio Guadalope south of the town of Alcañiz, Teruel. An older group of river terraces of Middle Pleistocene or earlier age are poorly preserved but a younger suite of Late Pleistocene terraces (12-23 m above present river level), that have been dated using luminescence techniques, can be traced along the entire reach. Middle and Upper Palaeolithic artefacts have been recovered from Late Pleistocene terrace deposits. Provisional dating of a major valley bottom alluviation in the Rio Guadalope to a cold but wet phase of the last interglacial complex, appears to run counter to conventional climate-river models in the western Mediterranean which considered cold and dry 'glacial' periods to favour river aggradation. It is argued, however, that in semi-arid northeast Spain vegetation is unlikely to have been a major moderator of runoff during the Late Pleistocene (prior to human disturbance) and that catchment water and sediment yields are likely to have been controlled more by changes in precipitation than temperature. Progressive incision of the Guadalope valley floor during the Pleistocene would appear to reflect long-term regional uplift with rates of channel downcutting increasing in Middle and Late Würm times, partly in response to a reduction of coarse sediment delivery to the main channel from tributary catchments.

KEYWORDS: Northeast Spain, Quaternary river terraces, alluvial archaeology, active tectonics, climate change, luminescence dating.

1 INTRODUCTION

Extensive suites of Quaternary alluvial deposits and well developed river terrace flights are a common and conspicuous feature of the Spanish landscape. High basin relief (34% of peninsula Spain lies between 800-2000 m above sea level), long-term regional uplift and extensive outcrops of sedimentary rock susceptible to mechanical breakdown and erosion have made Spanish rivers sensitive to Quaternary environmental perturbations (e.g. climate, vegetation, land use change) and provided conditions ideal for the formation, and preservation, of river terraces. River terraces in Spain and their underlying deposits have been the subject of scientific investigation since the early part of the century (see review by Freeman 1975), most notably by archaeologists on account of their frequent inclusion of prehistoric artefacts. Indeed, they constitute one of the most important sources of stratified archaeological material, particularly for the Palaeolithic period.

Quaternary fluvial research since the mid 1970's in mainland Spain has been concentrated in four main regions: the Ebro basin, northeast Spain, including its tributaries that drain the Pyrenees and Iberian mountains; the upper Tagus catchment centred on Madrid; the Turia, Júcar and Serpis rivers in the region of Valencia; and river systems in Almeria, Murcia and Alicante, southeast Spain, that issue from the Betic Cordillera. In the Ebro and upper Tagus, Holocene and Pleistocene alluvial archaeology and the impact of historic and prehistoric land-use change on hillslope and river channel dynamics have been central research themes (e.g. Burillo et al. 1986; Macklin et al. 1994; Peña et al. 1991; Sancho et al. 1988; van Zuidam 1975). While investigations in Valencia, Almeria and Murcia in the main have focused on evaluating the influence of long-term tectonics and Quaternary climatic oscillations on river (e.g. Harvey and Wells 1987; Payá and Walker 1986) and alluvial fan (e.g. Harvey 1984, 1988) aggradation and dissection sequences.

Although fluvial sedimentary sequences constitute the longest (albeit discontinuous) continental record of Quaternary environmental history their geochronologies in Spain, as elsewhere in southern Europe, are generally poorly documented. This is particularly true of Pleistocene alluvium beyond the effective dating range of [14]C (c. 45,000 years BP), which until recently relied solely on Palaeolithic tool typology and fossil mammal assemblages for a geochronological framework. Recently developed radiogenic and isotopic dating techniques, however, now offer the prospect of establishing fluvial geochronologies back to the Middle Pleistocene. Calcretes, commonly found in and on terraced alluvial sequences, particularly in eastern and southeastern Spain, have been dated using uranium series techniques (Radtke et al. 1988). Carbonate soil concretions, however, are very frequently multiphase and are often formed considerably later than the host sediment in which they are found (Smart 1991). These affects can give a considerable range of uranium concentrations for multiple analysis of coeval deposits, thus greatly limiting their use for river terrace dating and correlation. More satisfactory results have been obtained using luminescence techniques. For example, Prószynska-Bordas et al. (1992) successfully dated fossil soils in Late Pleistocene terraces of the Turia, Júcar and Serpis rivers near Valencia using thermoluminescence (TL). However, dated palaeosols, in common with uranium series dating of calcretes, provide only *terminus ante quems* or *terminus post quems* for associated alluvial units. To overcome this problem, Macklin et al. (1994) have recently applied infra-red stimulated luminescence (IRSL) techniques to date Holocene fluvial deposits in the Rio Regallo, Ebro basin, and found good agreement between luminescence ages of river sediment and AMS ^{14}C dating of charcoal incorporated within alluvium.

The inherent advantages of sediment based dating techniques, like luminescence, lies in being able to establish the age of the depositional event itself, as opposed to bracketing it between other dated material where it is difficult to determine precisely when alluviation began or ceased. This is critical for resolving true cause and effect relationships of river and drainage basin response to tectonics, climate and land-use change. For example, major periods of sediment generation in the Mediterranean during the Pleistocene are generally believed to relate to cold, dry 'glacial' phases (Payá and Walker 1976; Harvey 1988; Sabelberg 1977). In the absence of dating control, however, this hypotheses has not been adequately tested by direct comparison of Spanish Pleistocene river sequences with palaeoclimatic reconstructions from palustrine (e.g. Padul, Granada, Pons and Reille 1988) and marine (Turon 1984) pollen records in the region. The challenge that presently faces the fluvial earth scientist is how to relate river aggradation and incision episodes in the region to this climatostratigraphic framework.

In this paper we attempt to address this problem and report preliminary results from geomorphological and archaeological investigations of a series of river terraces in the Guadalope Valley near Alcañiz, Teruel, in northeast Spain (Fig. 1). Our study had three principal aims: to document the morphostratigraphy and sedimentology of Pleistocene fluvial deposits; to examine the relationships between the Palaeolithic archaeological record and river development, and (if possible) recover artefacts stratified within alluvium; and finally to develop an alluvial chronology, based on luminescence dating of fine-grained river sediments, and establish whether phases of aggradation and dissection in the Rio Guadalope were of climatic and/or tectonic origin.

2 STUDY AREA AND METHODOLOGY

The Rio Guadalope is a major south-bank tributary of the Ebro River and has a catchment area of c. 3000 km². Its headwaters drain the Montes de Maestrazgo which rise to 2019 m and form the most southeasterly part of the Iberian Cordillera. The Iberian massif is a typical 'basement and cover' structure with the basement in the Montes de Maestrazgo region formed by Palaeozoic rocks which are fractured into blocks and in some places thrust over Mesozoic and later cover material. Alcañiz lies within the Iberian Cordillera piedmont zone close to where the Rio Guadalope emerges from the highly deformed foothills of the Montes de Maestrazgo and flows across relatively undisturbed, gently northeasterly dipping Miocene conglomerates, sandstones and silts. The detailed chronology of post-Neogene tectonics in the Alcañiz region is unknown at present though deformed Quaternary fluvial deposits in contact with the Iberian Cordillera within Rioja province, in the southwest part of the Ebro basin (Palacios 1975), suggest that major faults in the Guadalope Valley have probably been active during the Quaternary. Furthermore, analysis of river terrace profiles of the Ebro in the Caspe area (Fig. 1) show evidence of regional uplift during the Middle-Late Pleistocene (Merino et al. 1992). Approximately 12-15 km downstream of Alcañiz the Guadalope traverses a folded and reverse faulted west-east striking anticline that brings Palaeozoic basement material to the surface at Puig Moreno (Fig. 1). Geomorphological and archaeological investigations have been focused within this inter-range basin bounded by the Puig Moreno anticlinal structure to the north and in the south by the Mesozoic limestone cuesta at Calanda.

Terrace mapping and levelling, sedimentological analyses and luminescence dating of Pleistocene alluvium was undertaken along a 12.5 km section of the Guadalope Valley immediately south of Alcañiz. (Fig. 1). Pleistocene river terraces and tributary stream alluvial fans are particularly well-developed in this reach which comprises a broad topographic embayment (locally more than 2 km wide) formed in relatively weak Miocene sandstones, marls and gypsum deposits. North of Alcañiz the valley floor narrows appreciably and the Guadalope cuts through a sandstone and conglomerate escarpment forming a series of highly sinuous, deeply entrenched bedrock-confined meanders. Along most of the study reach the present Guadalope River has a confined meandering pattern (cf. Lewin and Brindle 1977) with its channel almost entirely circumscribed by bedrock bluffs. It has a narrow (rarely attaining more than 100 m) shrub-covered floodplain comprised predominantly of vertically-accreted silts and sands.

Valley floor fluvial landforms in the study reach were mapped by field walking using 1:25,000 scale aerial photographs as base maps. All breaks of slope greater than 1 m were recorded and terrace heights were measured using an anaeroid barometer. Pleistocene alluvium is exposed in a number of gravel pits and road cuttings upstream of Castelseras and on the left bank of the

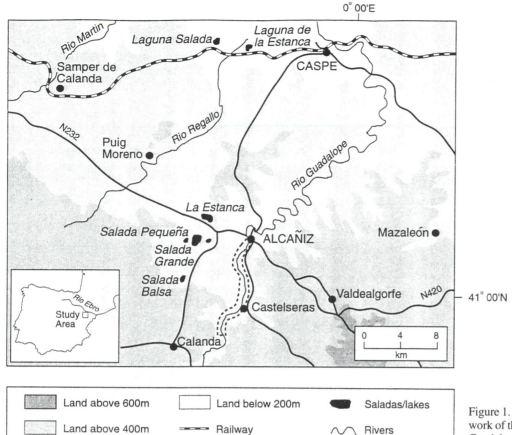

Figure 1. Relief and drainage network of the middle and lower Guadalope valley. Location of study reach, south of Alcañiz, is indicated.

Guadalope west of Alcañiz (Fig. 1). Sedimentary sequences were photographed and logged, and fine-grained material from two alluvial terraces (see below) were sampled for infra-red stimulated luminescence (IRSL) sediment dating following procedures outlined by Bailiff (1992).

3 GUADALOPE PLEISTOCENE ALLUVIAL MORPHOLOGY AND SEDIMENTARY SEQUENCE

3.1 *River and alluvial fan terrace sequence*

Eleven river terrace levels were identified in the study reach of which six, at c. 4, 9, 12, 17.5, 19 and 23 m above the present river, can be traced almost continuously between Castelseras and Alcañiz (Fig. 2). Five higher terraces (c. 29, 34.5, 45, 56 and 81 m above present river level) are also evident but are found only at the northern end of the reach where the valley floor is at its widest before the Guadalope enters a bedrock canyon at Alcañiz (Figs 2 and 3). Terraces T6 (23 m), T7 (19 m) and T8 (17.5 m) are the most extensive alluvial units and prominent relict alluvial fans developed at the mouths of a

number of tributary streams grade either to T6 or T7 (Fig. 3). Tributary valley alluvial fan deposition appears to have ceased or slowed shortly after the formation of T7 and was followed by tributary streams cutting narrow defiles or 'barrancas' through fan material. Incision of the Guadalope in post-T7 times has been more rapid than its tributaries leaving many hanging (e.g. Val de Carrascosa (F5), Fig. 3) above the level of the trunk river. T9 (12 m) and T10 (9 m) terraces are less extensive than earlier alluvial fills and lie partly (T9) or entirely (T10) within a bedrock trench that is between 150-500 m wide.

The longitudinal profiles of terraces T6, T7, T8, T9 and T10, and the present Guadalope River, are roughly parallel down to valley km 10 (Fig. 2). There is, however, a marked increase in the gradient of the Guadalope between km 10 and 11 which is also evident in T9, T8 and T7 but not in the downvalley profile of T6. Downstream of km 11, the gradients of T7, T8 and T9 decline and diverge from the present channel. The gradients of T3 and T6 between km 11 to km 13, in contrast, are very similar to the current river.

3.2 *Sedimentology of alluvial fills*

Road cuts and quarries provide sections in terraces T1,

T3, T6, T7, T8 and T9; T2, T4, T5 and T10 alluvium is presently not exposed. T1 alluvium underlies the castle at Alcañiz (Fig. 3) where it is poorly exposed in a small 1.5 by 3.0 m face and consists of clast supported and imbricated, subrounded pebble gravels.

Sections in T3 alluvium were examined in two quarries, c. 1 km west and 1.5 km southeast of Alcañiz (Fig. 3). The sedimentary sequence was similar at both sites, comprising two gravel lithofacies 7-10 m thick overlying Miocene sandstone. The lower unit consists of a series of vertically stacked tabular or gently lenticular concave-up gravel bodies 1.5-2.5 m thick that extend along the entire quarry face (c. 50 m). The dominant internal structure in this unit was Inclined Heterolithic Stratification (IHS, Thomas et al. 1987) comprising predominantly of clast-supported gravel alternating with sands and silts with dips ranging from 5 to 20°. Small channels (less than 0.5 m deep and between 3-8 m wide) are evident on the surface of a number of IHS sets and are infilled by fine-gravel and/or cross-bedded sands. Fine-grained overbank (interchannel) deposits were not observed. Prominent erosional channel form scours separate the two gravel units and are infilled by multi-storey, lenticular trough-cross stratified gravel and sands between 2-3 m thick.

IHS evident in the lower gravel member is most typically associated with lateral accretion on point or side bars separated or attached to a channel bank (Arche 1983; Bluck 1971). These bar forms are common in both high and low-sinuosity meandering gravel bed rivers in which there is one main active channel with occasional islands and subsidiary channels. The small sand and gravel-filled channels developed on the surfaces of some IHS units probably represent minor chute channels. The uppermost

and much larger channel forms with multi-storey fills (with each storey bounded by an erosion surface) would appear to have a rather different origin. They may represent major chute channels produced by the principal channel avulsing across the surface of a point or lateral bar, however, considering the relatively restricted outcrop it is unlikely that more than one channel would be evident in section. They are more likely to have been formed by continually shifting channels that braid and rejoin, and constantly change in position as the result of bank erosion and bar development. Similar sedimentary units have been described from Quaternary and older wandering or braided low-sinuosity rivers (Bluck 1979; Ori 1982). Thus there is evidence for a transformation of river pattern from a single thread meandering river to a multi-channel braided system during T3 times.

T6 sediment is exposed in a deep quarry at the southern end of the study reach, 3 km southwest of Castelseras. Alluvium here is more than 15 m thick (the depth to rockhead is unknown) and consist largely of flat-bedded massive matrix-rich gravels with minor horizontally laminated sand and silt beds. Their lower bounding surface is typically erosional and broadly concave in form with units frequently cutting each other producing vertically and laterally stacked trough-shaped bodies 25-30 m wide and 2-4 m deep. The internal structure of most gravel units is dominantly one of poorly defined or lenticular bodies 30-50 cm thick that are comprised both of open-work and matrix-rich gravel. In some sections these pass laterally into planar cross-bedded sandy gravels. The architecture of T6 alluvium, consisting of numerous multi-storey lenticular or tabular intersecting sheets, is typical of a low-sinuosity braided river that is aggrading its channel belt (Miall 1985; Ori 1982). In this river

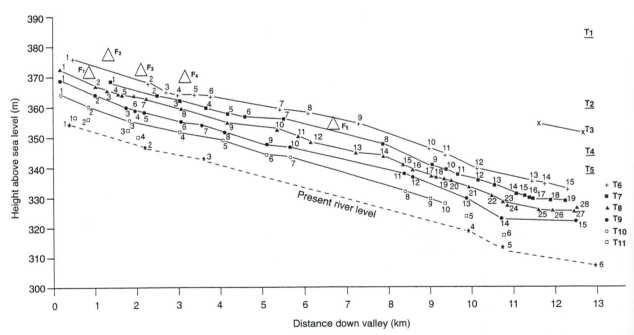

Figure 2. Longitudinal profiles of Quaternary river terraces (T1-T11) in the Guadalope Valley, Alcañiz. Heights of tributary stream alluvial fans (F1-F5) are also shown.

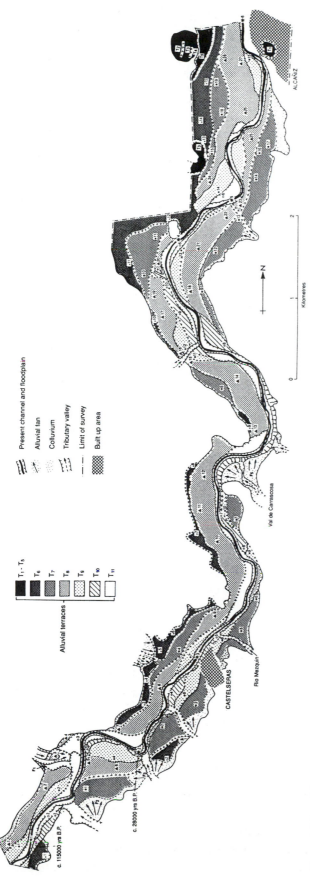

Figure 3. Map of Quaternary river terraces and alluvial fans in the Guadalope Valley, Alcañiz.

environment relatively thick laterally extensive units are generated as a result of the nucleation and lateral migration of complex bar forms (Bluck 1974; Dawson and Bryant 1987).

T7 is exposed at one site only in the study reach in a disused gravel pit 1 km southwest of Castelseras (Figs 1 and 3). where sections are degraded and relatively poor. Up to 10 m of sandy gravels was observed with a sedimentology and geometry similar to T6.

Tributary valley alluvial fan deposits, that are probably coeval with T6 and T7, consist of multi-storey sheets of flat-bedded, clast supported sub angular fine gravels, 0.3 to 0.5 m thick, overlain by silts (less than 0.2 m thick). Harvey (1984) has recognised similar 'sheet gravels' in Pleistocene alluvial fans in southeast Spain, and Bluck (1979) considers such deposits to typify braided streams with wide and shallow channels where bed relief is less than 0.3 m and there are no permanent channel banks. The vertical upward-fining sequence is thought to be produced over a single flood event.

T8 alluvium is particularly well exposed within a quarry 2 km southwest of Castelseras (Figs 1 and 3), in a 250 m long section running normal to the present Guadalope channel. Its sedimentary architecture contrasts markedly with those of T6 and T7, and consists of two main lithofacies: a lower gravel unit varying between 3-5 m in thickness lying on eroded gently dipping Miocene marls and sandstones, overlain by 2-3 m of silty sands. The gravel member is clast supported with a significantly lower matrix content than either T6 or T7. Gravel body thickness varies between 1-1.5 m and individual units can be traced up to 75 m along the quarry face. They have erosional, concave or quasi-planar lower bounding surfaces with an irregular, commonly convex top in which minor channels (3-5 m wide), infilled with crudely cross-stratified sands and fine gravels, have been cut. IHS stratification is particularly well developed within these gravel bodies and in the east part of the section inclined bounding surfaces downlap onto the floor of a major channel, 30 m wide. IHS, however, is not continuous from the top to the bottom of the sequence, and there is a gradational contact with, and no major discontinuities between the lower coarse and upper fine members. The latter unit has a sheet-like geometry and extends across the whole quarry. It consists of horizontal or gently inclined silty sands and massive silts, with occasional lenses or discontinuous layers of fine gravel. The architecture of T8 shares many features common to gravel deposits produced by the lateral migration of a meandering river (cf. Arche 1983; Forbes 1983), namely, horizontally extensive IHS and a thick associated fine member (fine-coarse member ratio > 0.6) which merge in major channels and appear to have accreted in the same direction. These architectural elements are consistent with meander-lobe development (Campbell and Hendry 1987) where fine-grained sediments were deposited either overbank, on a floodplain surface, or within channel, settling out of suspension in an area of slackwater.

Only the upper part of T9 fill has been observed in a road cutting 1.5 km southwest of Castelseras. The section consisted of 4 m of massive and horizontally-bedded sand and clayey silt which fined upward and overlay sandy gravels. Thin argillic horizons (with blocky to prismatic structure) were evident in the profile and indicate periods of soil development with negligible sediment deposition. It is unclear, from the limited exposure at the site, whether this sequence of fine-grained sediments represents overbank (interchannel) deposits or the infill of an abandoned channel. Preliminary quartz surface texture analysis by scanning electron microscopy, however, indicates that some of the units have been modified by wind action.

3.3 *Alluvial archaeology and geochronology*

Numerous chert and flint artefacts were recovered from terrace surfaces during the course of mapping but at two sites only were artefacts found stratified within alluvium. In a disused gravel pit 1 km southwest of Castelseras (Fig. 3) a faceted Levallois-struck flint core was discovered within sandy gravels 10 m below ground level. Recent uranium series dates from El Castillo Cave, Cantabria, northeast Spain (Bischoff et al. 1992) and thermoluminescence dates from the Asprochaliko rockshelter, Epirus, northwest Greece (Huxtable et al. 1992), have demonstrated a concentration of early Mousterian industries of this type both in Western Europe and around the Mediterranean during the last interglacial *sensu lato* (c. 130-80,000 years BP). This may therefore be an appropriate *terminus post quem* for the deposition of T7. Several tens of artefacts were also unearthed in T9 at the interface between the coarse and fine-grained lithofacies of this unit. Unfortunately, none of the artefacts were age diagnostic but all were in mint condition with no edge or face damage. Although they are clearly in a secondary, derived context their concentrated nature and excellent preservation suggests that they are unlikely to have moved far from their place of use or manufacture.

Preliminary numerical ages for silty sands infilling a minor channel within T6, and sands associated with lithic material in T9, have been obtained by Mr I.K. Bailiff (Department of Archaeology, University of Durham) using infra-red stimulated luminescence (IRSL) techniques. T6 sediments gave a luminescence age of 115,000 (±17,000) years BP and T9 material was dated to 28,000 (±4000) years BP. These equate, taking into consideration the error terms in these dates, with oxygen isotope boundaries substage 5e/5d and stage 3/2, respectively (Martinson et al. 1987).

4 RESPONSE OF THE RIO GUADALOPE TO PLEISTOCENE CLIMATE CHANGE AND TECTONICS

A provisional Pleistocene sequence for the Rio Guadalope is outlined in Table 1 and also in Figure 4 which shows a schematic cross-section of T6 and later alluvial terrace units. For comparison the heights, estimated ages

Table 1. Rio Guadalope Quaternary fluvial and archaeological records.

Alluvial terrace unit	Height (m) of terrace surface above present river	Maximum observed unit thickness (m)	Fluvial style	Archaeological record	Estimated age
T1	81	1.5	Gravel-bed river	Unknown	Middle Pleistocene or earlier
T2	56	Unknown	Unknown	Unknown	Middle Pleistocene or earlier
T3	45	> 10	Gravelly high-/low-sinuosity river	Unknown	Middle Pleistocene or earlier
T4	34.5	Unknown	Unknown	Unknown	Middle Pleistocene or earlier
T5	29	Unknown	Unknown	Unknown	Middle Pleistocene or earlier
T6	23	> 15	Gravelly low-sinuosity river	Unknown	Late Pleistocene (115 ka ±17)
T7	19	10	Gravelly low-sinuosity river	Levallois core	Late Pleistocene
T8	17.5	11	Gravelly high-sinuosity river	Unknown	Late Pleistocene
T9	12	> 4	Unknown	Debitage flakes	Late Pleistocene (28 ka ± 4)
T10	9	Unknown	Unknown	Unknown	Late Pleistocene/Holocene
T11	4	Unknown	Unknown	Unknown	Holocene

Table 2. Pleistocene fluvial and archaeological records of the Jarama, Manzanares, Júcar, Serpis and Turia river basins.

River system	Height (m) of terrace surface above present river	Estimated age	Archaeological record
Jarama and Manzanares (Freeman, 1975)	100-110	Middle Pleistocene or earlier	Culturally sterile
	85	Middle Pleistocene or earlier	Culturally sterile
	60-70	Middle Pleistocene or earlier	Culturally sterile
	40	Middle Pleistocene or earlier	Early or Early Middle Acheulean
	20-25	Late Pleistocene (Early Würm?)	Mousterian of Acheulean tradition
	15-18	Late Pleistocene (Early Würm?)	Mousterian of Acheulean tradition, rolled
	4-10	Late Pleistocene (Late Würm)	Aurignacian, rolled
Júcar, Serpis and Turia (Payá and Walker, 1976)	100	Middle Pleistocene	Unknown
	70-80	Middle Pleistocene	Unknown
	30-40	Late Pleistocene (Early Würm?)	'Levallois' struck platform and Mousterian disc core
	15	Late Pleistocene (46 ka ±7)	Unknown
	10	Late Pleistocene (post 29 ka)	Unknown

and archaeological records of Pleistocene river units in the upper Tagus, Júcar and Turia basins, which share a common watershed with the Ebro catchment, are summarised in Table 2. It highlights very clearly how few Pleistocene river terrace sequences in northern and eastern Spain are reliably dated and have independent chronological control, and the heavy reliance on incorporated artefacts for estimating alluvial unit age. However, one interesting point to emerge is the similar height of Late Pleistocene river terraces in the upper Tagus and Guadalope basins, showing between 25-40 m of valley floor incision in piedmont regions of the Iberian massif (possibly in response to long-term uplift) during the Würm and Holocene. Although tentative 'late' and 'early' Würm aggradation phases have been identified in a number of catchments in northern and eastern Spain, dating control is currently inadequate to establish firm inter-basin correlation of Pleistocene fluvial units. Even in the Rio Guadalope where dating provision by comparison is relatively good, it is only Late Pleistocene fluvial events that can be reconstructed with some degree of confidence; Middle Pleistocene and earlier river development remains entirely conjectural.

T6 represents a period of major valley floor aggradation (> 15 m) during which tributary streams constructed alluvial fans at their mouths and delivered large quantities of sediment to the Rio Guadalope. T6 is dated to around 115,000 (±17,000) years BP and would appear to have been deposited at the 5e/5d oxygen isotope substage boundary (Mangerud 1989). Although at present there are no pollen sequences in the Iberian Peninsula that cover this period (the last interglacial complex at Padul, Granada, begins at substage 5c; Pons and Reille 1988), palaeoclimatic reconstructions are available from deep sea pollen records 100 km off the Iberian coast near the Spanish-Portuguese border (Turon 1984) and from a number of mires and lakes in southwest France (see review by Pons et al. 1992). These both show a marked decline in arboreal pollen at the 5e/5d boundary with 5d characterised by low mean temperatures and high annual total precipitation in the early part of the substage. At the Cova Negra Palaeolithic site, Valencia, a phase of cooler and much wetter climate (dated to 117,000 ±17,000 years BP) is also evident at the time (Prószynska-Bordas et al. 1992).

Plant cover, as it is today, was probably rather sparse in the Alcañiz region throughout the Late Pleistocene. Vegetation and soil in this environment would have had a very limited buffering affect on increased precipitation with extra rainfall likely to create a net increase in over-

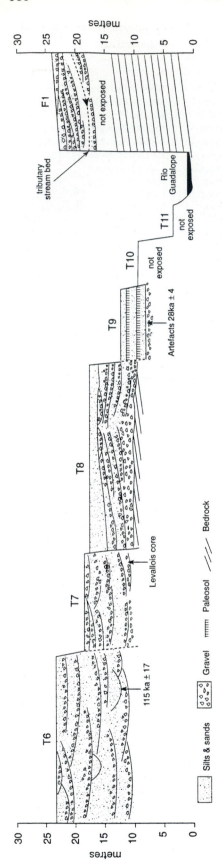

Figure 4. Schematic cross-section of the Guadalope Valley, Alcañiz, showing sedimentary architectures of Late Pleistocene river terrace units and tributary stream alluvial fan.

land flow and erosion significantly augmenting alluviation rates and flood levels along valley bottoms. Reduced evapotranspiration, as a consequence of a cooler climate, is almost certain to have been less important in a semi-arid environment like Alcañiz than in a more humid area because actual losses are limited more by available water (Kirkby 1989). This would appear to contradict the more generally held view that Pleistocene river aggradation in Spain occurred only during dry and cold 'pleniglacial' periods (cf. Payá and Walker 1976). A key factor, however, is likely to have been changes in the seasonality of climate which has been shown to be a critical control of sediment yields and run-off in Mediterranean environments (Kirkby and Neale 1987). Prentice et al. (1992) have recently proposed that the climate of the Last Glacial Maximum in the Mediterranean region was characterised by cold winters and a pronounced summer drought (as at present) but with intense winter precipitation and high runoff. Such conditions would have undoubtedly promoted tributary catchment erosion, hillslope stripping and aggradation of the Rio Guadalope in the middle and lower parts of the basin.

At present, because of poor exposure, it is difficult to determine whether T7 represents a discrete aggradation event or a fill-cut terrace (*sensu* Bull 1991) in T6 (Fig. 4). The former scenario is considered more likely in that, similar to T6, T7 merges and appears to be coeval with a number of tributary alluvial fans. The phase of incision separating alluvial units T6 and T7 may therefore reflect a temporary reduction in sediment supply, although downcutting could have also been induced by increased rates of uplift. The only evidence however for base level change in T7 times caused by tectonics is found in the lower part of the study reach (km 11.5-12.5) where there is a marked flattening of the longitudinal profile of T7 in comparison to T6 (Fig. 2). This anomalous reach is not related to tributary influences and valley-slope deformation could reflect reactivation of the Puig Moreno anticline which may have acted as a dam causing sediment deposition and reduction of channel gradients upstream of the axis of uplift (cf. Ouchi 1985). The relatively low gradients also of T8 and T9 within this reach, particularly in comparison to the present channel longitudinal profile, suggests episodic uplift downstream of Alcañiz continued until well after 28,000 years BP. No geochronometric dates are yet available for T7 but the Levallois core recovered from this unit suggests it was deposited during or after the last interglacial complex (oxygen isotope stage 5).

T8 is the most extensive alluvial unit in the study reach (in places occupying more than half of the valley bottom, Fig. 3) and, as discussed earlier, has a very different sedimentary architecture to T6 or T7; it is presently undated. Its lithofacies assemblages and architectural elements are typical of an actively laterally migrating coarse-grained meandering river (Miall 1985) and it overlies a prominent strath or bevelled-bedrock surface. The predominance of lateral erosion points towards the Rio Guadalope being close to static or dynamic equilibrium (cf. Bull 1991) during this period with rates of channel downcutting broadly matching rates of uplift. All major tributary valley alluvial fans lie above and are truncated by T8 indicating that very little coarse sediment was being delivered to the trunk river by tributary streams. Lower catchment sediment yields could account for the differences in fluvial style (a meandering river replacing a braided channel pattern) between T8 and earlier Late Pleistocene fluvial units (T6 and T7).

The upper fine-grained member of T9 has been dated to 28,000 (±4000) years BP and shows trenching of T8 and refilling of the valley bottom (T9), occurred before the end of oxygen isotope stage 3. Palaeoclimatic reconstructions for this period are, however, problematic as it lies at the limit of the dating range of ^{14}C and also because it was characterised by frequent but relatively small amplitude climatic oscillations. At Padul a clear climate signal cannot be discerned (Pons and Reille 1988) but at Les Echets in southwest France (Guiot et al. 1989) the latter part of stage 3 was cool with relatively low precipitation. Recent analysis of growth frequency variations of secondary carbonate deposits in northwest Europe by Baker et al. (1993) shows a very slight climatic improvement (warm and/or wet conditions ca. 28-31,000 years BP). Evidence of human activity in the Guadalope Valley around this time would also suggest relatively favourable environmental conditions in northeast Spain during this period.

Two alluvial units, T10 and T11, have been deposited in the Guadalope Valley over the last 28,000 years though poor exposure has prevented the recording of their sedimentology or collection of material for geochronometric dating. The extent to which they have been affected by active tectonics has also not yet been fully resolved. Although the relatively steep gradient of the present channel compared with the longitudinal profiles of T7, T8 and T9 would suggest that uplift of the Puig Moreno anticline may have slowed, or even ceased, in recent times allowing the Guadalope to fully adjust to a new base level. It is likely that T11 is of Holocene age.

5 CONCLUSIONS

Geomorphological and archaeological investigations of Quaternary fluvial deposits in the Guadalope Valley, northeast Spain, have revealed an extended record of river response to long-term environmental change. Eleven Quaternary alluvial terraces, ranging from 4 to 81 m above present river level, have been identified along a 13 km reach of the Rio Guadalope south of the town of Alcañiz, Teruel. An older group of river terraces (T1-T5) of Middle Pleistocene age or earlier are poorly preserved but a younger suite of Late Pleistocene terraces (T6-T10) can be traced along the entire study reach. Preliminary luminescence dating of two alluvial units (T6, T9) shows that the formation of the younger group of terraces (T6-T10) spans the last interglacial-glacial cycle. Fill-terrace units range from 1.5-15 m thick and are comprised of gravel-sized material with minor sand members. Progressive incision of the Guadalope valley floor during the Pleistocene would appear to reflect long-term regional uplift with rates of channel downcutting increasing in Middle and Late Würm times, partly in response to declining coarse sediment delivery to the main channel from tributary catchments.

Middle and Upper Palaeolithic artefacts have been recovered from two alluvial units (T7, T9). Early Palaeolithic material, if present in the Guadalope Valley, is likely only to be preserved in older and more fragmentary terraces (T1-T5) set above the more extensive Late Pleistocene alluvium. It is recommended that future archaeological surveys should focus on these deposits.

In common with most river systems in the Iberian Peninsula the chronology of Late Pleistocene river development in the Guadalope basin is relatively poorly controlled and it is difficult to relate individual aggradation and trenching episodes to specific climatic and/or tectonic controls. Provisional dating of major valley bottom alluviation in the Rio Guadalope to a cold but wet phase of the last interglacial complex (oxygen isotope stage 5) would appear to run counter to conventional climate-river response models in the western Mediterranean which considered cold and dry 'glacial' periods to favour river aggradation, principally as the result of reduced vegetation cover and evaporation (Payá & Walker 1976; Sabelberg 1977). It is argued, however, that in semi-arid environments (such as northeast Spain) vegetation is unlikely to have been a major moderator of runoff during the Late Pleistocene (prior to human disturbance) and that catchment water and sediment yields are likely to have been controlled more by changes in precipitation (seasonality and annual totals) than temperature. Nevertheless, geomorphic responses to Pleistocene climate change in the Guadalope basin are likely to have been complex not least because of the additional affect of active tectonics.

Differences between climatic vegetation zones in mountain headwater and piedmont parts of the catchment may have been more accentuated during cooler phases of the Pleistocene with the Iberian Massif contributing a larger proportion of runoff, in the form of snowmelt, than in warmer periods. As a consequence rain-fed tributary streams in the Alcañiz area would have had a markedly different hydrological and sediment regime to that of the Rio Guadalope. This is most clearly shown by the much higher rates of trunk river downcutting after the deposition of the T7 fill which resulted in most extra montane tributary streams being decoupled from the main channel. This may have been caused by a sharp decline in precipitation and runoff in the Iberian piedmont sometime during Würm times. Detailed reconstruction of Late Pleistocene alluvial histories and catchment hydrology in the Guadalope basin, however, must await further dating assays and improved pollen-based vegetation and climate records of the last interglacial-glacial cycle in the northern Iberian Peninsula. Sediment based dating techniques like thermoluminescence (TL) and infra-red stimulated luminescence (IRSL) are likely to be integral to future studies as they offer, for the first time, the possibility of establishing high resolution alluvial geochronologies for the entire Late Pleistocene period.

6 ACKNOWLEDGEMENTS

We wish to thank Tony Stevenson, Basil Davis, José Antonio Benavente and Carlos Navarro for assistance in the field and Ian Bailiff for luminescence dating. MGM is grateful to Newcastle University Research Committee for supporting his investigations in northeast Spain through a research grant.

REFERENCES

Arche, A. 1983. Coarse-grained meander lobe deposits in the Jarama River, Madrid, Spain. In J.D. Collinson & J. Lewin (eds), *Modern and Ancient Fluvial Systems: Sedimentology and Processes.* International Association of Sedimentologists Special Publication 6: 313-321.

Bailiff, I.K. 1992. Luminescence dating of alluvial deposits. In S. Needham & M.G. Macklin (eds), *Alluvial Archaeology in Britain.* Oxbow Monograph 27: 27-35. Oxford: Oxbow Press.

Baker, A., P.L. Smart & D.C. Ford 1993. Northwest European palaeoclimate as indicated by growth frequency variations of secondary calcite deposits. *Palaeogeography, Palaeoclimatology, Palaeoecology* 100: 291-301.

Bischoff, J.L., J.F. Garcia & L.G. Straus 1992. Uranium-series isochron dating at El Castillo Cave (Cantabria, Spain): the 'Acheulean'/'Mousterian' question. *Journal of Archaeological Science* 19: 49-62.

Bluck, B.J. 1971. Sedimentation in the meandering River Endrick. *Scottish Journal of Geology* 7: 93-138.

Bluck, B.J. 1974. Structure and directional properties of some valley sandur deposits in southern Iceland. *Sedimentology* 21: 533-554.

Bluck, B.J. 1979. Structure of coarse grained braided stream alluvium. *Transactions of the Royal Society of Edinburgh* 70: 181-221.

Bull, W.B. 1991. *Geomorphic Responses to Climatic Change.* Oxford: Oxford University Press.

Burillo, F., M. Gutierrez, J.L. Peña & C. Sancho 1986. Geomorphological processes as indicators of climate change during the Holocene in northeast Spain. In V.F. Lopez (ed), *Quaternary climate in the western Mediterranean*: 31-44. Madrid: Universidad Autónoma.

Campbell, J.E. & H.E. Hendry 1987. Anatomy of a gravelly meander lobe in the Saskatchewan River, near Nipawin, Canada. In F.G. Etheridge, R.M. Flores & M.D. Harvey (eds), *Recent developments in fluvial sedimentology.* Society of Economic Paleontologist and Mineralogist Special Publication No. 39: 179-189.

Dawson, M.R. & I.D. Bryant 1987. Three-dimension facies geometry in Pleistocene outwash sediments, Worcestershire, U.K. In F.G. Ethridge, R.M. Flores & M.D. Harvey (eds), *Recent developments in fluvial sedimentology.* Society of Economic Paleontologists and Mineralogists Special Publication No. 39: 191-196.

Forbes, D.L. 1983. Morphology and sedimentology of a sinuous gravel-bed channel system: Lower Babbage River, Yukon coastal plain, Canada. In J.D. Collinson & J. Lewin (eds), *Modern and Ancient Fluvial Systems: Sedimentology and Processes* International Association of Sedimentologists Special Publication 6: 195-206.

Freeman, L.G. 1975. Acheulean sites and stratigraphy in Iberia and the Maghreb. In K.W. Butzer & G.L. Isaac (eds), *After the Australopithecines; stratigraphy, ecology and culture change in the Middle Pleistocene*: 661-743. Chicago: Aldine.

Guiot, J., A. Pons, J.L. Beaulieu & M. Reille 1989. A 140 000-year continental climate reconstruction from two European pollen records. *Nature* 338: 309-313.

Harvey, A.M. 1984. Aggradation and dissection sequences on Spanish alluvial fans: influence on morphological development. *Catena* 11: 289-304.

Harvey, A.M. 1988. Controls of alluvial fan development: the alluvial fans of the Sierra de Carrascoy, Murcia, Spain. *Catena Supplement* 13: 123-137.

Harvey, A.M. & S.G. Wells 1987. Response of Quaternary fluvial systems to differential epeirogenic uplift: Aguas and Feos river systems, southeast Spain *Geology* 15: 689-693.

Huxtable, J., J.A.J. Gowlett, G.N. Bailey, P.L. Carter & V. Papaconstantinou 1992. Thermoluminescence dates and new analysis of the early Mousterian from Asprochaliko. *Current Anthropology* 33 (1): 109-114,

Kirkby, M.J. 1989. Forecast changes in sediment yield due to expected global warming for Mediterranean Spain based on a slope evolution model. In A.C. Imeson & R.S. de Groot (eds), *Landscape ecological impact of climate change on the Mediterranean region*: 1-18. Lunteren.

Kirkby, M.J. & R.H. Neale 1987. A soil erosion model incorporating seasonal factors. In V. Gardiner (ed), *International Geomorphology 1986, Part II*: 189-210. Chichester: John Wiley.

Lewin, J. & B.J. Brindle 1977. Confined meanders. In K.J. Gregory (ed), *River Channel Changes*: 221-233. Chichester: Wiley.

Macklin, M.G., D.G. Passmore, A.C. Stevenson, B.A. Davis & J.A. Benavente 1994. Responses of rivers and lakes to Holocene environmental change in the Alcañiz region, Teruel, northeast Spain. In A.C. Millington & K. Pye (eds), *Effects of Environmental Change in Drylands*: 113-130. Chichester: John Wiley.

Mangerud, J. 1989. Correlation of the Eemian and the Weichselian with the deep sea oxygen isotope stratigraphy. *Quaternary International* 3/4: 1-4.

Martinson, D., N.Pisias, J. Hays, J. Imbrie, T. Moore & N.J. Shackleton 1987. Age dating and the orbital theory of the Ice Ages: Development of a high-resolution 0 to 3 000 000-year chronostratigraphy. *Quaternary Research* 27: 1-29.

Merino, J.K., A. Davo & M.H. Muñoz 1992. Evidencias geomorfologicas de la existencia de actividad neotectonica durante el Pleistoceno en un sector de la zone central de la cuenca del Ebrol. *Estudias de Geomorfologia en España*. 643-651.

Miall, A.D. 1985. Architectural-element analysis: a new method of facies analysis applied to fluvial deposits. *Earth-Science Reviews* 22: 261-308.

Ori, G.G. 1982. Braided to meandering channel patterns in humid-region alluvial fan deposits, River Reno, Po plain (northern Italy). *Sedimentary Geology* 31: 231-248.

Ouchi, S. 1985. Response of alluvial rivers to slow active tectonic movement. *Geological Society of America Bulletin* 96: 504-515.

Palacios, J.L. 1975. Nota Sobre las relaciones de la red fluvial camerana y las tectonica del borde septentional del sistema Iberico. *Berceo* 88: 93-99.

Payá, A.C. & M.J. Walker 1986. Palaeoclimatological oscillations in continental Upper Pleistocene and Holocene formations in Alicante and Murcia. In V.F. Lopez (ed), *Quaternary climate in the western Mediterranean*: 365-376. Madrid: Universidad Autónoma.

Peña, J.L., A. Julian & J. Chueca 1991. Séquences évolutines des accumulations Holocènes a la Hoya de Huesca dans le contexte général du Basin de L'Ebre (Espagne). *Physio-Geo* 22/3: 55-60.

Pons, A., J.L. Guiot & M. Reille 1992. Recent contributions to the climatology of the last glacial-interglacial cycles based on French pollen sequences. *Quaternary Science Reviews* 11: 439-448.

Pons, A. & M. Reille 1988. The Holocene and Upper Pleistocene pollen record from Padul (Granada, Spain): A new study. *Palaeogeography, Palaeoclimatology, Palaeoecology* 66: 243-263.

Prentice, I.C., J. Guiot & S.P. Harrison 1992. Mediterranean vegetation, lake levels and palaeoclimate at the Last Glacial Maximum. *Nature* 360: 658-660.

Prószynska-Bordas, H., W. Stanska-Prószynska & M. Proszynski 1992. TL dating of river terraces with fossil soils in the Mediterranean region. *Quaternary Science Reviews* 11: 53-60.

Radtke, V., H. Brückner, A. Mangini & R. Hausmann 1988. Problems encountered with absolute dating (U-series, ESR) of Spanish calcretes. *Quaternary Science Reviews* 7: 439-445.

Sabelberg, U. 1977. The statigraphic record of late Quaternary accumulation series in southwest Morocco and its consequences concerning the pluvial hypothesis. *Catena* 4: 209-214.

Sancho, C., M. Gutierrez, J.L. Peña & F. Burillo 1988. A quantitative approach to scarp retreat starting from triangular slope facets: Central Ebro basin, Spain. *Catena Supplement* 13: 139-146.

Smart, P.L. 1991. Uranium series dating. In P.L. Smart and P.D. Frances (eds), *Quaternary dating methods – a user's guide* 45-83. Cambridge: Quaternary Research Association.

Thomas, R.G., D.G. Smith, J.M. Wood, J. Visier, E.A. Calverley-Range & E.H. Koster 1987. Inclined heterolithic stratification – terminology, description, interpretation and significance. *Sedimentary Geology* 53: 123-179.

Turon, J.L. 1984 Direct land/sea correlations in the last interglacial complex. *Nature* 309: 673-676.

van Zuidam, R.A. 1975. Geomorphology and archaeology. Evidence of interrelation at historical sites in the Zaragoza region, Spain. *Zeitschrift für Geomorphologie* 19(3): 319-328.

CHAPTER 11

Glaciation, river behaviour and Palaeolithic settlement in upland northwest Greece

JAMIE C. WOODWARD
Department of Environmental and Geographical Sciences, Manchester Metropolitan University, Manchester, UK

JOHN LEWIN
Institute of Earth Studies, UCW Aberystwyth, Aberystwyth, Dyfed, UK

MARK G. MACKLIN
School of Geography, University of Leeds, Leeds, UK

ABSTRACT: Despite long-standing recognition of the widespread occurrence of glacial features and sediments in the headwaters of many Mediterranean river basins, the downstream impact of former glacial activity upon Mediterranean Quaternary river environments is poorly known. The Voidomatis River basin of northwest Greece, however, provides a notable exception as a substantial body of geomorphological data has recently been assembled. A large headwater basin of this river system lies in the high karst terrain of the Pindus Mountains and was glaciated during the Pleistocene on at least two occasions. Evidence is presented from Pleistocene river sediments preserved in the lower reaches of the Voidomatis basin which suggests that the cold stage river maintained a braided planform with a sediment load dominated by glacially-derived material. During the full-glacial period, riverine sediment fluxes were very high and meltwater discharges accelerated sediment delivery from headwater basins to bedrock-confined downstream reaches. At this time extensive spreads of predominantly coarse-grained glacio-fluvial materials built up across the valley floor. During phases of incision following the last glacial maximum, these sediments were terraced and now form prominent landscape features. Glacial activity was a *major* agent of landscape modification in parts of the Mediterranean region and this influence extended well beyond glacier margins by effecting wholesale changes in river regime and depositional environments. In addition, there is also a growing body of geoarchaeological evidence to suggest that, in marginal upland environments, glacial activity may have also influenced the timing of Palaeolithic settlement.

KEYWORDS: Greece, glaciation, Quaternary river environment, sediment sources, aggradation, Palaeolithic settlement, Klithi rockshelter.

1 INTRODUCTION

Wherever European uplands lay above the snowline during glacial stages of the Pleistocene, ice formed upon them and valley and cirque glaciers developed (Flint 1971). This was also the case in many of the high mountains of the Mediterranean lands as evidence for Pleistocene glacial activity is widespread in each of the countries which share the northshore coastline of the Mediterranean Sea (Fig. 1). Glacial landforms and sediments have been identified in northern and central Spain and southern France, along the length of both the Italian and Balkan peninsulas as well as in the uplands of Corsica (Conchon 1986) and Turkey (Messerli 1967). More recently some workers have also reported the presence of glacial sediments from as far south as Crete (e.g. Stewart 1993). The review by Messerli (1967) also documents several locations in Mediterranean northwest Africa where evidence for Late Pleistocene valley glaciation has been found (Fig. 1). To date, however, apart from the Alps themselves, detailed information about the nature, extent and age of the glacial sediments of southern and Mediterranean Europe is not yet available for many areas. Indeed, in comparison to our understanding of the glacial sedimentary sequences in northern and northwest Europe (see Ehlers 1983; Ehlers et al. 1991), the chronology and geomorphological significance of Pleistocene glacial activity in the Mediterranean region is poorly known and recent major reviews have highlighted the scarcity of basic field data (see Denton and Hughes 1981; Sibrava et al. 1986). Existing information is largely restricted to small scale maps which were compiled simply to illustrate the location of ancient glacial deposits in the mountain chains of Europe (cf. Messerli 1967; Flint 1971) (Fig. 1).

Despite long-standing recognition of the widespread occurrence of such glacial features in the headwaters of many river basins, the impact of former glacial activity upon Mediterranean Quaternary river environments is similarly poorly known. One notable exception, however, is provided by the Voidomatis River basin of northwest Greece where a substantial body of geological and geomorphological data has recently been assembled (Woodward 1990; Lewin et al. 1991; Woodward et al. 1992; Woodward et al. 1994). A large headwater basin of this river system lies in the high karst terrain of the Pindus Range and was glaciated during the Pleistocene on at least two occasions.

Using the example of the Voidomatis River this chapter aims to illustrate how glacial activity constituted a *major* agent of landscape change in certain parts of the Mediterranean region and to show that this influence

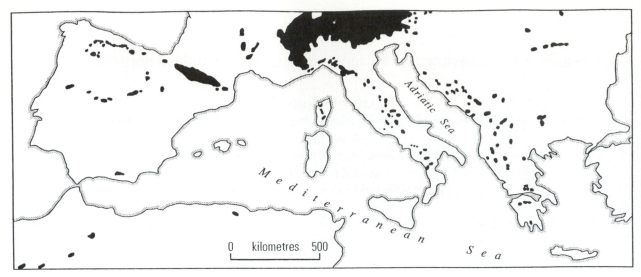

Figure 1. The distribution of Pleistocene glacial sediments and landforms in the Mediterranean region (after Messerli 1967 and Denton & Hughes 1981).

extended far beyond the glacial margins of the upland zone by forcing major changes in river regime and downstream depositional environments. These changes included large increases in both suspended and bed sediment loads, shifts in channel planform and sedimentation style, and substantial increases in rates of valley floor accretion. In addition, there is also a growing body of geoarchaeological evidence to suggest that, in marginal upland environments, glacial activity may have also influenced the timing of Palaeolithic settlement (cf. Bailey et al. 1990). Palaeolithic investigations in the Voidomatis basin began in 1983 and interest has mainly been focused upon the rockshelter site of Klithi located in the lower reaches of the Voidomatis River (see Bailey et al. 1984; Bailey and Gamble, 1990). Two principal archaeological research goals have been identified: (1) to provide a detailed definition of the Palaeolithic sequence at Klithi and the surrounding area and (2) to establish the relationship of this sequence to the environmental changes associated with the last glacial cycle (cf. Bailey et al. 1984).

2 THE GLACIAL DEPOSITS OF GREECE

In the Pindus Mountains, as in most of the world's major mountain ranges, glaciers under average Quaternary conditions extended well beyond cirques at valley heads and during times of glacial maxima, equilibrium-line altitudes (ELAs) fell about 1000 m (Porter 1989). Between Mount Gramos (2520 m) on the Albanian border and Mount Taygetos (2407 m) southwest of Sparta, many peaks in the Pindus Range exceed 2000 m and glacial deposits have been reported from several locations (Figs 1 and 2). One of the earliest accounts reporting evidence for former glacial activity in Greece was published by Niculescu in 1915. This paper describes the glacial features around Mount Smolikas (2637 m), the second highest peak in Greece, which lies within the extensive

ophiolite terrain of northern Epirus in the middle reaches of the Aoos River basin. Two decades later Sestini (1933) published his '*Tracce glaciali nel Pindo epirota*' which described glacial landforms at a number of sites and included an estimate of the position of ice age snowlines across Greece. More recently Pechoux (1970) has reported the occurrence of several small cirques and glacial sediments in the uplands of central Greece in the vicinity of Mount Parnassus (2450 m), Mount Ghiona (2510 m) and Mount Vardhousua (2495 m). In a brief discussion of the Quaternary geology in the Metsovitikos area of central Epirus, Lorsong (1979) has reported the occurrence of debris flow deposits and glacial moraines in the catchment of the Metsovitikos River.

The distribution of glacial features in Greece closely matches the present distribution of average annual rainfall which itself is closely related to topography (Fig. 2). Figure 2 also serves to further emphasise the marked contrasts in river regime which exist across Greece and the rest of the Balkan Peninsula. There is a tenfold increase in annual rainfall moving northwest from southeast Greece (< 400 mm) to parts of upland Epirus and Albania (2000-4000 mm). With such pronounced differences in present and probably past river regime and annual runoff it not surprising that inter-regional correlations between episodes of Quaternary alluviation within Greece have so far proved elusive (see Lewin et al. 1991). In addition, inter-basin comparisons are hampered by the overall paucity and often poor quality of dating control for Quaternary river deposits.

3 THE VOIDOMATIS RIVER BASIN

3.1 *Geological and geomorphological setting*

The Voidomatis basin (384 km²) is located in the Epirus region of northwest Greece and drains part of the high-

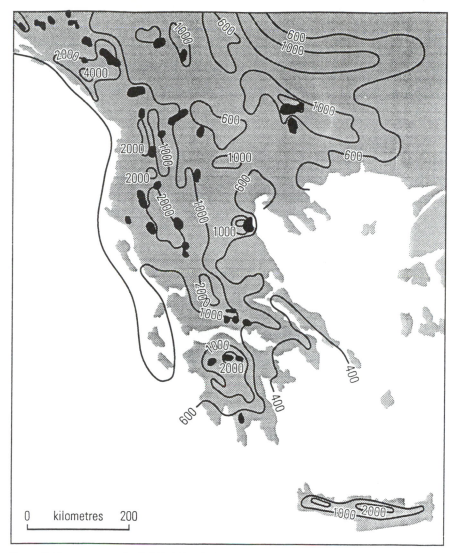

Figure 2. Average annual rainfall (after Osborne 1987) and the distribution of Pleistocene glacial features and sediments in Greece (after Denton & Hughes 1981).

relief, block-faulted karst of the Pindus Mountain Range (Figs 3 and 4). Elevations within the catchment range from c. 450 m on the Konitsa Plain to over 2400 m along the watershed of the Voidomatis and Aoos Rivers. The Voidomatis River is a 4th order, gravel-bed stream of steep average gradient (0.016 m m^{-1}) whose catchment is developed in resistant Jurassic and Eocene limestones which are capped in places by thick flysch beds of Late Eocene to Miocene age. The hard limestone rocks support the development of deep gorges and steep-sided tributary ravines (Vikos Gorge and Lower Vikos Gorge, Figs 3 and 4) whilst, in marked contrast, the erodible flysch terrains are characterised by sub-catchments of lower relief with greater drainage densities and higher present day suspended sediment yields (see Lewin et al. 1991).

3.2 Climatic regime

Epirus forms a zone of transitional climate between the Mediterranean region and central Europe and the heavy snowfalls and freezing temperatures of the Epirus winter are well documented (see Hammond 1967; Furlan 1977). The study area falls within the 'Mountain climate' zone of Walter and Lieth (1960) and mean annual rainfall frequently exceeds 2000 mm (Furlan 1977) (Fig. 2). The summer months are generally hot and dry although heavy thunderstorms are not uncommon in July and August.

4 THE GLACIATED HEADWATERS OF THE VOIDOMATIS BASIN

The limestone massif in the central and eastern part of the basin forms the highest part of the catchment where elevations locally exceed 2400 m. In the district surrounding the village of Tsepelovon the topography is dominated by an extensive range of morphological and sedimentological features indicating recent Pleistocene glaciation (Figs 4 and 5). Large-scale erosional features (glacial troughs and cirques) are carved into the hard limestone bedrock and the widespread deposition of glacially-modified sediments has produced a series of morainic lobes with associated glacio-fluvial landforms. These moraine ridges are capped by boulder-strewn surfaces and form a highly distinctive landform assemblage which blankets almost the entire area north of the modern stream channel (Figs 3 and 4). Large-scale ice-scoured

troughs are present in the high limestone plateau and these features can be seen in the SPOT satellite image shown in Figure 5. All these features serve to demonstrate that ice accumulation in the high Pindus Mountains was *not* merely confined to a few isolated cirque hollows. The major centre of ice accumulation in the Voidomatis basin lay due north of Tsepelovon in the high plateau of the Gamilla Massif.

4.1 *Sedimentology of the Tsepelovon glacial sediments*

When exposed in section, the most striking feature of

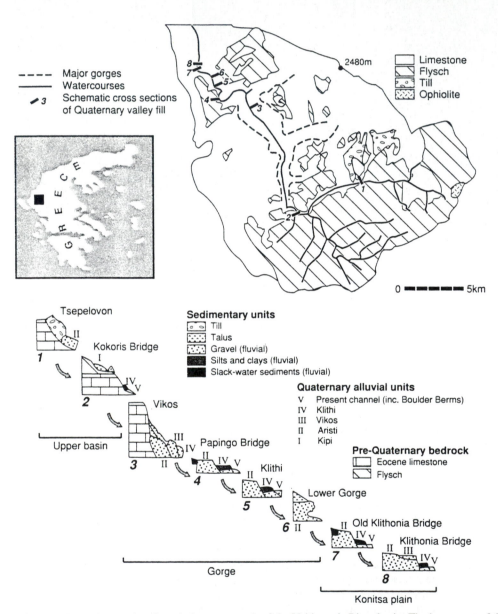

Figure 3. The geology and surface drainage network of the Voidomatis River basin. The lower part of the diagram shows schematic sections of the alluvial valley-fill sequence and the river terrace surfaces at each of these sites. Site 1 is located approximately 1.5 km south of the village of Tsepelovon in the glaciated part of the catchment. The main Vikos Gorge lies between Kokoris Bridge (2) and Vikos (3), and the Lower Vikos Gorge lies between Vikos (3) and the Old Klithonia Bridge (7). The Late Upper Palaeolithic rockshelter site of Klithi is located in the middle reaches of the Lower Vikos Gorge approximately 100 m upstream of site 5. The Voidomatis River joins the Aoos River approximately 5 km north of site 7 in the centre of the Konitsa Plain.

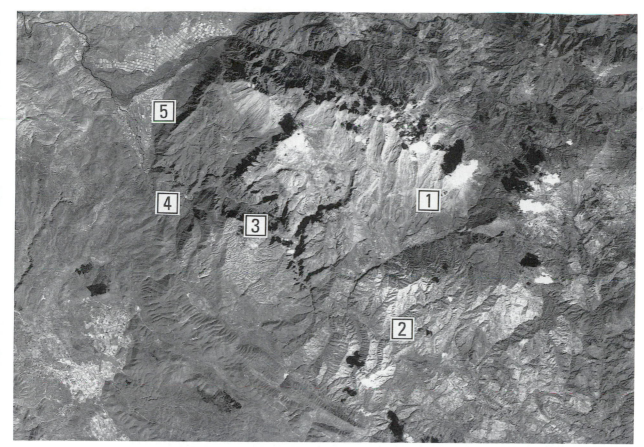

Figure 4. SPOT satellite image of the Voidomatis River basin highlighting the five major physiographic units described by Lewin et al. (1991). This scene provides a graphic illustration of the high-relief, block-faulted terrain of the study region. The scene incorporates all the area shown in Figure 1 and further highlights the distinctive, comparatively high-density pinnate drainage network of the flysch terrain upstream of Kokoris Bridge (site 2). This scene covers an area of approximately 40×28.5 km. Both the Voidomatis and Aoos River systems drain sub-catchments containing Pleistocene glacial sediments and both streams exit fault-bounded limestone gorges onto the Konitsa Plain. The Voidomatis-Aoos confluence is approximately 10 km from the Albanian border. Physiographic units: (1) Glaciated Tsepelovon district; (2) Headwater flysch terrain; (3) Vikos Gorge; (4) Lower Vikos Gorge; (5) Konitsa Plain.

these glacial sediments is their brilliant creamy-white colour which reflects their almost exclusively limestone origin (Fig. 6). These sediments are massive diamictons containing boulders in excess of 1 m in diameter and all size grades finer. The coarse fraction is dominated by limestone clasts (> 95%) with a minor flint and flysch gravel component. Most clasts are sub-angular to sub-rounded in form, unweathered, and frequently have scratched and striated surfaces. The glacial sediments are compact, unsorted and largely unstratified with a closed matrix-supported fabric.

Processes of glacial crushing and grinding have reduced the hard limestone source rock into a chemically-fresh rock flour. This fine matrix is composed largely of calcium carbonate (up to 95%) with quartz and plagioclase accounting for most of the acid insoluble residue – part of which can be accounted for by the presence of a minor flysch component. The mineralogical composition of the two main rock types in the catchment and the till sediments is shown in Figure 7. These peak-height data

from X-Ray Diffraction (XRD) traces indicate a limestone source for the rock flour component of the glacial sediments. The coarse and fine elements of the glacial sediments also contain a very minor flysch component.

In Figure 8 the particle size characteristics of the silt and clay fraction (< 63 μm) of samples of till sediment (taken from the sections in Figure 6) are shown. This component is dominated by silt-grade $CaCO_3$ with a clay content of approximately 20% and median grain size (D_{50}) ranging from 12 to 15 microns. The resistant limestone rocks of the Voidomatis basin can only liberate significant amounts of material of this calibre when subjected to physical comminution processes in a glacial environment (cf. Woodward et al. 1992; Lautridou, 1988).

The glacial sediments are not strongly weathered and soil profiles on moraine crests are generally less than 50 cm thick. This probably reflects the fact that much of the winter precipitation at this altitude (> 1500 m) falls as

Figure 5. SPOT satellite image of the glaciated headwaters of the Voidomatis basin and the distinctive flysch terrain south of the contemporary river channel. This image highlights the marked contrast in slope form and drainage pattern and density in the headwater area. The Voidomatis River flows from right to left in the centre of this image. This image covers an area of approximately 18 × 11.5 km. The glacially-scoured limestone troughs of the Gamilla Massif are clearly visible at the top of this scene.

Figure 6. Top- A section in the Tsepelovon moraine complex exposing limestone-derived till sediments. Bottom- The terminus of a large moraine of Late Würm age adjacent to the bed of the modern Voidomatis River (view looking downstream) due south of the village of Tsepelovon.

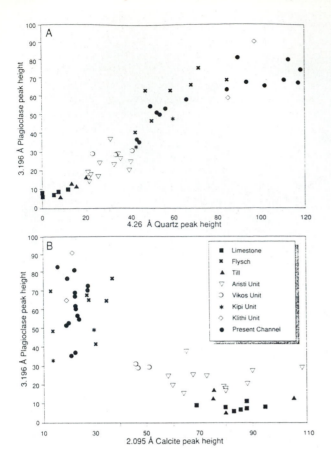

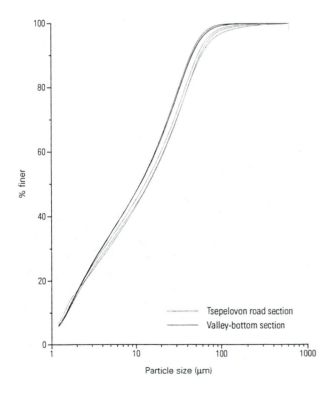

Figure 7. Peak height data from X-ray diffraction (XRD) traces showing the broad mineralogical composition of the < 63 μm component of the Voidomatis glacial and alluvial sediments. A) Plot showing the strong positive correlation between quartz and plagioclase in the basin sediments ($r^2 = 0.86$). B) Plot showing calcite and plagioclase relationships. The Aristi unit sediments are poor in plagioclase and quartz and rich in calcite reflecting their derivation from limestone-rich glacial sediments. In contrast, the Klithi unit sediments are rich in non-carbonate (flysch-derived) materials. The Vikos unit sediments are intermediate between these two groups (see Woodward et al. 1992).

Figure 8. Particle size characteristics of the fine fraction of till sediments exposed in the Tsepelovon road section and the valley-bottom section shown in Figure 6. Sample preparation involved screening through a 63 μm sieve and chemical dispersion in sodium hexametaphosphate. Size distributions were determined using a Malvern Mastersizer Laser Diffraction Particle size analyzer (see Agrawal et al. 1991).

snow and infiltration rates (effective precipitation) are thus much reduced in comparison to conditions in the Lower Vikos Gorge and Konitsa Plain at lower altitude (< 500 m) where strongly leached soils are present on age-equivalent Pleistocene alluvial sediments (Woodward et al. 1994).

5 THE IMPACT OF GLACIAL ACTIVITY UPON THE RIVER ENVIRONMENT

This section discusses the impact of the headwater glaciers upon the Late Pleistocene Voidomatis River environment. Table 1 shows that two alluvial units (Aristi and Vikos) have been dated to this period. Most attention is directed to the properties and significance of the Aristi unit sediments as these are the most extensive and have the largest volume of all the Voidomatis river deposits.

5.1 *The Quaternary alluvial sequence in the Voidomatis basin*

The Quaternary alluvial succession has been investigated through a combination of detailed geomorphological survey and both field- and laboratory-based lithological analyses of valley fill deposits allied to a multi-method (ESR, TL and ^{14}C) dating programme (see Lewin et al. 1991). This sequence embraces at least five major episodes of Quaternary alluvial sedimentation prior to the development of the modern floodplain system and each is followed by a period of incision and soil development. The lithological and sedimentary properties, altitudinal relationships, ages and depositional environments of these units are summarised in Table 1 (see also Lewin et al. 1991; Woodward et al. 1992).

Recent work based on terrace surface soil development in the Lower Vikos Gorge and Konitsa Plain has suggested that the Aristi unit may be subdivided into two distinct alluvial units which are difficult to differentiate using only primary lithological criteria (see Woodward et al. 1994). At present this pre-Late Würm episode of 'Aristi-type' sedimentation has only been identified with certainty at a single location in the catchment. While this chronosequence largely corroborates the model of alluvial history reported by Lewin et al. (1991), there is now some evidence to suggest that the northern Pindus Mountains were also glaciated before the last major (Late Würm) ice advance and that at least two phases of glacial activity can now be recognised in the Greek Pleistocene. Only isolated remnants of this earlier glacio-fluvial aggradation remain and it is the Aristi unit sediments of Late Würm age which dominate the Pleistocene fluvial geomorphology of the basin and provide the main focus for the following discussion.

5.2 *The Late Würm pro-glacial Voidomatis River*

The Aristi unit is believed to be the terraced remnant of a formerly extensive glacio-fluvial outwash train that was deposited by an aggrading, low-sinuosity proglacial stream which drained a series of valley glaciers in the Tsepelovon region (Lewin et al. 1991). This unit forms a striking alluvial terrace averaging 12.4 m above the level of the contemporary stream bed which can be traced almost continuously from the former glacier margins in the Tsepelovon district to the Aoos River junction on the Konitsa Plain more than 40 km downstream. During the Late Würm, channel morphology was influenced largely by meltwater discharges with high sediment loads (Lewin et al. 1991). The Aristi unit sediments consist almost exclusively of massive and flat-bedded, matrix-rich cobble gravels with occasional boulder-sized clasts. Clast imbrication is locally well developed where there is a lower proportion of creamy-white silty matrix.

Figure 6 shows the terminus of a large moraine in the bottom of the present valley system adjacent to the contemporary stream bed of the Voidomatis River. The basal sections of this moraine are being actively eroded by the modern stream where it is possible to observe in section the transition from fully-glacial to fully-fluvial sedimentary facies of the Aristi unit.

5.3 *Lithology of the Aristi unit sediments*

The lithological composition of both the coarse and fine elements of this alluvial unit are similar to the till sediments of the Tsepelovon district. The coarse (8-256 mm) component is dominated by limestone (95.4%) gravels with a significant minor proportion of flint (Table 1) and the fine matrix (< 63 μm) contains up to 85% $CaCO_3$. XRD analysis of the < 63 μm fraction of this unit shows these sediments to be characterised by a large proportion of limestone-derived (calcite-rich) sediment (Woodward et al. 1992).

During meltwater flood discharge events the Tsepelovon glaciers would have provided the major source of alluvial material for the cold stage river system, whereas today, the fine sediment load of the modern Voidomatis River is derived almost exclusively from erosion of local flysch rocks and soils. During the full-glacial period, however, limestone-rich rock flour provided the main source of fine alluvial sediment. Limestone material dominates the *gravel* fractions of the modern river (72.7%) and the late Holocene Klithi unit (69.3%), but the limestone rocks do not supply significant amounts of sediment in the < 63 μm range under modern climatic conditions. The fine matrix of the Late Pleistocene Aristi unit sediments is dominated by limestone-derived sediment as indicated by plagioclase-quartz and plagioclase-calcite relationships (Fig. 7). Solutional processes are important in the contemporary weathering environment and material < 63 μm is not a major constituent of recent (limestone) breakdown products. The limestones supply large amounts of rudaceous (> 2 mm) sediment while any limestone-derived material < 63 μm is largely removed in solution.

The marked contrast between the modern river system and the cold-stage glacial Voidomatis is shown in Figure 7. This simple mineralogical comparison serves to high-

Table 1. Altitudinal relationships, lithological properties, depositional environments and ages of the Quaternary alluvial units in the Voidomatis basin. The dates were obtained by the following methods: a) AMS ^{14}C; b) ESR; and c) TL (see Lewin et al. 1991). Recent work reported by Woodward et al. (1994) – based upon terrace surface soil profile development – has shown that the Aristi unit sediments represent two distinct phases of glacio-fluvial sedimentation (see text).

Alluvial unit	Height of terrace surface above river bed level (m)	Maximum observed thickness of unit	Clast lithological (bedload) composition (8-256 mm)						Coarse (C)/ fine (F) sediment member ratio	Munsell colour of <63 μm fraction	Fluvial sedimentation style and (in brackets) the dominant source of the suspended sediment load	Age of unit (years BP)
			Clasts N	Samples N	% Lime-stone	% Flysch	% Flint	% Ophio-lite				
Present channel			8388	7	72.7	26.6	0.5	0.2	C > F	Yellowish brown 10YR 5/8	Incising, confined meandering gravel bed river. Low suspended sediment load. (Flysch sediments)	<30
Klithi unit	x = 3.2, s = 0.7, Range = 1.8–4.5	4.5	1139	2	69.3	29.6	1.0	0.1	C ⪇ F	Yellowish brown 10YR 5/8	Aggrading, high sinuosity gravel bed river. High suspended sediment load. (Flysch sediments)	1000 (±50)[a] – 30
Vikos unit	x = 6.8, s = 1.7, Range = 3.9–9.7	8.3	695	2	82.3	12.8	0.6	4.3	C ⪈ F	Brownish yellow 10YR 6/8	Incising wandering gravel bed river. Low suspended sediment load. (Flysch & Glacial sediments)	24,300 (±2600)[b] – 19,600 (±3000)[c]
Aristi unit	x = 12.4, s = 3.9, Range = 6.7–25.9	25.9	5680	9	94.6	3.1	2.2	0.1	C ⪈ F	Very pale brown 10YR 8/7-4	Aggrading, low sinuosity, coarse sediment river system. High suspended sediment load. (Glacial sediments)	28,200 (±7000)[c] – 24,300 (±2600)[b]
Kipi unit	56	22.9	361	1	18.7	36.7	0.9	44.0	C > F	Yellowish brown 10YR 5/8	Aggrading (?) low sinuosity, coarse sediment river system. (Flysch sediments)	>150,000[c]

light the dramatic influence of glaciation upon this Mediterranean river system.

5.4 *Dating the glacio-fluvial sediments*

Below an exposure in Aristi unit gravels underlying slackwater sediments at Old Klithonia Bridge (Fig. 3 site 7), part of a red deer jaw bone and a number of Palaeolithic flint flakes were discovered in a small palaeochannel infilled with sandy silts. The sharp, unabraded edges of the lithic material and the fragile jaw bone (which still contained several teeth) indicated that this material was probably *in situ*. The jaw bone was submitted for ^{14}C dating but was devoid of collagen and therefore unsuitable. However, teeth from the mandible were submitted for enamel dating by ESR (using the linear, continuous U-uptake model; R. Grün, pers. comm. 1989) and yielded ages of 24,300 ±2600 (571c), 25,000 ±500 (571a), and 26,000 ±1900 (571b) years BP from three separate dating assays. These ages, together with a TL date of 28,000 ±7100 years BP (VOI23) also obtained from fine sediments at Old Klithonia Bridge, indicate that the Aristi unit, and the glaciation to which it relates, is of Late Würm age.

5.5 *Slackwater sedimentation*

The large proportion of silt- and sand-size matrix in the Aristi gravels indicate that suspended sediment loads were high during the full-glacial period. A closed clast-supported fabric in which all the voids are filled with fine materials is typical of the Aristi unit sediments and thin (up to 20 cm) lenses of exclusively fine-grained (sand- and silt-rich) sediment are occasionally present. The combination of high suspended sediment loads and the ponding of tributary ravines resulted in the development of thick sequences of fine-grained slackwater sediments (cf. Baker et al. 1983) at a number of tributary junctions in the Lower Vikos Gorge. These sediments were laid down in low-energy conditions and individual laminae may contain up to 40% clay and fine silt (Woodward 1990).

5.6 *River regime*

At present, from early summer to late autumn, stream flow is maintained only in the reaches downstream of a major exsurgence in the Vikos Gorge. In the central and upper reaches of the basin main channel flows are ephemeral and controlled by late autumn and winter precipita-

Figure 9. The Klithi rockshelter in the Lower Vikos Gorge. The site is approximately 30 m wide and 10 m deep. It is the largest rockshelter site known in Epirus and was discovered in 1979. Excavations commenced in 1983 led by Geoff Bailey and his co-workers. The site has proved to be immensely rich in both lithic and faunal material with a high density of artefacts in the upper part of the rockshelter sedimentary fill. The large faunal assemblage is dominated by ibex and chamois and the stone tools comprise a typical Late Upper Palaeolithic microlithic flint industry (see Bailey et al. 1986).

tion and snowmelt through the spring. During the Late Pleistocene, however, it is likely that peak flows would have been associated with spring and summer meltwater discharges.

6 PALAEOLITHIC SETTLEMENT AND ROCKSHELTER SEDIMENTATION AT KLITHI

6.1 *Geoarchaeological objectives*

The archaeological research objectives in the Voidomatis basin were broadly twofold: (1) to establish the nature of the Palaeolithic sequence at the Klithi rockshelter (Fig. 9) and (2) to determine the relationship of this sequence to the environmental changes associated with the last glacial cycle (see Bailey et al. 1986). At the beginning of the Klithi project in the early 1980s, it soon became apparent that the sedimentary fill at the Klithi rockshelter would play a central role in any attempt to correlate the environmental ('off-site') and archaeological ('on-site') records. The rockshelter sedimentary sequence thus provides an important bridge between the Palaeolithic record at Klithi and the climatically-driven changes recognised in the Pleistocene river environment (Woodward 1990).

6.2 *Excavation, coring and Palaeolithic occupation*

Since 1983 excavations have focused upon the rockshelter site of Klithi in the Lower Vikos Gorge (Figs 3 and 9). This site contains at least 7 m of Pleistocene sediments – the upper two metres of which are extremely rich in lithic and faunal material (Bailey et al. 1986; Bailey and Thomas 1987). This archaeological assemblage and a

series of radiocarbon dates indicate that occupation of the site took place during the Late Upper Palaeolithic between c. 16,000 and 10,000 years BP, with no evidence of Palaeolithic occupation at this site before or after this period (Bailey et al. 1986; Bailey and Gamble 1990). Evidence for both earlier and later human activity has, however, been found on the Konitsa Plain but *not* in the rockshelter and cave sites of the Lower Vikos Gorge (see Figs 3 and 4).

In 1986, during an experimental drilling programme, a 7 m-long sediment core was recovered from Klithi which provided a valuable window into the nature of the pre-occupation sediments (below 2 m) in the rockshelter sequence (Bailey and Thomas 1987). A detailed study of this core was undertaken by the first author in order to place the rockshelter sediments within an appropriate local environmental context and to establish linkages with the climatically-driven changes recorded in the 'off-site' Pleistocene sedimentary records (Woodward 1995). A range of sediment analyses including detailed particle size analysis, $CaCO_3$ content, mineralogy, and magnetic susceptibility were performed on over 50 samples throughout this sequence and some of the results are shown in Figure 10 and Table 2. The rockshelter sedimentary sequence is described fully in Woodward (1995) and only the main features of the sequence are presented here.

6.3 *The Klithi rockshelter sediments*

The rockshelter deposits are composed largely of unconsolidated angular limestone clasts of various sizes within a predominantly silt-grade calcareous fine matrix. These materials are typically roughly stratified and very poorly sorted.

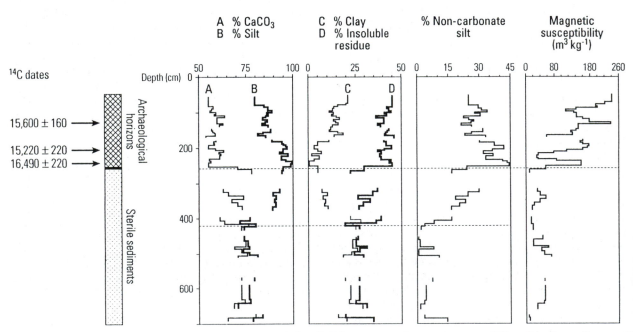

Figure 10. Downcore changes in the particle size characteristics, $CaCO_3$ content and magnetic susceptibility of the fine sediments of core Y25. Depth is in centimetres below the surface. Below the upper dashed line the sediments are archaeologically sterile. Laboratory numbers for radiocarbon dates (top to bottom): OxA – 1155, OxA – 1091, OxA – 1092.

Table 2. Summary of selected sediment property values for Klithi rockshelter core Y25. Ranges and mean values for each core section are shown. The proportion of *non-carbonate silt* has been estimated as the difference between the total insoluble residue fraction and the clay content of each sample. All values refer to the sediment fraction < 63 μm. Bulk magnetic susceptibility measurements (low frequency) were carried out on a mass specific basis on the < 1 mm sediment fraction using a standard Bartington system.

Sediment property	Upper section (0-2.5 m)	Central section (2.5-4.2 m)	Lower section (4.2-7 m)	Whole core (0-7 m)
$CaCO_3$ (%)	54.0-64.0 (58.5)	61.4-78.1 (69.0)	65.2-80.6 (72.9)	54.0-80.6 (64.0)
Clay (%)	0.28-20.5 (10.4)	5.2-27.8 (11.4)	15.7-28.7 (22.2)	0.28-28.7 (13.5)
Non-carbonate silt (%)	21.1-43.9 (31.1)	8.2-29.4 (19.7)	0.9-14.8 (4.9)	0.9-43.9 (22.5)
16-32 μm (%)	26.3-37.9 (31.3)	23.8-30.7 (27.3)	13.3-25.9 (20.7)	13.3-37.9 (27.9)
Magnetic susceptibility ($m^3 kg^{-1}$)	28.1-239.6 (143.3)	8.7-54.5 (27.3)	5.5-68.7 (39.0)	5.5-239.6 (95.7)

Downcore changes in fine sediment lithology within the Klithi rockshelter sequence reflect the changing supply of three main fine sediment sources – namely materials ultimately derived from the limestone and flysch rocks of the catchment and aeolian dust of uncertain provenance. Sedimentological and geomorphological evidence suggests that the basal and archaeologically-sterile portion of the core sequence (below 4.2 m) correlates with the Late Würm Aristi unit of the Voidomatis alluvial sequence (Fig. 10) and thus represents full-glacial conditions in the Voidomatis basin before c. 24,000 years BP (Lewin et al. 1991). The fine matrix of the basal core material is lithologically similar to the fine matrix of the Aristi alluvial unit and is composed largely of silt-grade $CaCO_3$ (up to 80%). This fine material is not a by-product of *in situ* limestone breakdown. Much of this material was blown into the site during the full-glacial period and was locally derived. Fine-grained alluvial materials, deflated from the adjacent Voidomatis floodplain immediately below the rockshelter site, provided the dominant source.

The central and upper sections (above 4.2 m) of the rockshelter core sequence record a considerable change in fine sediment character (Fig. 10 and Table 2). This change in lithology was a response to a decrease in the availability and supply of limestone-derived fines from the floodplain surface of the pro-glacial Voidomatis River. During the Late Würm, as the supply of glacio-fluvial fine sediment waned, the delivery of flysch-derived (quartz-rich) silts increased in importance. A proportion of these quartz-rich silts was also locally derived from local riverine/aeolian sources. The fine fraction of the Vikos alluvial unit contains a significant flysch component (Fig. 7 and Table 1). In addition, a significant proportion of this latter (flysch-derived) non-carbonate fine silt material in the rockshelter deposits was washed down through fissures in the host limestone bedrock (Woodward 1990, 1995).

The Palaeolithic occupation horizons begin in the upper part of the Klithi rockshelter sedimentary record (post c. 16,000 years BP – above 2.2 m in core Y25: Fig. 10). The deposits in this part of the sequence contain a significant amount of non-carbonate silt (Fig. 10) and show evidence of modification by human activity. This human impact is reflected in the magnetic enhancement of the fine fraction due to burning (Fig. 10) and the abundance of human habitation debris and organic material (Bailey and Thomas 1987).

The Vikos alluvial unit was deposited following incision of Aristi unit sediments around 19,600 ±3000 years BP (Table 1). During the deposition of the Vikos alluvial unit and the subsequent phase of incision, the central and archaeologically-rich upper parts of the rockshelter fill were deposited (Fig. 10). It seems likely that rockshelter occupation at Klithi began during incision of the Vikos unit sediments during the climatic improvement which accompanied at least partial deglaciation of the headwater region. At this time the Vikos unit river was flowing at a level up to c. 7 m above present river level, but was probably reworking the extensive spread of Aristi unit sediments and incising into the alluvial valley floor. The presence of glaciers in the basin headwaters appears to have created conditions severe enough to deter any human use of the upland interior at the Last Glacial Maximum (cf. Bailey et al. 1993). Throughout this period attempts to access the Lower Vikos Gorge during the spring and summer may have been thwarted by peak stream flows associated with glacial meltwater discharges.

In summary, the lithological and environmental changes recorded in the Late Pleistocene rockshelter sequence broadly mirror the major changes recognised in the alluvial sedimentary record of the adjacent river environment. The sedimentary sequence at the Klithi rockshelter has provided an important link between the environmental and archaeological records.

7 CONCLUSIONS

The profound impact of glacial activity on Quaternary river behaviour has been well documented in many areas outside of the Mediterranean region (e.g. Rose et al. 1980; Baker 1983; Church and Ryder 1972; Macklin and Lewin 1986). However, despite long-standing recognition of the former presence of glaciers in the headwaters of many Mediterranean river basins, the impact of glacial activity upon Mediterranean Quaternary river environments has received relatively little attention. The example of the Voidomatis River presented in this chapter demonstrates that glacial activity was a *major* agent of landscape modification in parts of the Mediterranean region and that this influence extended well beyond glacier margins by effecting wholesale changes in river regime and depositional environments.

Evidence from Pleistocene river sediments preserved in the lower reaches of the Voidomatis basin suggests that the cold stage river maintained a braided planform with a sediment load dominated by glacially-derived material. Compared to present conditions, sediment fluxes were very high as large suspended and bed sediment loads caused shifts in channel planform and sedimentation style and substantial increases in rates of valley floor accretion. Meltwater discharges accelerated sediment delivery from the headwater basins to bedrock-confined downstream reaches and extensive spreads of predominantly coarse-grained, glacio-fluvial materials built up across the valley floor. During phases of incision following the Last Glacial Maximum, these sediments were terraced and now form prominent landscape features.

In comparison to present conditions, it is clear from the glacial geomorphological evidence that the headwaters of the Voidomatis basin were subjected to a more severe climatic regime for at least part of the last glacial period. In a recent paper Prentice et al. (1992) suggest that the ice age climate of the Mediterranean region was characterised by cold winters, intense winter precipitation and summer drought, and central to this hypothesis is an *increased seasonality* of precipitation as a key to the full-glacial Mediterranean palaeoclimate. Such a scenario is not incompatible with the glacial evidence from the Pindus Mountains. Summer temperatures were probably cooler allowing snowfields to persist and thicken and valley glaciers to develop. A plentiful precipitation supply must have been available to feed the snowfields of the Gamilla Massif.

As far as the archaeological record of the Late Würm is concerned, more information is needed from sites at intermediate altitude (500-1000 m) in the upland interior of the Mediterranean zone. The available geoarchaeological evidence from Epirus suggests that glacial activity not only effected major landscape changes, but may have also influenced the timing of Palaeolithic settlement in marginal upland locations such as the Lower Vikos Gorge of the Voidomatis basin (cf. Bailey et al. 1990). We await with interest the opportunity to compare these findings with data from other glaciated basins in the Mediterranean lands.

8 ACKNOWLEDGEMENTS

Much of the work reported here was carried out while JCW held a SERC Ph.D. studentship at Darwin College and the Subdepartment of Quaternary Research (Godwin Laboratory) at the University of Cambridge. We would especially like to thank Geoff Bailey and all the members of the Klithi Project for their support and also IGME (Athens) for permission to undertake our field research in Epirus. We also thank the sedimentology laboratory in the Department of Earth Sciences in Cambridge for generously allowing access to the SediGraph and XRD facilities. Andrew Teed and Terry Bacon of the Department of Geography at the University of Exeter kindly prepared the photographs and diagrams. We are also grateful to the British Geomorphological Research Group's Research and Publication Fund for providing a research grant to JCW for work on SPOT satellite imagery of NW Greece.

REFERENCES

Agrawal, Y.C., I.N. McCave & J.B. Riley 1991. Laser diffraction size analysis. In J.P.M. Syvitski (ed.), *Principles, methods, and application of particle size analysis*. Cambridge University Press. 119-128.

Bailey, G.N., P.L. Carter, C.S. Gamble, H.P. Higgs & C. Roubet 1984. Palaeolithic investigations in Epirus: the results of the first season's excavations at Klithi, 1983. *Annual of the British School of Archaeology at Athens* 79:7-22.

Bailey, G.N., C.S. Gamble, H.P. Higgs, C. Roubet, D.A. Sturdy & D.P. Webley 1986. Palaeolithic investigations at Klithi: preliminary results of the 1984-1985 field seasons. *Annual of the British School of Archaeology at Athens* 81:7-35.

Bailey, G.N. & G. Thomas 1987. The use of percussion drilling to obtain core samples from rockshelter deposits. *Antiquity* 61:433-439.

Bailey, G.N., J. Lewin, M.G. Macklin & J.C. Woodward 1990. The 'Older Fill' of the Voidomatis Valley Northwest Greece and its relationship to the Palaeolithic Archaeology and Glacial History of the Region. *Journal of Archaeological Science* 17:145-150.

Bailey, G.N. & C.S. Gamble 1990. The Balkans at 18,000 BP: the view from Epirus. In O. Soffer & C.S. Gamble (eds), *The World at 18,000 BP. Volume One: High Latitudes*. London: Unwin Hyman, 149-167.

Bailey, G.N., G. King & D. Sturdy 1993. Active tectonics and land-use strategies: a Palaeolithic example from northwest Greece. *Antiquity* 67:292-312.

Baker, V.R. 1983. Late Pleistocene fluvial systems. In S.C. Porter (ed.), *Late Quaternary Environments of the United States. Volume One: The Late Pleistocene*. London: Longman, 115-129.

Baker, V.R., R.C. Kochel, P.C. Patton & G. Pickup 1983. Palaeohydrologic analysis of Holocene flood slack-water sediments. In J. Collinson &J. Lewin (eds), *Modern and ancient fluvial systems. International Association of Sedimentologists Special Publication* 6:229-239.

Church, M. & J.M. Ryder 1972. Paraglacial sedimentation: a consideration of fluvial processes conditioned by glaciation. *Bulletin of the Geological Society of America* 83:3059-67.

Conchon, O. 1986. Quaternary Glaciations in Corsica. In V. Sibrava, D.Q. Bowen & G.M. Richmond (eds), *Quaternary glaciations in the northern hemisphere.* Quaternary Science Reviews 5:429-432.

Denton, G.H & J.T. Hughes 1981. *The Last Great Ice Sheets.* New York: John Wiley and Sons.

Ehlers, J. 1983. *Glacial deposits in North-west Europe.* Rotterdam: Balkema.

Ehlers, J., P.L. Gibbard & J. Rose 1991. *Glacial Deposits in Great Britain and Ireland.* Rotterdam: Balkema.

Flint 1971. *Glacial and Quaternary Geology.* New York: John Wiley and Sons.

Furlan, D. 1977. The Climate of Southeast Europe. In C.C. Wallen (ed.), *Climates of Central and Southern Europe.* Elsevier Scientific Publishing Company. 185-223.

Grün, R., H.P. Schwarcz & S. Zymela 1987. ESR dating of tooth enamel. *Canadian Journal of Earth Sciences* 24:1022-1037.

Hammond, N.G.L. 1967. *Epirus.* Oxford: Clarendon Press.

Lautridou, J.P. 1988. Recent Advances in Cryogenic Weathering. In M.G. Clark (ed.), *Advances in Periglacial Geomorphology.* Chichester: John Wiley and Sons, 33-47.

Lewin, J., M.G. Macklin & J.C. Woodward 1991. Late Quaternary fluvial sedimentation in the Voidomatis Basin, Epirus, northwest Greece. *Quaternary Research* 35:103-115.

Lorsong, J.A. 1979. Sedimentation and deformation of the Pindos and Ionian Flysches, northwestern Greece. Ph.D. Thesis, University of Cambridge.

Macklin, M.G. & J. Lewin 1986. Terraced fills of Pleistocene and Holocene age in the Rheidol valley, Wales. *Journal of Quaternary Science* 1:21-34.

Messerli, B. 1967. Die Eiszeitliche und die gegenwartige Vergletscherung im Mittelmeeraum. *Geographica Helvetica* 22:105-228.

Niculescu, C. 1915. Sur les traces de glaciations dans le massif du Smolika (chîne du Pinde). *Bull. Sect. Sc. Ac. Roumaine* 4.

Osborne, R. 1987. *Classical landscape with figures: the ancient Greek city and its countryside.* London: George Philip.

Pechoux, P. 1970. Traces d'activité glaciaire dans les montagnes de Grèce centrale. *Rev. Géographie Alpine* 58:211-224.

Porter, S.C. 1989. Some geological implications of average Quaternary glacial conditions. *Quaternary Research* 32:245-261.

Prentice, I.C, J. Guiot & S.P. Harrison 1992. Mediterranean vegetation, lake levels and palaeoclimate at the Last Glacial Maximum. *Nature* 360:658-660.

Rose, J., C. Turner, C.R. Coope & M.D. Bryan 1980. Channel changes in a lowland river catchment over the last 13,000 years. In R.A. Cullingford, D.A. Davidson & J. Lewin (eds), *Timescales in Geomorphology.* Chichester: John Wiley and Sons. 159-175.

Sestini, A. 1933. Tracce glaciali nel Pindo epirota. *Boll. Soc. geog. ital.* 10:136-156.

Sibrava, V., D.Q. Bowen & G.M. Richmond 1986. Quaternary glaciations in the northern hemisphere. *Quaternary Science Reviews* 5:1-511.

Stewart, I.S. 1993. Sensitivity of fault-generated scarps as indicators of active tectonism: some constraints from the Aegean Region. In D.S.G. Thomas & R. Allison (eds), *Landscape Sensitivity.* Chichester: John Wiley and Sons. 129-147.

Walter, H. & H. Leith 1960. *Klimadiagramm-Weltatlas.* Jena.

Woodward, J.C. 1990. Late Quaternary Sedimentary Environments in the Voidomatis Basin, northwest Greece. Ph.D. Thesis, University of Cambridge.

Woodward, J.C., J. Lewin & M.G. Macklin 1992. Alluvial sediment sources in a glaciated catchment: the Voidomatis Basin, northwest Greece. *Earth Surface Processes and Landforms* 17:205-216.

Woodward, J.C., M.G. Macklin & J. Lewin 1994. Pedogenic weathering and relative-age dating of Quaternary alluvial sediments in the Pindus Mountains of northwest Greece. In D.A. Robinson & R.B.G. Williams (eds), *Rock Weathering and Landform Evolution.* Chichester: John Wiley and Sons 259-283.

Woodward, J.C. 1995. The Klithi Sediments. In: G.N. Bailey (ed.) *Excavations at Klithi 1983 to 1988: Palaeolithic Archaeology and Landscape in Epirus, Northwest Greece.* McDonald Institute of Archaeological Research. Cambridge (in press).

CHAPTER 12

Early Neolithic farming in a Thessalian river landscape, Greece

TJEERD H. VAN ANDEL
Department of Earth Sciences, University of Cambridge, Cambridge, UK

K. GALLIS and G. TOUFEXIS
Archaeological Museum, Larisa, Greece

ABSTRACT: The Peneios River, flowing east from the Pindos Ranges, crosses the subsiding Trikala and Larisa basins before entering the Aegean Sea. In both basins it built wide Late Pleistocene floodplains that were settled some 8500 years ago by Neolithic farmers. It has been generally assumed that they farmed small, rain-fed clearings on the elevated river terraces, using the hills for hunting and grazing. In reality the first settlement took place in the floodplain during a phase of aggradation that was probably due to soil erosion. The incision of the river to its present level happened no earlier than 6000 years ago, and only then did the settlements on the old floodplain acquire their present position 3-15 m above the river. A study of the deposits underlying and surrounding Neolithic mounds in the northeastern Trikala basin shows that occupation began on the bank of a creek in an active floodplain, presumably to practice farming on annually freshened silt. Only late in the Middle Neolithic, when the river began to incise and the human accumulation rate overtook natural sedimentation, did the mounds rise above the plain. Neolithic mounds in similar settings exist in the Larisa basin and probably also in Early Neolithic settlements of the southeastern European river plains.

KEYWORDS: Neolithic agriculture, Greece, floodplain farming, soils.

1 INTRODUCTION

The plains of the Trikala and Larisa basins in Thessaly (Fig. 1) are dotted with over three hundred mounds known as magoules (s. magoula) marking the sites of Neolithic and Bronze Age villages. The vast majority rest on the late Quaternary alluvium of the Peneios River and its tributaries (Halstead 1984: Fig. 6.2) now elevated 3-15 m above the present floodplains. Halstead (1981, 1987, 1989) has suggested that the Neolithic settlers farmed light, arable soils watered by rain in small woodland clearings on the elevated river terraces while using the surrounding hill country for hunting, pastoralism and wood-cutting. This fits the widely-held view that Anatolian farmers moving into southeastern Europe successfully applied the dry-land farming practices of their country of origin. The Neolithic pollen record, however, although not easily interpreted in terms of human interference (Bottema 1982), shows no evidence of substantial clearing until about 4000 years ago and can be explained as a result of climatic change, perhaps enhanced by increasing grazing (Bottema 1979; Willis 1992).

During the Late Pleistocene and earliest Holocene, the Peneios and its tributaries created in two steps, represented by the Agia Sophia and Mikrolithos Alluvia, the large alluvial plains that today form Schneider's Niederterrasse (1968). The Niederterrasse was settled more than 8000 years ago, but afterwards the rivers aggraded once more to the level of the old floodplain (Demitrack 1986), possibly because of soil erosion resulting from land clearing (van Andel and Zangger 1990). This middle Hol-

ocene Girtoni Alluvium covers the edges of some Early and Middle Neolithic magoules, showing that they were built on an active floodplain (Demitrack 1986). Subsequently the rivers abandoned the Niederterrasse surface (Fig. 2), and at a much lower level formed the present floodplain (older and younger Peneios Alluvia: Demitrack 1986).

This left the earlier Neolithic settlements stranded well above the river in a setting that was drastically different from that in which they were established, in contrast to the opinion of Jarman et al. (1982) that 'the stratigraphic evidence of the relationship between Neolithic settlements and the modern land surface does not suggest that there has been much change in the relevant period, and the modern situation can probably be taken as a rough indication of pedological conditions at the time'.

Therefore the environment of the Neolithic settlements of the Thessalian plain discussed so extensively by Halstead (1984) should be reconsidered in terms of the type, distribution and age of the deposits surrounding and underlying a representative set of Neolithic magoules. We report here a test of the feasibility of such a study. To render this enterprise practical it must not only be informative in terms of soil conditions, but also so efficient that it permits the study of a large set of mounds without having to incur major costs over a long time. Because the Netherlands Soil Survey has dealt successfully with similar problems in the prehistoric settlement of the lower Rhine Valley (e.g. Edelman 1950; Louwe Kooijmans 1974), we adopted their method of drilling shallow holes (1-3 m) with a Dutch soil auger (Steur 1961). This tool

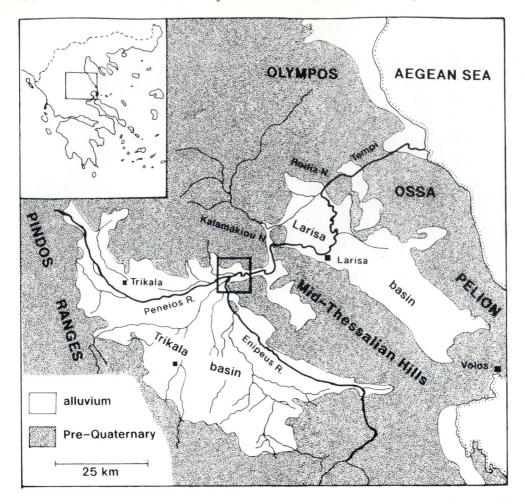

Figure 1. Trikala and Larisa basins in Thessaly, Greece. Small square marks location of study area of Figure 5.

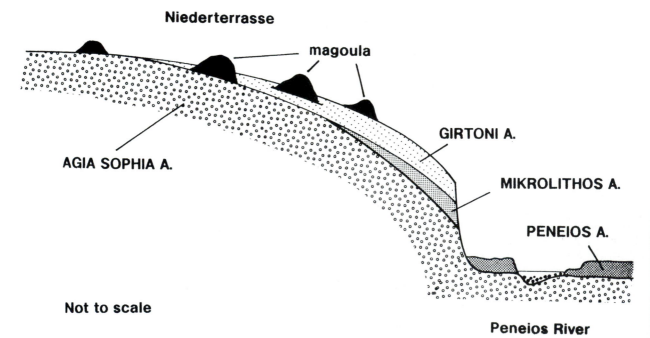

Figure 2. Diagram of the Late Pleistocene and Holocene stratigraphy of the Peneios floodplain in the Larisa basin. Not to scale. Note different stratigraphic positions of magoula bases.

Figure 3. Dutch soil auger with core bits, handle, extension rod, core-cutting spatula, colour chart and notebook.

Figure 4. Coring with the Dutch soil auger.

(Figs 3 and 4) permits convenient coring to about 8 m with bits for various soil conditions. In dry soils, a little water is added to improve cohesion and ease penetration. Stones and clean sand may present difficulties, but in finer-grained deposits much detail can be seen. The cores are laid out at the site, sliced lengthwise with a metal spatula, described and if desired sampled. The 20 cm long core samples are somewhat twisted but little disturbed and are free of internal contamination. However, because of the nature of the coring process (the cores expand when extruded), sediment boundaries tend to have an error of up to ±5 cm and sometimes more.

On the spot, the lithological characteristics (grain size, colour by Munsell Colour Chart, components identifiable with the naked eye or a hand lens, cohesion, bedding and bed contacts, features such as root channels, and the presence of artefacts) were noted on a depth scale in centimetres, and soil horizons described according to Birkeland's scheme (1984).

The dating of the Holocene deposits in Thessaly is not yet robust. Demitrack (1986) dated Larisa basin alluvia and soils with U/Th disequilibrium and uncalibrated ^{14}C data and Halstead's (1984) Neolithic chronology, also based on uncalibrated ^{14}C dates. Recently, Gimbutas et al. (1989) have revised the Early-Middle Neolithic chronology with 'calibrated' ^{14}C dates from the Achilleion mound, but their calibration is unconventional. Because recalibration with the now widely used CALIB programme (Stuiver and Reimer 1986) reveals problems with the phasing (Runnels 1990), we have used only broad definitions of Neolithic boundaries and relied otherwise on Demitrack's data and our own radiocarbon dates for Platia Magoula Zarkou, both calibrated with Stuiver and Reimer (1986). The latter place the Early/Middle Neolithic boundary just before 7700 BP, and not at 8000 BP as Gimbutas et al. (1989) do.

2 THE ARCHAEOLOGY OF THE STUDY SITES

This study centred on a well-preserved mound, the Platia Magoula Zarkou (PMZ) which rises 6-7 m above its surroundings to 94.6 masl and covers 2 ha. It and the smaller Koutsaki Magoula are situated south of the village of Zarkos, where a gently south-sloping alluvial fan merges into a late Quaternary Peneios terrace just downstream from the confluence of this river with the Enipeus, and not far above the gorge through which the Peneios flows into the Larisa basin (Fig. 5).

In 1974 a cremation cemetery was discovered 300 m north of PMZ (Figs 5 and 6). Excavation gave it a Late Neolithic age and showed (Gallis 1982) that in the burials the well-known Tsangli grey ware of late Early Neolithic age co-existed with black Larisa ware then thought to belong to the end of that period.

The burial urns were found dug 30 cm or more into a stiff, greyish black, pebbly loam. The top of this loam was the land surface at that time, because the overlying c. 50 cm of yellowish brown (Munsell 10YR 4/4) sand with fine gravel were undisturbed below a 50 cm thick plough

zone. On the west side of two trenches the black loam itself rested on a sterile, yellowish grey brown (10YR 3/2-4/6) gravelly loam and sandy gravel.

To examine the stratigraphic position of the two wares, an excavation of the Platia Magoula was begun in 1976 (Gallis 1983). It lasted until 1990 when the sterile base of sand and gravel was reached at −10.5 m below a late Early Neolithic (Pre-Sesklo) stratum (Fig. 7) that also filled a shallow ditch in the northwest corner of the trench. Middle Neolithic habitation layers occurred from −9.00 to −5.10 m, followed without a break by the earliest Late Neolithic. As in the cemetery, it was marked by the co-existence of Tsangli and Larisa pottery, thus defining the validity and position of a Tsangli-Larisa phase (Gallis 1985, 1987; Demoule et al. 1988).

After the Tsangli-Larisa phase the site remained unoccupied until the Early Bronze Age. The upper four metres of the section belong to the not yet differentiated Early and Middle Bronze Age. The site was abandoned before Mycenaean times.

It is worth noting here that all strata between −10.00 and −8.50 m, including the living floors, dip 3-6° eastward, implying that the settlement was established on the slope of a terrain rise or on the bank of a gully.

In addition to a 3 m core (PMZ-1) taken near the bottom of the trench for comparison with the excavation profile (Fig. 7), we cored 9 sites near the magoula (Fig. 6), one close to the Peneios River south of the mound, and one near the cemetery. All site locations and their elevations were surveyed.

About 1.5 km southeast of Platia Magoula Zarkou and 300 m west of a large meander of the Peneios River lies Koutsaki Magoula or Zarko 3 (Fig. 5). It is an oval hillock of 1.5 ha and 2 m high, sloping gently north and west and more steeply south and east. The top has been recently levelled. The site has not been excavated, but sherds, stone tools, and figurines profusely litter the surface that belong exclusively to the Early Neolithic, more precisely to phases EN II (Protosesklo) and III (Presesklo). Apparently the site was abandoned earlier than PMZ and failed to develop the high profile of the latter. On a NS line across the mound four cores, their positions and elevations estimated rather than surveyed, were taken.

3 PLATIA MAGOULA ZARKOU IN THE LANDSCAPE

3.1 *The geological setting*

The Quaternary of the Trikala basin is less well known than that of the Larisa basin. Schneider (1968) regarded most deposits as Holocene in age, laid down by the spring floods of the Peneios River and its tributaries which still occur today where they are not controlled by humans (Sivignon 1975). Late Pleistocene and early Holocene deposits are found mainly in the southern half of the basin (Piket 1959). Generally, the Peneios and its tributaries have shallow, braided channels, but where the river ap-

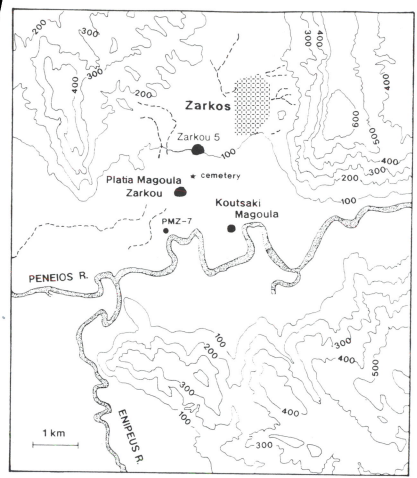

Figure 5. Platia Magoula Zarkou and Koutsaki and the Late Neolithic cemetery on the bank of the Peneios River. Channels of the Peneios and Enipeus rivers stippled. Dashed lines mark seasonal streams draining fans and low wetlands; the one near hole PMZ-7 may connect with the subsurface channel of Figure 10. Contours in metres.

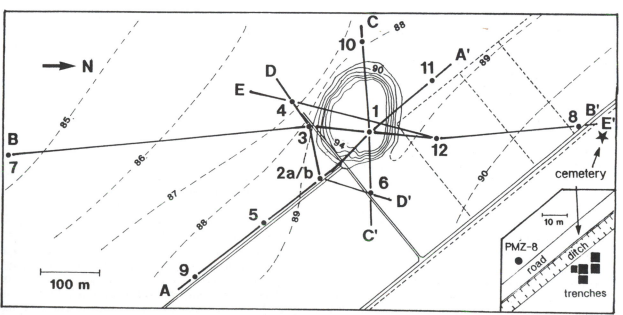

Figure 6. Platia Magoula Zarkou: topography (in metres asl) and core locations. Labeled lines are sections of Figures 8 and 9. Inset: details of Late Neolithic cemetery.

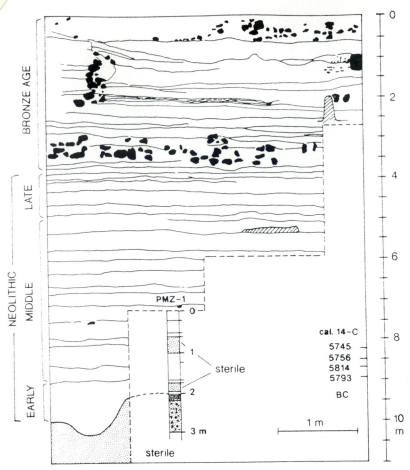

Figure 7. North face of excavation trench on Platia Magoula Zarkou with core hole PMZ-1. Black: Bronze Age structures and stones; stippled: sterile base (thinly bedded, silty river clay over gravelly sand in PMZ-1). Depth in metres below mound surface on right side. Lower right: radiocarbon dates (BC) for Early/Middle Neolithic boundary, calibrated after Stuiver and Reimer (1986).

proaches the Kalamakiou Narrows leading to the Larisa basin, its course and that of the Enipeus, both incised a few metres into a former floodplain, begin to meander (Fig. 5).

The Peneios exits to the sea through three bedrock gorges of which the middle one, the Rodia Narrows, lies in the tectonically most active part of the northern Thessalian basin. The deposition of the Niederterrasse and its later abandonment thus depended on temporal changes of the sediment supply rate, the rate of bedrock erosion in the gorges, and the rate of subsidence of the Larisa basin (Demitrack 1986). Their impact is also evident above the Kalamakiou Narrows and the late Quaternary deposits of the northeastern corner of the Trikala basin reflect a history of floodplain deposition and incision closely related to that of the Larisa basin.

4 THE LATE QUATERNARY DEPOSITS AROUND PLATIA MAGOULA ZARKOU

The auger core descriptions include two separate aspects: (i) depositional features (lithology), and (ii) alteration due to subsequent soil formation. Sediment and soil are different terms relating to different parts of the depositional history, and must not be used synonymously. The lithology depends on environmental conditions during aggradation, whereas soils form when the land surface is stable or subject only to slow or intermittent erosion and deposition. Absence of soils implies a young age (less than a few millennia) or an unstable landscape of rapidly alternating deposition and erosion (human-induced or otherwise).

Quaternary Mediterranean soils have evolved to form a characteristic chronosequence, and their maturity stages tell their relative and, when calibrated, absolute ages (summary in Birkeland 1984; examples in Ajmone Marsan et al. 1988; Busacca 1987; Gile et al. 1966; Harden 1982; Leeder 1975; MacFadden and Weldon 1987; Pope and van Andel 1984; Wieder and Yaalon 1982).

Lithologically, the PMZ sediments form four groups (Table 2). Unit A, at the base of most cores, is the oldest and Unit D, observed only near the modern Peneios channel (core PMZ-7), the youngest. Unit B is lithologically closer to A but much finer-grained, while Unit C resembles the late Holocene loam of Unit D.

Many archaeologists think of the alluvium of large rivers as a heavy clay not suited to primitive farming methods. This is not true. Besides the gravel and sand of former channels, sometimes rather fine and loamy as in Unit B, there are two basic types of river alluvium, the natural levée and overwash fan deposits and those of the backswamps. Natural levées and overwash fans tend to

Table 1. Mid-Holocene chronology (Agia Sophia, Mikrolithos and Girtoni alluvia together form the Niederterrasse).

Alluvia and Soils*	Archaeology**	Alluvia, this paper
New Peneios alluvium		
—— modern ——		
—— historical ——		Unit D
Deleria soil		
—— historical ——		
Old Peneios alluvium		
—— historical ——		
	Bronze Age	
—— < 6000 bp ——	--- c. 6000 bp----	
Girtoni soil		
——6000 bp——	Late Neolithic	?
Girtoni alluvium		
——7000 bp——	---------? ---------	?
	Middle Neolithic	< 7500 bp
	-- c. 7700 bp$ ---	
	Early Neolithic	Unit C (=Girtoni)
Non-calcareous brown soil	--> 8400 bp$$---	> 8200 bp
		?
—— 10,000 bp ——		
Mikrolithos alluvium		Unit B
—— 14,000 bp ——		
—— 18,000 bp ——		
Agia Sophia soil		
—— < 27,000 bp ——		
Agia Sophium alluvium		Unit A
—— 42,000 bp ——		

*After Demitrack (1986, Table 3).

**Generalised from Gimbutas et al. (1986) after provisional calibration with Stuiver and Reimer (1986).

$Based on radiocarbon dates from Platia Magoula Zarkou, calibrated with Stuiver and Reimer (1986); using Kromer et al. (1986) for the interval; range 6843-6723 BP (1σ); 6882-6669 BP (2σ).

$$From an uncalibrated ^{14}C date of 5900 ±45 BP (Demitrack 1986) on the base of an overlying buried Late Neolithic site, calibrated with Stuiver and Reimer (1986) as 6747 BC, range (1σ): 6882-6696.

Table 2. Description of lithological units in Platia Magoula Zarkou cores.

Unit A	Clayey coarse sand and gravel and, more rarely, sandy, gravelly loam; reddish brown (5YR 5/3-4) to brown (10YR 4/3); no or very little carbonate
Unit B	Sandy, gravelly loam and, more rarely, clayey, sandy gravel; dark to very dark greyish or yellowish brown (10YR 3/2-3, 10YR 4/2-6); not calcareous
Unit C	Plastic silty clay or loam; very dark to dark brown or greyish brown (10YR 2/2, 10YR 3/2-3, 10YR 4/2-3); slightly to moderately calcareous
Unit D	Clayey, silty loam; very dark greyish to dark yellow-ish brown (10YR 3/2, 10YR 4/4); moderately to strongly calcareous

Note: Where a soil profile is present, clay content, colour and carbonate content are based on the A, E, or C horizons. Detailed core descriptions available from the senior author on request.

consist of sandy to silty loams, often marked by fine laminations due to episodic flooding and deposition. Such laminations can be seen in Unit D, and in Unit C in a roadside ditch near the cemetery (Fig. 6). They, and Unit B as well, are overwash fan or levée deposits. These are well suited to cultivation even with primitive methods (e.g. Edelman 1950; Buringh and Edelman 1955). In contrast, the heavy, water-logged clays of the low, wet backswamps with their herbaceous, alder and willow cover are only useful for grazing and have only recently yielded to advanced agricultural technology.

Many cores display soil profiles, mainly B horizons because the fragile A and E horizons are usually missing. In different units these possess different characteristics (Table 3) that cannot be attributed to lithological differences alone. The soils of Unit A (best developed in PMZ-4) are the most mature; their colour range of 5YR to 7.5YR, the medium-strong clay development of the Bt horizon, and the abundance of small to medium-sized CaCO₃ nodules in the Bca resemble Demitrack's Agia Sophia Soil (1986) which formed on the Late Pleistocene Agia Sophia Alluvium of the Larisa basin.

The soils of Unit B have a less mature Bt horizon and, being carbonate-free throughout, lack a Bca. In PMZ-8 an upper immature Bt resembling the Girtoni Soil of the Larisa basin overlies a buried A-E-Bt profile close in maturity to the early Holocene Non-calcareous Brown Soil there (Demitrack 1986). The hole bottomed in Unit A deposits with a well-developed, mature A-Bt profile.

The Unit C soil has developed on a finer-grained, more clay-rich substrate than the others. Its Bt horizon has a massive structure and a reddish tinge. A Bca was seen only in PMZ-10 and it is weak. This low CaCO₃ content reminds one of Demitrack's Non-calcareous Brown Soil, but may be due merely to formation in a low-lying part of the alluvial plain in the same manner as described by Piket (1959) for non-calcareous soil profiles in the young backswamps of the northern Trikala basin. In all other ways the Unit C soil corresponds to the Girtoni Soil on Middle Neolithic alluvium (Demitrack 1986).

Unit D, which is calcareous throughout, contains many buried immature A-B horizon couplets spaced 50-100 cm apart. They stand out as slightly more resistant ledges in the low river cliffs of the present Peneios channel, but their weak development suggests that the implied intervals of reduced river flooding were brief.

Within the mound the sediments are mainly, although in the Early Neolithic not exclusively (Fig. 7), anthropogenic. Farther away, the anthropogenic component is reduced to scattered sherds, charcoal fragments and organic matter (e.g. PMZ-4, PMZ-10) that are more probably due to cultivation or refuse disposal than to occupation (Bintliff and Snodgrass 1988; Wilkinson 1982, 1988, 1989). We are not able to draw the limits of true occupation precisely, but they lie close to the mound.

Table 3. Soil horizons in Platia Magoula Zarkou cores.

Unit	Horizon	CaCO₃ content	Colour
A	Bt	None	red (2.5YR 4/6) to yellowish red (5YR 4-5/6, 7.5YR 5/4)
	Bca	Moderate (nodules)	Pale to light brown (7.5YR 6/4, 10YR 6/3)
B	Bt	None	Dark brown to yellowish brown (10YR 4/6 to 4/3)
C	Bt	None/small	Reddish brown (5YR 3/2, 5YR 5/4) yellowish brown (10YR 3/2-4) to brown (7.5YR 4/2-4, 7YR 3/4)
	Bca	Small/moderate	Reddish brown (10YR 5/4), brown (7.5YR 5/4) to dark yellow brown (10YR 4/6)
D	Bt	Moderate/large	Dark reddish brown (5YR 3/4) to brown (7.5YR 4/4)

Note: Clay films, bridges, and alluvial clay are common in the Bt horizons of Units A, B and C. Calcareous nodules occur only in Unit A.

5 DEVELOPMENT OF PLATIA MAGOULA ZARKOU

In the featureless plain sloping gently toward the Peneios River the magoula is a prominent feature. At its deepest point the base lies 4-5 m below the surrounding surface and at least 1-2 m below the top of Unit A. Only the latest Middle and Late Neolithic and the Bronze Age strata rise above the present surface. The mound is steeper now than when it was young, its slopes having been truncated by ploughing and by removal of fertile anthropogenic deposits that are now spread across the adjacent fields. This cover is generally 25-30 cm thick and can be recognized in some borings (e.g. PMZ-5, 6 and 11). The scars of borrow pits dug for this purpose can still be seen on the periphery of the mound.

Cross sections connecting the borings (Figs 8 and 9) reveal complex relationships among the four stratigraphic units defined above. The top of Unit A forms a gently SW-sloping terrace (Fig. 10). Under the eastern half of the mound (Figs 8 and 9) it contains a gully some 75 m wide that trends north towards the cemetery. The present surface shows no trace of it except near PMZ-12 (Fig. 6), because it has been obliterated by the magoula itself and by the deposits of Units B and C.

Near the cemetery the upper soil horizon of Unit B in PMZ-8 (Fig. 8, B-B'), set on the west bank of the gully, is equivalent to the Girtoni Soil of the Larisa basin and probably corresponds to the land surface when the cemetery was in use. The cemetery itself is situated on the fill of the gully near its west bank (Fig. 11: Top) and its burials were cut in a loam which, where it is exposed in the road-side ditch, closely resembles Unit C. Their late Early Neolithic age (Tsangli-Larisa phase) defines the end of the floodplain stage in the area.

The gully was probably part of a channel system connecting the alluvial fan and the backswamps along the foot of the mountains (Piket 1959) to the Peneios; some of it is still visible on aerial photos southwest of the mound and west of PMZ-7 (Fig. 5).

It is obvious that the Early and Middle Neolithic occupation and buildup of the site were contemporaneous with the deposition of several metres of river sand, loam and silty clay of Units B and C. In the vicinity of the magoula, these deposits document the contemporaneity of flooding and occupation in the form of disseminated charcoal, artefacts and disturbed bedding (PMZ-2b, 4, 6 and 10).

In this context the temporal relation between units A, B, and C is of interest (Table 1). Lithology and soil maturity leave little doubt that Unit A was deposited in the latest Pleistocene; subsequently a mature soil was formed, then truncated and covered by Units B and C. Unit B could be a finer-grained, late phase of A, but its two palaeosols are younger; unlike A it was deposited in stages alternating with incipient soil development, and in PMZ-11 and possibly PMZ-2b it interfingers with Unit C. Thus, it probably represents deposition on the toe of the Zarkos alluvial fan and in channels crossing the lower alluvial plain, long after the deposition of Unit A had ceased. Its lower part may be roughly equivalent to Demitrack's (1986) Mikrolithos Alluvium and Non-calcareous Brown Soil.

Unit C is more fine-grained, more plastic and darker than the finest strata of B. Its soil is much less mature than Unit A, but may be the equivalent of the Unit B soils in PMZ-8 (upper) and PMZ-11. It is contemporaneous with the Early and early Middle Neolithic of the magoula (Figs 8 and 9) and we regard it as the equivalent of the Girtoni aggradation phase that terminated deposition of the Niederterrasse in the Larisa basin (Demitrack 1986).

Thus the Early Neolithic settlement appears to have occupied the west bank of the gully at a time when it and its surroundings were still subject to frequent flooding. Flooding is also confirmed by the occurrence within the mound of sterile beds at 60-100 cm and 175-200 cm in PMZ-1 that closely resemble Unit C. We suspect that more flood deposits were once intercalated within the anthropogenic layers, but that all thin ones were destroyed by the occupation of the site. It is therefore not possible to estimate the frequency of local flooding. Underneath the site itself Unit A, which elsewhere has a well preserved Bt horizon (in PMZ-2b, 3, 11 and 12), lacks any trace of soil, so implying removal of as much as one metre of sediment by human beings or by human-induced erosion.

We conclude either that permanent occupation began elsewhere on the Unit A terrace or that it was intermittent and took place only outside the flood season. One needed not go far to find a dry site; the 100 m contour, about 18 m above the present channel (Piket 1959), is only about 1 km away, and the low magoula of Zarkou 5, which was

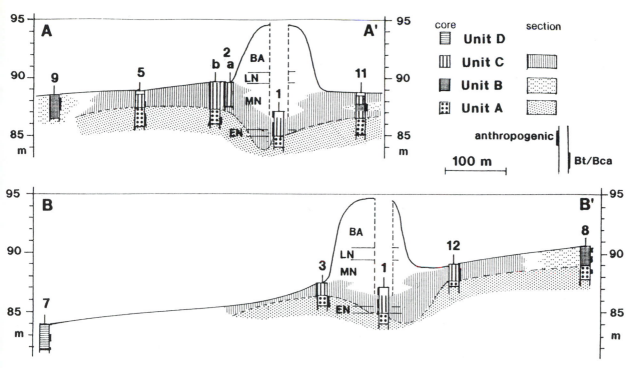

Figure 8. Platia Magoula Zarkou: sections A-A' and B-B'. Elevations in metres asl. Holes 2a and b are adjacent. EN = Early Neolithic; MN = Middle Neolithic; LN = Late Neolithic; BA = Bronze Age. Anthropogenic part of mound in white. Core locations on Figure 6.

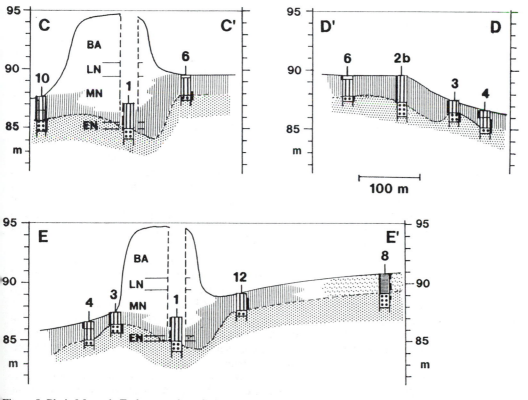

Figure 9. Platia Magoula Zarkou, sections C-C', D-D' and E-E'. Key on Figure 8. Core locations on Figure 6.

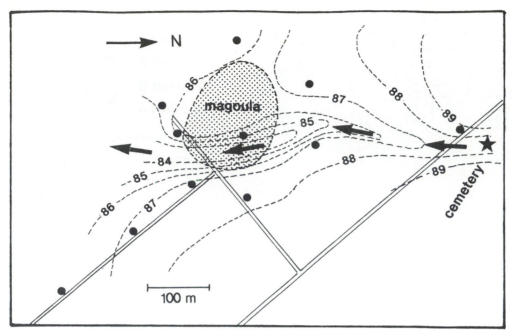

Figure 10. Contour map (metres asl) of Late Pleistocene Unit A at Platia Magoula Zarkou (stippled). Arrows show buried gully; star is Neolithic cemetery. Black dots are core locations.

also inhabited in the Early Neolithic, lies 800 m north of PMZ (Fig. 5). As in the 19th and 20th centuries, the flood season probably lasted from December through May (Sivignon 1975). The most recent Peneios flood took place in March-April of 1982, when a large area in the vicinity of PMZ turned into a shallow lake.

6 KOUTSAKI MAGOULA

Koutsaki Magoula was explored briefly to furnish comparative material to test the validity of the stratigraphy at Platia Magoula Zarkou. The mound proved to be thin: KM-2 (Fig. 11: Bottom) passed through a mere three metres of anthropogenic deposits of Neolithic age, including a hearth with carbonised cereal grains. It ended in a sandy gravel with a well developed red Bt horizon of unquestionably Late Pleistocene age and closely resembling Unit A in lithology and soil maturity. North of the magoula this unit occurred at the same or a slightly lower level, but on the south slope KM-4, less than one metre below KM-2, after passing through 150 cm of anthropogenic deposits, ended in 150 cm of plastic, silty loam identical with Unit C. The same loam occurs in hole KM-1 100 m south of the mound and close to the river. It also forms the upper portion of KM-3 north of the mound (Fig. 11).

The small number of holes and lack of precise elevations do prevent firm conclusions, but it is evident that the depositional units and ages of PMZ are valid at Koutsaki Magoula also and that here also the Early Neolithic deposits formed within an active floodplain. As at PMZ, the floodplain sediments top out well above the earliest human deposits.

7 DISCUSSION

Our observations of early floodplain farming in Thessaly were predicted long ago by Sherratt (1980) for Nea Nikomedia. His view was rejected by Barker (1985) based on a geomorphic study by Bintliff (1976) which, however, rested on a model of Holocene alluviation in Greece that has since proved unworkable.

Clearly, our data set is small and we must resist drawing too many conclusions. Still, the evidence we have from the position of other Early Neolithic sites in Thessaly indicates that some Early Neolithic farmers, accepting some serious drawbacks, selected the sites because they permitted post-flood cultivation of fresh, moist silt. The idea of floodwater farming of wheat, a main staple of the Early Neolithic in Greece, has been unpopular because of its assumed poor yield on wet ground, but Davies and Hillman (1988) have shown that the tolerance of wheat species or of populations within species varies widely. The so-called Pontus population of emmer (*Triticum dicoccum*), for example, is well adapted to wet, heavy soils, and emmer happens to be the dominant cereal cultivated in Early Neolithic Thessaly and in Greece as a whole (Hansen 1988).

These are not the only Neolithic sites in Thessaly that were started on active floodplains. Demitrack (1986) cites eighteen sites which, although set like ours on Pleistocene alluvium, were occupied during the deposition of the middle Holocene Girtoni Alluvium and partly buried under it. One Late Neolithic site was built upon the Girtoni while deposition was still going on (Fig. 2). Also, the cultivation of spring-watered fields has been suggested by van Andel and Runnels (1987, 1988) for the earliest Neolithic at Franchthi Cave and elsewhere in the

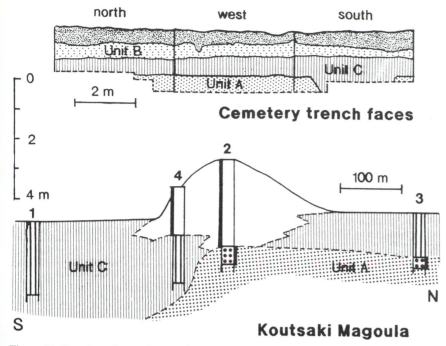

Figure 11. Top: three faces of a trench in Late Neolithic cemetery north of Platia Magoula Zarkou. Uppermost dark shading is the plough zone. Units A, B and C as in Figures 8 and 9. Bottom: NS section; core KM-1 is approximately 200 m from the river. Core key in Figure 8. Distances and elevations estimated. Vertical scale bar valid for both sections.

Peloponnese and is documented from the Levant (Henry 1985; Hopf 1969; Moore 1982).

A floodplain setting may have been chosen elsewhere in southeastern Europe also (Whittle 1987), such as for the levée-based Körös settlements of the middle Danube basin (Barker 1975). In the Great Hungarian Plain the Neolithic sites of the 6th-5th millennia BC were aligned along terrace edges slightly above the present floodplain similar to the sites described above (Sherratt 1982, 1983a, b; Chapman 1989).

These parallels remain speculative, but they add to the attraction of our hypothesis that floodplain rather than dryland farming was practiced in the Early Neolithic at many sites in southeastern Europe. Its corollary is that the early settlements must have been occupied intermittently, perhaps even seasonally, thus casting new doubt on the connection between early farming and sedentism (see also Henry 1985; Edwards 1989). We expect to come back to the archaeological issues raised here in another paper.

8 ACKNOWLEDGEMENTS

This exploratory study was made possible by the resources of the Ephoria of Antiquities in Larisa and by private gifts to Stanford University on behalf of the geoarchaeological research of the senior author. We are grateful to Professor Curtis N. Runnels of Boston University and his field party, in particular Stavros Zabetas and Tom Tartaron, and to members of the staff of the Ephoria for their cheerful field assistance in the hot June days of 1991. Curtis Runnels, Andrew Sherratt and John Chapman commented helpfully on the broader aspects of the subject.

REFERENCES

Ajmone Marsan, F., E. Barberis & E. Arduina 1988. A soil chronosequence in Northwestern Italy: Morphological, physical and chemical characteristics. *Geoderma* 42:51-64.

Barker, G. 1975. Early Neolithic land use in Yugoslavia. *Proceedings of the Prehistoric Society* 41:85-104.

Barker, G. 1985. *Prehistoric Farming in Europe*. Cambridge: Cambridge University Press.

Bintliff, J. 1976. The plain of western Macedonia and the Neolithic site of Nea Nikomedia. *Proceedings of the Prehistoric Society* 42:241-261.

Bintliff, J. & A. Snodgrass 1988. Off-site pottery distribution: A regional and interregional perspective. *Current Anthropology* 29:506-13.

Birkeland, P. W. 1984. *Soils and Geomorphology*. New York: Oxford University Press.

Bottema, S. 1979. Pollenanalytical investigations in Thessaly, Greece. *Palaeohistoria* 21:19-40.

Bottema, S. 1982. Palynological investigations in Greece with special reference to pollen as an indicator of human activity. *Palaeohistoria* 24: 237-289.

Buringh, P. & C.H. Edelman 1955. Some remarks about the soils of the alluvial plain of Iraq south of Bagdad. *Netherlands Journal of Agricultural Science* 3:40-49.

Busacca, A.J. 1987. Pedogenesis of a chronosequence in the Sacramento Valley, California: I. Applications of a soil development index. *Geoderma* 41:123-148.

Chapman, J.C. 1989. Neolithic of southeastern Europe and its Near Eastern connections. *Varia Archaeologica Hungarica* 2:33-53.

Davies, M.S. & G.C. Hillman 1988, Effects of soil flooding on growth and grain yield of populations of tetraploid and hexaploid species of wheat. *Annals of Botany* 62:597-604.

Demitrack, A. 1986. The Late Quaternary Geologic History of the Larissa Plain, Thessaly, Greece: Tectonic, Climatic, and Human Impact on the Landscape. Ph.D. dissertation, Stanford University. Ann Arbor, Michigan: University Microfilms.

Demoule, J.-P., K. Gallis & L. Manolakakis 1988. Transition entre les cultures néolithiques de Sesklo et de Dimini: Les catégories céramiques. *Bulletin de Correspondance Hellénique* 112, Etudes:1-58.

Edelman, C.H. 1950. *Soils of the Netherlands*. Amsterdam: North Holland.

Edwards, P.C. 1989. Problems of recognizing earliest sedentism: The Natufian example. *Journal of Mediterranean Archaeology* 2:5-48.

Gallis, K. 1982. Kauseis Nekroon apo ti Neolithiki Epochi sti Thessalia. Athens.

Gallis, K. 1983. Platia Magoula Zarkou. *Archaiologikon Deltion* 38: Chronika: 201-203 (in Greek).

Gallis, K. 1985. I sosti stromatographiki thesi tis neolithikis keramikis tis gnostis os politismou tis Larisas. *Acts of the First Historical and Archaeological Symposium in Larisa, Larisa: Past and Future, April 26-28, 1985:37-55.*

Gallis, K. 1987. Die stratigraphische Einordnung der Larisa Kultur: eine Richtigstellung. Praehistorische Zeitschrift 62:147-163.

Gile, L.H., F.F. Peterson & R.B. Grossman 1966. Morphological and genetic sequences of carbonate accumulation in desert soils. *Soil Science* 101:347-360.

Gimbutas, M., S. Winn & D. Shimabuku (eds) 1989. *Achilleion: A Neolithic Settlement in Central Greece: 6400-5600 B.C.* Monumenta Archaeologica 14. Los Angeles: Institute of Archaeology, University of California.

Halstead, P.L.J. 1981. Counting Sheep in Neolithic and Bronze Age Greece. In I. Hodder, G. Isaac & N. Hammond (eds), *Pattern of the Past: Studies in Honour of David Clarke*:307-339. Cambridge: Cambridge University Press.

Halstead, P.L.J. 1984. Strategies for Survival: An Ecological Approach to Social and Economic Change in the Early Farming Communities of Thessaly, Northern Greece. Unpublished Ph.D. dissertation. Cambridge: Cambridge University.

Halstead, P.L.J. 1987. Traditional and ancient rural economy in Mediterranean Europe: Plus ca change? *Journal of Hellenic Studies* 107:77-87.

Halstead, P.L.J. 1989. The economy has a normal surplus: Economic stability and social change among early farming communities of Thessaly, Greece. In P. Halstead & J. O'Shea (eds), *Bad Year Economics: Cultural Responses to Risk and Uncertainty*:68-80. Cambridge: Cambridge University Press.

Hansen, J.M. 1988, Agriculture in the prehistoric Aegean: Data versus speculation. *American Journal of Archaeology* 92:39-52.

Harden, J.W. 1982. A quantitative index of soil development from field descriptions: Examples from a chronosequence in Central California. *Geoderma* 28:1-28.

Henry, D.O. 1985. Pre-agricultural sedentism: The Natufian example. In T.D. Price & J.A. Brown (eds), *Prehistoric hunter-gatherers: the emergence of cultural complexity*: 365-384. Orlando, Florida: Academic Press.

Hopf, M. 1969. Plant remains and early farming in Jericho. In P.J. Ucko and G. Dimbleby (eds.), *The domestication and exploitation of plants and animals*:355-360. London: Duckworth.

Jarman, M.R., G.N. Bailey & H.N. Jarman 1982. *Early European Agriculture: Its Foundation and Development.* Cambridge: Cambridge University Press.

Kromer, B., M. Rhein, M. Bruns, H. Schoch-Fisher, K.O. Munnich, M. Stuiver & B. Becker 1986. Radiocarbon calibration data for the 6th and 8th millennia B.C. *Radiocarbon* 28:954-960.

Leeder, M.R. 1975. Pedogenic carbonates and flood sediment accretion rates: A quantitative model for alluvial arid zone lithofacies. *Geological Magazine* 112:257-270.

Louwe Kooijmans, L.P. 1974. The Rhine/ Meuse Delta: Four studies on its Prehistoric Occupation and Holocene Geology. *Analecta Praehistorica Leidensia* 7:1-420.

MacFadden, L.D. & R.J. Weldon 1987. Rates and processes of soil development in Quaternary terraces in Cajon Pass. *Bulletin of the Geological Society of America* 98:280-293.

Moore, A.M.T. 1982. Agricultural origins in the Near East: A model for the 1980s. *World Archaeology* 14:224-236.

Piket, J.J.C. 1959. Een physisch-geografische onderzoeking van het Trikala bekken (west Thessalie). *Tijdschrift van het Koninklijk Nederlandsch Aardrijkskundig Genootschap* 76:379-396.

Pope, K.O. & T.H. van Andel 1984. Late Quaternary alluviation and soil formation in the southern Argolid: Its history, causes and archaeological implications. *Journal of Archaeological Science* 11:281-306.

Runnels, C.N. 1990. Review of: *Achilleion: A Neolithic Settlement in Thessaly, Greece, 6500-5600 B.C.*, by M. Gimbutas, S. Winn & D. Shimabuku. *Journal of Field Archaeology* 17:341-344.

Schneider, H.E. 1968. *Zur quartärgeologischen Entwicklungsgeschichte Thessaliens (Griechenland)*. Bonn: Habelt.

Sherratt, A.G. 1980, Water, soil and seasonality in early cereal cultivation. *World Archaeology* 11:313-330.

Sherratt, A.G. 1982. The development of Neolithic and Copper Age settlement in the Great Hungarian Plain. I The regional setting. *Oxford Journal of Archaeology* 1:287-316.

Sherratt, A.G. 1983a. Early agrarian settlement in the Körös region of the Great Hungarian Plain. *Acta Archaeologica, Academia Scientiae Hungarica* 35:155-169.

Sherratt, A.G. 1983b. The development of Neolithic and Copper Age settlement in the Great Hungarian Plain. II. Site survey and settlement dynamics. *Oxford Journal of Archaeology* 2:13-41.

Sivignon, M. 1975. La Thessalie: Analyse géographique d'une province grecque. *Institut des Etudes Rhodaniennes de l'Université de Lyon: Mémoires et Documents* 17.

Steur, G.G.L. 1961. Methods of soil surveying in use at the Netherlands Soil Survey Institute. *Boor en Spade* 11:59-77.

Stuiver, M. & J. Reimer 1986. A computer program for radiocarbon age calibration; In M. Stuiver & R.S. Kra (eds), *Radiocarbon Calibration Issue: Proceedings of the 12th International Radiocarbon Conference, Throndhjem, Norway. Radiocarbon* 28:1022-1030.

van Andel, T.H. & C.N. Runnels 1987. *Beyond the Acropolis: A Rural Greek Past*. Stanford, California: Stanford University Press.

van Andel, T.H. & C.N. Runnels 1988. An essay on the 'Emergence of Civilisation' in Greece and the Aegean. *Antiquity* 62:234-247.

van Andel, T.H. & E. Zangger 1990. Landscape stability and destabilisation in the prehistory of Greece. In S. Bottema, G. Entjes-Nieborg & W. van Zeist (eds), *Man's Role in the Shaping of the Eastern Mediterranean Landscape*:159-182. Rotterdam: Balkema.

Whittle, A. 1987. Neolithic settlement patterns in temperate Europe: Progress and problems. *Journal of World Prehistory* 1:5-51.

Wieder, M. & D.H. Yaalon 1982. Micromorphological fabrics and developmental stages of carbonate nodular forms related to soil characteristics. *Geoderma* 28:203-220.

Wilkinson, T.J. 1982. The definition of ancient manured zones by means of extensive sherd-sampling techniques. *Journal of Field Archaeology* 9:323-333.

Wilkinson, T.J. 1988. The archaeological component of agricultural soils in the Middle East: The effects of manuring in Antiquity. In W. Groenman van Waateringe & M. Robinson (eds), *Man-made Soils*:93-114. Oxford: British Archaeological Reports, International Series 410.

Wilkinson, T.J. 1989. Extensive sherd scatters and land-use intensity: Some recent results. *Journal of Field Archaeology* 16:31-46.

Willis, K.J. 1992. The late Quaternary vegetational history of northwest Greece: III. A comparative study of two contrasting sites. *New Phytologist* 121:139-155.

CHAPTER 13

Quaternary valley floor erosion and alluviation in the Biferno Valley, Molise, Italy: The role of tectonics, climate, sea level change, and human activity

G.W. BARKER
School of Archaeological Studies, University of Leicester, Leicester, UK

C.O. HUNT
Department of Geographical and Environmental Sciences, University of Huddersfield, Huddersfield, UK

ABSTRACT: This paper describes the results of an integrated programme of archaeological and geomorphological fieldwork in the Biferno Valley in the Italian region of Molise. The archaeological survey was one of the largest thus far conducted in the Mediterranean, and was combined with the excavation of a number of prehistoric, Roman, and Medieval settlements, documentary research, and palaeoenvironmental studies. In the Pleistocene, valley alluviation responded to both tectonic and climatic change. In the Holocene, major episodes of valley sedimentation occurred in prehistoric, classical, medieval, and post-medieval times. Whereas for the Pleistocene the survey found only sparse traces of Palaeolithic settlement, during the Holocene all the episodes of alluviation coincided with significant intensifications in settlement forms and land use. The repeated correlations between human activity and valley sedimentation are one of the most striking results of the Biferno Valley study.

KEYWORDS: Agriculture, alluviation, archaeology, Biferno Valley, climatic change, erosion, field survey, Italy, land use, sea level change, tectonics.

1 INTRODUCTION

Ever since the publication of Vita-Finzi's classic study of Mediterranean valley alluviation (Vita-Finzi 1969), the possible roles of climatic change and human action in shaping the Holocene Mediterranean environment have been strongly debated. Summaries of the arguments are given by Bell (1982), Bintliff (1992), Boardman and Bell (1992) and Hunt et al. (1992). Vita-Finzi concluded that climatic change was the primary factor responsible for the major phase of Holocene alluviation which he identified throughout the Mediterranean and termed the Younger Fill, dating it from its association with archaeological features and inclusion of Roman potsherds to the early-to-mid centuries of the first millennium AD. Since then, a variety of earlier and later episodes has been identified by geomorphologists in various parts of the Mediterranean basin, and, like the classical Younger Fill itself, these are mostly ascribed by the fieldworkers to human impact on the environment (in terms of deliberate deforestation for cultivation or as a result of overgrazing), rather than to climatic change (e.g. Bell 1982; Chester and James 1991; Davidson 1980; Hunt et al. 1992; Pope and van Andel 1984; van Andel and Zangger 1990). Proponents of the original thesis, however, point out that the aggradation episodes dated generally to classical times are significantly different in their scale and characteristics from earlier and later aggradations, suggesting that, whilst land use systems may well have been a contributory factor in their genesis, climatic change was probably still the prime mover (e.g. Bintliff 1976; Gutierrez-Elorza and Pena-Monne 1990).

In order to evaluate the respective roles of climate and people in shaping the Mediterranean landscape, we need to investigate Mediterranean valleys with integrated methodologies linking geomorphology, archaeology and history, so that we can compare reliable evidence for environmental change with similar evidence for settlement history in the same area. This paper presents the results of a project that was planned with these goals, a study of the Biferno Valley in central-southern Italy. The Biferno River is the principal river system of the Italian region of Molise (Fig. 1). With a total catchment measuring some 100 km in length by 30 km in width, a topography rising from sea level to almost 2000 m, and a geological structure that is typical of the eastern side of the Italian Peninsula, the valley seemed an ideal study area for the investigation of the long-term relationship between a Mediterranean people and their landscape (Barker, 1991, 1994).

The study area for the project (Fig. 2) was principally defined by the watershed of the Biferno Valley and its main tributary stream, the Cigno, though in the lower valley we also included the catchment area of the small stream to the north of the Biferno, the Sinarca, and limited survey took place on the floors of the high basins of the Matese Mountains just beyond the Biferno watershed. The form of the Matese is typical of the limestone Apennines: steep ridges rising to some 2000 m above sea level enclose a series of karstic basins (*altipiani*), the floors of which are at about 1000 m. The tributary streams of the Biferno gather in the large intermontane basin north of the Matese named after its principal settlement, Boiano. The floor of this, covered by fine alluvial

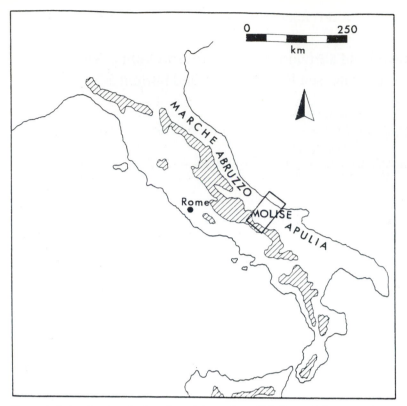

Figure 1. Italy, showing the location of the region of Molise and the approximate area of the Biferno Valley survey. The hatching indicates land over 1000 m above sea level.

sediments, is at about 500 m. The river flows northeast of here for some 60 km to the sea at Termoli. The geology of the middle section of the valley consists of soft Tertiary sands and clays interspersed with limestone outcrops, an extremely unstable landscape prone to massive landslips. The topography on either side is steep, the watershed being 200-300 m higher than the river in the upper section and 500 m or more downstream. The bottom of the valley gets increasingly narrow in this section, and some 20 km from the sea the river passes through a narrow gap in the hills that is now dammed. Below the dam the river meanders across a wide floodplain. The topography of the lower valley, underlain by Plio-Pliocene marine siltstones, is much gentler and more rolling than upriver, the land rising generally to 200-300 m above sea level.

The main field programme took place between 1974 and 1978, though material studies and other fieldwork continued through the 1980s. Some 350 km^2 were investigated by systematic field-walking, almost a third of the total catchment (Fig. 2), making this project the second largest systematic archaeological survey in Italy (and we think the Mediterranean) after the South Etruria survey conducted by the British School at Rome in the 1950s and 1960s (Potter 1979; Ward-Perkins et al. 1968). To complement the survey record of site distributions and densities, a representative series of occupation sites found by the survey, of each major period of settlement from Neolithic to Medieval, was investigated by a combination of geophysical survey and excavation. In part, the

excavations were to seek stratified artefact sequences to enable us to build up control typologies as yardsticks against which to classify the surface data. The second priority was to recover samples of animal bones and plant remains to inform on former economic, particularly agricultural, practices.

The goal of the geomorphological fieldwork was twofold: to assist in understanding the modern distributions of the surface archaeological materials and sites, and more particularly to contribute towards an explanation of changing settlement patterns in antiquity. Regarding the first problem, geomorphology was essential to understand whether blank areas of the archaeological map of a particular period were blank because the people of the time chose not to put their settlements there, or because those settlements, though once there, could no longer be found as surface traces because of later processes of erosion, colluviation, and alluviation. Regarding the second, we hoped that studies of the valley sediment stratigraphies would generate models not only of past climates and environments, but also of the relative impact of human settlement on the landscape in terms of the size of the cultivation area and the intensity of systems of land use. Preliminary geomorphological studies were carried out during the archaeological survey by Derrick Webley, followed by a more detailed study in the mid 1980s by Chris Hunt. The results are summarised in the following section. Dates are given as radiocarbon years bc (uncalibrated), or historical dates BC/AD.

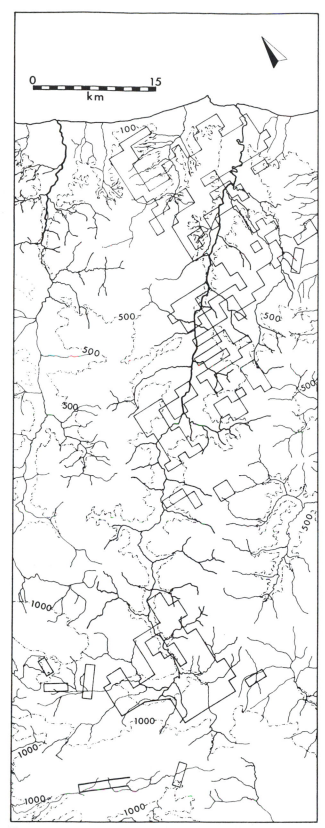

Figure 2. The Biferno Valley, showing topography and the zones selected for the archaeological survey (boxes). Contours in metres.

2 THE GEOMORPHOLOGICAL SURVEY: THE PLEISTOCENE SUCCESSION

The Pleistocene and Holocene succession described below has been established from the investigation of some 40 exposures of sediments throughout the Biferno Valley, combined with geomorphological mapping (Fig. 3). The exposures were selected to be representative of the succession, but it is acknowledged that, given the relatively short period of time in the field and the chance nature of the exposures in a largely vegetated landscape, the selection may be incomplete or subject to some form of sampling bias. The exposures are described in detail in the final report on the Biferno Valley project (Barker 1994).

A number of sediment facies can be distinguished in the deposits examined in the Biferno Valley. These are briefly described and explained in Table 1. The Quaternary stratigraphy proposed for the Biferno Valley on the basis of these facies is summarised in Table 2.

In the landforms and Pleistocene and Holocene deposits of the Biferno Valley lies the evidence of the palaeoenvironments and processes which shaped the landscape we see today. The broad form of the present landscape, with the Matese Mountains overlooking the Boiano basin and with hilly upland between the Boiano basin and the coast, was established by the end of the Pliocene, although relief was probably less pronounced than today and the coastal lowlands were under the sea at that time.

The oldest terrestrial Pleistocene deposits in the Biferno Valley are the Quirino gravels and de Francesco beds of the Boiano basin, the Gentile gravels of the middle valley, and the Guglionesi beds of the lower valley (see Table 2 and Fig. 3). It is impossible to correlate these deposits accurately, but they reflect alluvial fan and lacustrine sedimentation in the Boiano basin, the presence of the middle valley gorge, and a short-lived phase of

Table 1. Sedimentary facies distinguished in the Biferno Valley; terminology after Miall (1977); and Rust and Koster (1984).

Abbreviation	Description
Dm	Diamicton – an unsorted deposit with clasts of all sizes 'floating' in a muddy matrix
Gm	Clast-supported, commonly imbricated gravel with subhorizontal gravel
Gms	Muddy matrix-supported gravel without imbrication or internal stratification
Gt	Trough cross-bedded clast-supported gravel
Gp	Planar cross-bedded clast-supported or matrix-supported gravel
Ge	Epsilon cross-bedded, imbricated sheets
GFp	Planar gravel and sand/mud sheet
Sh	Horizontally stratified sand
Sm	Massive sands
St	Trough cross-stratified sands
Sp	Planar cross-bedded sand
Fm	Massive mud or fine sandy mud
Fl	Laminated silt or mud
P	Pedogenic concretionary carbonate
C	Cryoturbation and other frost disturbance

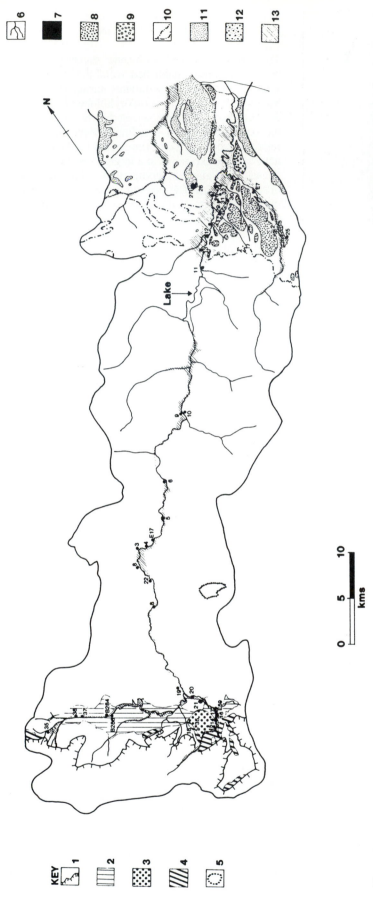

Figure 3. The Biferno Valley, showing the location of the sedimentary deposits identified in the text (see Table 2). The numbered key is as follows: 1. front of Matese limestone massif; 2. ancient basin fill – Quirino and De Francesco beds; 3. Campochiaro gravels; 4. talus, scree; 5. Campobasso lake deposits; 6. rivers; 7. Guglionesi beds; 8. S. Columba gravels; 9. Ripetello gravels; 10. Paledri gravels; 11. marine beds, Pleistocene; 12. Castello gravels; 13. valley floor deposits. The small numbers refer to the locations of the main sedimentary sections examined.

Table 2. Quaternary succession in the Biferno Valley.

Lower Biferno	Middle Biferno	Boiano basin
Late Pliocene		
Marine deposition		Termine silts Fm
Early Pleistocene		
Marine deposition	Gentile gravels Gp Ge Fm	Quirino gravels Gp Fm
Guglionesi beds GdP FmP	De Francesco beds Gp Ge Fm	
Marine/dune deposition		
major uplift/incision		
S.Columba gravels GeP FmP GdP	Campochiaro gravel Gm Gp Ge Fm	
major uplift/incision		
?Middle Pleistocene		
Ripetello gravels, Castello gravels GpC GeC Ge GFp		
uplift/incision		
Paledri gravels GpC		
uplift/incision	incision	
Valley floor 10 GtCP	Moline gravels GdP Gt	
Last interglacial		
Valley floor 9 FlP	Moline gravels FmP	
Interstadial during last glacial period		
Valley floor 8 GtP		
?Glacial maximum		
Valley floor 7 FmC		
?Neolithic/Bronze Age		
Valley floor 6, 5 Ge Fm	Valley floor 5 Ge Fm	Valley floor 8, 7 GFp Gp
Late Samnite/Roman		
Valley floor 4 Gp Ge GFp	Valley floor 4 Ge Sh Fm	Valley floor 6 Gp Ge GFp
	Tre Archi beds GFp	
Medieval		
Valley floor 3 Ge Fm	Valley floor 3 Ge	Valley floor 5 soil
Early post-medieval		
		Valley floor 4 Ge
Late 19th century		
		Valley floor 3 Gp GFp Fm
1930-1950		
	Valley floor 2 Ge Sp Fl	
1960-1975		
Valley floor 2 Gp Fm		Valley floor 2 Gp GFp
Modern		
Valley floor 1 Gp Gt Fm	Valley floor 1 Gp Fm Fl	Valley floor 1 Gp Gt Fm

marine regression and river gravel deposition in what is now the lower valley. Deposition probably took place over a considerable period of time, and the climate probably varied from temperate to cold-arid.

After a substantial phase of earth movements and erosion (these early deposits are tectonised), the sea retreated finally from the lower valley, and alluvial fan gravels – the S. Columba gravels – was laid down. The Matese was also substantially raised by these earth movements and a great volume of alluvial fan gravels – the Campochiaro gravels – were laid down. The Campochiaro gravels contain pollen of extinct Fagaceae (*Tricolpopollenites liblarensis fallax*) and Aquifoliaceae (*Tricolpopollenites margaritatus*) and exotic taxa such as *Scia-*

dopitys, Rhus, Cedrus, Liquidambar, Juglans and *Pterocarya*, as well as extinct terrestrial molluscs, and are thus assumed to be of Lower Pleistocene age. They were laid down during a temperate phase and then a phase of subarctic climate, and are overlain by two substantial fossil soils.

During the Middle Pleistocene, at least three further phases of uplift occurred, and each time the Biferno incised its bed. Further terrace gravels – the Ripitello, Castello and Paledri gravels – were laid down in the lower valley and Cigno Valley and coarse gravels – part of the Moline gravels – in the Biferno gorge. Most are coarse planar and trough cross-bedded gravels with occasional *Pupilla muscorum* probably laid down in cold, arid climates and some of the terraces have a cover of loess-like silts which are also likely to be the product of cold-stage conditions. The Castello gravels also include a warm stage unit laid down by a meandering channel and with a temperate scrubby woodland and marsh mollusc assemblage including *Pomatias* spp., Clausiliidae, *Rumina decollata*, *Trichia striolata* and *Helicella itala*.

The Biferno had incised almost to its current altitude in the lower valley by the cold stage preceding the last interglacial, but further earth movements occurred at this time upstream, probably around the top of the middle valley gorge. Gravels and channel fills of this age are widespread in the gorge (part of the Moline gravels) and are also found in the lower valley (Valley floor 10), but although artefacts of Middle Palaeolithic type were found, the deposits are otherwise without fossils. However, the sedimentary style – epsilon cross-bedded fine gravels and silty palaeochannel fills – is consistent with fully temperate conditions and a well-vegetated landscape during the last interglacial. Following deposition of fluvial sediments, pedogenic carbonates indicate a period of warm-arid climate. The last glacial period is represented only in the lower valley by gravels with an interstadial pollen flora (Valley floor 8) of grasses, sedges, herbs, birch and buckthorn, and trough cross-bedded cryoturbated gravels of the ?glacial maximum (Valley floor 7).

3 THE GEOMORPHOLOGICAL SURVEY: THE HOLOCENE SUCCESSION

The complexity and the poor dating control of the Holocene sediments, relying largely on derived artefacts, limits the interpretation of the valley floor deposits. Nevertheless, at least seven sedimentary units can be recognised, with the following chronology: 1, (broadly) Neolithic/Bronze Age; 2, Classical; 3, Medieval; 4, early post-Medieval; 5, 19th century; 6, early 20th century; and 7, late 20th century. These are distinguished firstly on stratigraphical grounds, since units overlie or lie adjacent to one another, and secondly by the artefact assemblages found within them.

The fluvial units are characterised by two main facies-groups. The first of these is in-channel gravels, characterised by a variety of bedding styles. Epsilon cross-bedding is common. This most probably relates to a meandering single channel stream. Also common is plane bedding, probably associated with deposition in a low-sinuosity single channel or multiple-channel river with large tabular lateral or medial bars. In a few places trough cross-bedding, sometimes with colluvial lenses, was present. This lithofacies is characteristically laid down in multi-channel rivers (Rust and Koster 1982). The colluvial lenses result from mass-flows. The second main facies-group consists of usually plane-bedded sheets of silts, sands and gravels, laid down in an overbank or back-basin location by floodwaters. The aggradation of significant epsilon cross-bedded gravels is a slightly unusual feature of the sedimentation pattern of Holocene

rivers in Italy. Experience suggests that aggradation is typically associated with low sinuosity single thread or multiple channel streams (Hunt et al. 1992). It is suggested that the Biferno aggraded from a single meandering channel most probably as the result of high sediment input but cohesive (most likely well-vegetated) riverbanks.

3.1 Mid-Holocene deposits

In the lower valley, about 2 m of silts, clays and soil profiles (Valley floor 5) rest upon epsilon cross-bedded gravels (Valley floor 6) which contain epipalaeolithic/Mesolithic artefacts. Pollen analysis of the silts gave spectra dominated by oak pollen with a variety of other trees including hornbeam, fir, yew, elm, hazel, pine, pistachio, buckthorn, hawthorn, sycamore, lime and ash. Herbaceous taxa are also present including grasses, sedges, black bindweed, dandelion group, daisy group, wormwood, plantain, and others. These assemblages reflect species-rich mixed oak woodland and are comparable with mid-Holocene spectra from elsewhere in Italy (Alessio et al. 1986; Hunt and Eisner 1991). Mollusc assemblages contain woodland taxa such as *Pomatias elegans*, *Oxychilus cellarius*, and Clausiliidae, together with the open ground species *Helicella itala* and *Vallonia excentrica*. The plant macrofossil assemblages contain charcoal, leaves of trees and seeds of 'weeds' such as Polygonaceae and Chenopodiaceae. The presence of significant numbers of 'weed' pollen and plant fossils and open ground molluscs may reflect areas of naturally open ground such as gravel bars in river courses, landslide scars, and tree fall scars, but it is tempting to equate them with clearances resulting from early agriculture, particularly given the presence of charcoal.

In the gorge, about a metre thickness of river gravels and slackwater deposits, undated but most probably of middle Holocene age, is known from two sites (Valley floor 5). An extremely small pollen assemblage, containing a mixture of trees and open-ground species, is possible evidence for contemporaneity with an early clearance. In the Boiano basin, coarse river gravels containing Bronze Age potsherds and overbank alluvium of probably mid Holocene age are known from two sites (Valley floor 7, 8).

3.2 Samnite/Roman deposits

In the lower valley, plane-bedded and epsilon cross-bedded gravels and significant overbank deposits (sometimes over 3 m thick) containing classical-age potsherds are very widespread (Valley floor 4). A major aggradation of 3 m of mostly epsilon cross-bedded gravels containing potsherds, brick and tile of classical age is also known from the gorge (Valley floor 4). A small mollusc assemblage from the gravels contained only open-ground taxa. Colluvial sequences up to 3m thick are also known from the gorge. Tree stump casts are common in the colluvial deposits and the mollusc faunas are a mixture of sheltered habitat taxa such as Clausiliidae, *Pomatias ele-*

gans, *Oxychilus* spp. and open ground species such as *Helicella itala*. The associated potsherds range up to the 1st century AD. In the Boiano basin, gravels and over-bank alluvium containing classical artefacts are wide-spread and up to 2 m thick (Valley floor 6). At high altitude, trough cross-bedded gravels with colluvial lenses are the dominant lithofacies. Lower in the basin, plane bedded gravels and overbank silt and gravel sheets are common, and by the entrance to the gorge are epsilon cross-bedded gravels.

3.3 *Medieval to 19th century deposits*

Sediments of medieval to 19th century age are rare in the Biferno Valley. In the lower valley, over 2 m of epsilon cross-bedded gravels and significant overbank sands and silts postdate alluvium with classical ceramics and pre-date sediments with late 20th century artefacts, but con-tain no distinctive artefacts at the sites visited (Valley floor 3). At one site in the gorge, sherds of the 13th century and a little later were recovered from around 2 m of epsilon cross-bedded gravels (Valley floor 3). In the Boiano basin, soils containing Medieval potsherds (Val-ley floor 5) overlay gravels containing Roman artefacts and at another site (Valley floor 4) epsilon cross-bedded gravels contained sherds of early post-Medieval age. Also in the Boiano basin planar gravels and overbank alluvium contained sherds of 19th century age.

3.4 *Early 20th century deposits*

In the gorge, over 4m of lacustrine sandy silts and epsilon cross-bedded gravels (Valley floor 2) relate to the con-struction of an industrial weir at Santa Elena and the rapid infilling of the lake behind it with sediment. The weir is most probably of interwar date: a piece of rubber tyre was preserved in the gravels below it and the construction style is typical of the earlier part of the 20th century. Robust tin cans of typical interwar type were found in fluvial gravels upstream from the lacustrine deposits. The deposits that accumulated behind the dam are richly fossiliferous, containing a variety of woodland molluscs (*Oxychilus*, *Neaovitrea*, *Pomatias*), open ground species (*Hellicella*, *Truncatellina*), marsh taxa from the edge of the lake (*Vertigo*, *Cochlicopa*) and aquatic species (*Val-vata*, *Acroloxus*, *Lymnaea*). Plant macrofossils reflect a similar range of habitat with trees represented by oak and other charcoal and a poplar seed, weeds such as *Cheno-podium* and marsh and aquatic plants like *Schoenoplec-tus*, Cyperaceae, *Najas*, and *Ranunculus*.

3.5 *Later 20th century deposits*

In the lower valley, up to 5 m of mostly plane-bedded gravels and overbank alluvium (Valley floor 2) are wide-spread. The deposits contain abundant artefacts from the 1960s and 1970s: polythene bags, tin cans, bottle and so on. Gabions and other flood control structures were buried by this aggradation at two sites, at one by over 3 m of gravels and overbank sediments. Similarly in the Boia-no basin, gravels with planar and trough cross-bedding, and significant overbank alluvium, have overwhelmed flood-control structures to a depth of over 1 m (Valley floor 2). The deposits again contain artefacts typical of the period since 1960.

3.6 *Sediments of the active floodplain*

In most places in the Biferno Valley, the river is a low-sinuosity single channel flanked by large tabular lateral bars, but on the higher slopes of some alluvial fans and occasional reaches downstream it is a multi-channel river with tabular medial bars. The modern floodplain (Valley floor 1) contains much late 20th century rubbish and clearly accumulated over approximately the last 10-20 years. The Biferno has recently incised into its bed, but the causes and chronology of this appear complex. The river has incised 2-5 m in the lower valley since the construction of the Guardialfiera dam in 1975. By the mid 1980s, aggradation was beginning to extend upstream from the lake that has formed behind the dam. In the Biferno gorge, about 4 m of incision has occurred after the breaching of the industrial weir at Santa Elena. The abandoned in-channel sediments appear to have stopped accumulating some time around the middle of the 20th century, because no modern plastic or other artefacts were found over several hundred metres of exposure. In the Boiano basin, 2-4 m of incision has occurred and in-channel sediments that contain artefacts dating to the 1960s and 1970s have been abandoned.

In the following section, we summarise the survey record for site distributions and land use, and in the concluding discussion we review the evidence for corre-lations between the geomorphological record for periods of aggradation identified above and the archaeological record for expansions in settlement and/or inten-sifications in land use in the valley.

4 THE ARCHAEOLOGICAL SURVEY: SETTLE-MENT AND LAND USE TRENDS

4.1 *Palaeolithic settlement*

Although one of the earliest Palaeolithic sites in Europe, dated to some three quarters of a million years ago, lies just beyond the Biferno Valley watershed at Isernia la Pineta (Peretto et al. 1983), the first evidence for system-atic utilisation of the valley is Middle Palaeolithic. The most striking feature of the distribution of chert and flint artefacts identified as Middle Palaeolithic is its clustering in the lower valley – over 200 findspots, compared with about 40 in the middle valley and less than 20 in the upper valley. Another characteristic was the difference between the lower and middle valley material: abundant and wide-spread 'off-site' distributions in the lower valley, com-pared with fewer but much larger and more discrete concentrations – 'sites' – in the middle valley.

The material suggests a pattern of consistent utilization of the lower elevations and a variety of subsistence tasks being practised there, with forays into the middle valley creating more task-specific sites. It is noteworthy that the differential distributions of artefacts in the Biferno Valley are like those observed by Turq (1978) in the Lot and Garonne valleys in France, with chopping tools and debitage clustered on the river terraces and flake tool assemblages more common on the intervening plateaus. It is impossible to translate the differences in the survey material into different kinds of subsistence activities, though faunal assemblages from Middle Palaeolithic sites elsewhere in peninsular Italy indicate generalised systems of hunting and scavenging (Barker 1981). As mentioned in the preceding section, Middle Palaeolithic artefacts were recovered from stratified river sediments identified as belonging to the cold stage before the last interglacial.

Upper Palaeolithic settlement also concentrated in the lower valley, but forays were made at that time not only to the middle but also to the upper valley, particularly in the latter part of the Upper Palaeolithic. In 1981 one of us put forward a model of Upper Palaeolithic subsistence in central Italy, arguing that it was mobile and specialised in response to the glacial environment, which was characterized by severe winters in the mountains, more extensive coastal plains, and a generally open landscape (Barker 1981). Whereas Middle Palaeolithic sites were mostly at lower elevations, Upper Palaeolithic sites were both here and in the mountains, and the faunal samples tended to be dominated by two species, red deer and the steppe horse *Equus hydruntinus*. The model was undoubtedly too simplistic, but the basic distinction between Middle and Upper Palaeolithic subsistence patterns still seems to hold good. It does seem clear that Upper Palaeolithic subsistence, in contrast with earlier patterns of settlement, included the use of the Apennine intermontane basins, presumably in the summer (and/or during interstadial phases). The Biferno Valley survey data conform with this model. From recent studies of Upper Palaeolithic settlement in this part of Italy (Donahue 1988), it seems increasingly likely that Upper Palaeolithic hunter-gatherers were organised in systems of logistical mobility, using base camps or residential sites at the hub of sets of satellite kill sites and other kinds of special purpose camps.

4.2 *Later prehistoric settlement*

The paucity of typical Mesolithic artefacts, and the fact that the little such material we found was always near the coast, tend to support the hypothesis that little use was made of the main part of the valley by foraging groups in the early Holocene. Systematic farming began in the late fifth millennium BC, following centuries of its practice immediately south of the valley on the Tavoliere plain.

The main earlier Neolithic (c. 4500-3750 BC) settlements were situated in both the lower and middle valley, in a number of topographical situations, and off-site activities that created spreads of lithic material – presumably hunting and herding - ranged throughout the valley up to the Boiano basin (Fig. 4a). We obtained good

botanical and faunal samples from one site, Monte Maulo, which indicated a subsistence system dominated by mixed farming. The main later Neolithic (c. 3750-3000 BC) settlements were in the lower valley, in more restricted topographical situations, and off-site activities in the middle and upper valley were generally near the river. There was also a general trend towards an increase in settlement size. These people also practised mixed farming, but one significant difference compared with the earlier Neolithic suggested by the faunal samples and strainer sherds was the increasing importance of animal secondary products, particularly milk for cheese. The trend is reflected in the use of a small rockshelter excavated at Ponte Regio in the upper valley, used as a temporary campsite (Barker 1974). The intensification in the subsistence base was paralleled by a variety of evidence for social elaboration.

The absence of Neolithic lithic material from the Matese Mountains suggests that subsistence activities did not yet make use of the high Apennines. Pollen analyses from the Matese (Hunt, unpublished), as elsewhere in central Italy (Alessio et al. 1986; Kelly and Huntley 1991; Watts 1985), in fact indicate that, in the relatively humid climate of this period, the upland landscape was generally heavily wooded. The evidence for exchange systems also emphasises that contacts between Neolithic communities were along the Adriatic littoral rather than over the Apennines.

The lower valley remained the major zone of permanent settlement until the 3rd millennium BC. During the 2nd millennium (Fig. 4b), Bronze Age settlement extended into the middle valley as well and to a limited extent into the upper valley, and off-site activities for the first time reached into the Matese Mountains. On the evidence of excavation and geophysical survey, the main settlements consisted of several huts, social units of perhaps three to five family groups. The agricultural system was fundamentally the same as that practised by the later Neolithic communities. Although most pottery was locally produced, the decoration shows that there were now well established communication systems from the valley not only up and down the coastal zone but also across the Apennines to Campania. The Biferno Valley data correlate well with the general settlement trends observed throughout central Italy in the 2nd millennium of a filling out of the landscape and the first systematic use of the Apennine Mountains, the latter generally thought to reflect the development of some kind of transhumant pastoralism, albeit on a small scale (Barker 1981; Barker and Grant 1991).

The process of settlement expansion continued through the first half of the 1st millennium BC, the Iron Age (Fig. 4c), but there were substantial changes in settlement forms, social systems, and land use. A clear settlement hierarchy developed, with substantial nucleated settlements, presumably villages, established for the first time. Cemeteries provide as clear signs of a markedly stratified society (Di Niro 1981). The agricultural base continued as before, but in the lower valley there is evidence for wine production for the elites to

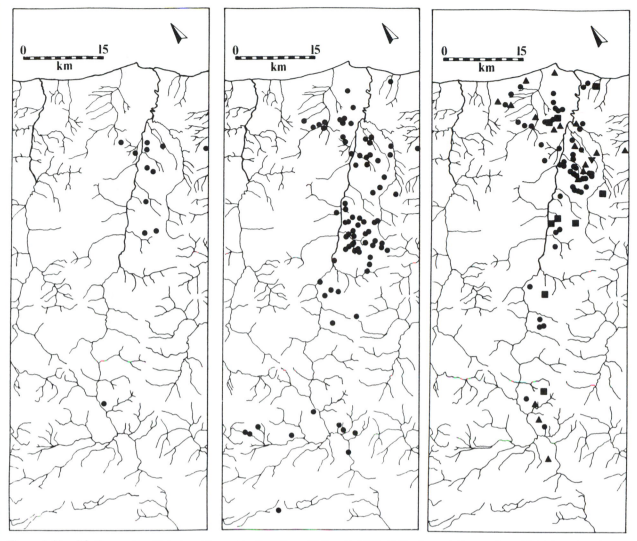

Figure 4. Simplified map of prehistoric settlement in the Biferno Valley: (left) Neolithic; (centre) Bronze Age; (right) Iron Age. In the Neolithic and Bronze Age maps the circles mark the principal habitation sites found by the survey; in the Iron Age map squares denote villages, circles denote smaller settlements, and triangles denotecemeteries.

indulge in Greek-style *symposia*.

4.3 *Samnite and Roman settlement*

The Biferno Valley in the second half of the 1st millennium BC was part of ancient Samnium, the Pentri tribe of the Samnites living in the upper valley and the Frentani living on the lowlands (Salmon 1967). The ancient sources record how the rising population of Samnium led to emigration over the mountains to the western side of the Italian Peninsula, where the Samnites came into conflict with Rome, losing three bitter wars with the Romans between 343 and 290 BC. Although urbanization on the Graeco-Roman model did not develop until the 1st century BC, after the devastations by the Romans following the Social War (91-82 BC), the archaeological survey demonstrated a huge rise in rural settlement in the Biferno Valley during the period of black glaze pottery production (the late 4th to the end of the 1st centuries

BC), filling out the countryside to a level unparalleled in the valley's history perhaps until the early modern period (Lloyd 1991; Fig. 5). Cereals, legumes and vines were grown in the upper valley, and olive cultivation was introduced in the lower valley. Sheep, goats, and pigs were the principal stock, cattle kept mainly as plough animals. The major settlements such as the pre-Roman *vicus* at Saepinum were thriving centres of trade and manufacture by the 2nd century BC. Whilst many farmers probably operated at a subsistence level, the wealthier landowners were producing surplus commodities, particularly stock, which they used to trade for exotic luxuries or as benefactions for religious sanctuaries to legitimate their authority over the peasantry.

With full Romanization in the later 1st century BC, urban centres were established in the lower, middle and upper valley. The survey indicates a major restructuring of the landscape by the early imperial period: throughout the valley, sites with Italian *terra sigillata* pottery number

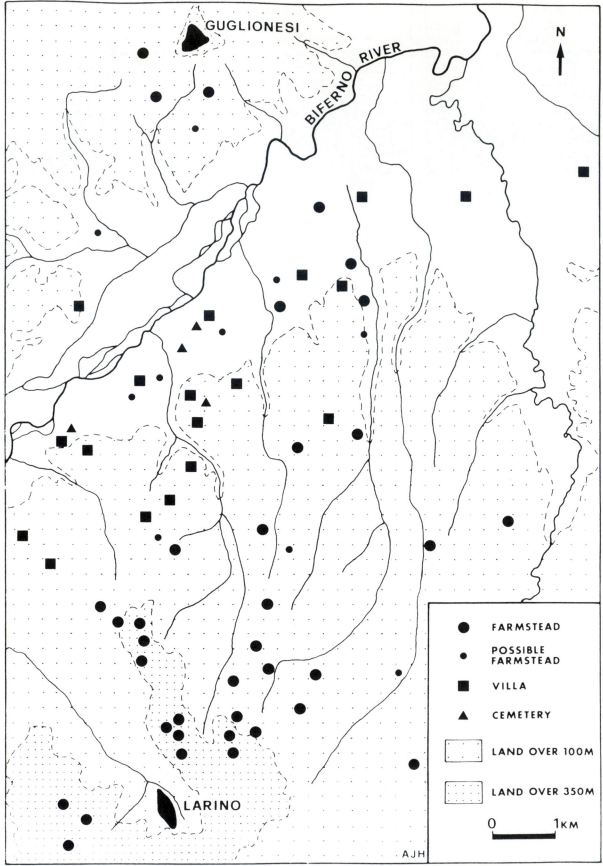

Figure 5. An example of Roman settlement in the lower Biferno Valley.

only a third of those with black glazed pottery. The trend indicates in part a decline in the rural population but more the development of large estates, as noted elsewhere in Italy at this time. The Samnite farm of Matrice was rebuilt in the early 1st century AD, not only substantially enlarged but also elaborately refurbished both in the living quarters but also the working area, where pressing equipment, probably for wine, was installed. The documentary sources for the period show the emergence of increasingly powerful families in the valley, and the faunal and botanical residues and material culture from the excavated sites make it clear that the villa estates were involved in intensive agricultural production for external markets. In the upper valley the surplus was largely produced by animal husbandry, in the lower valley by cereal, olive and vine cultivation. The trend to more land becoming concentrated in fewer hands had dire consequences for the ordinary rural population: by the late 1st century AD the Emperor Hadrian had to institute food relief schemes for the dispossessed of the region. After these 'boom and bust' centuries of the late Republic and early empire in the valley, the later Roman empire was characterised by a relatively stable, if more contracted, settlement system.

4.4 Medieval/Post-medieval settlement

As in the rest of Italy, the middle centuries of the 1st millennium AD were characterized by a massive decline in population. The major towns were deserted by about AD 550, except as homesteads for the elite. The countryside was almost deserted, with rural settlements being small, defended, and widely scattered. The survey found only one site for the 6th to 8th centuries AD (Hodges et al. 1980), and scarcely half a dozen for the 9th. It was not until the 10th century AD that the basic pattern of the modern hill villages was established. The 11th and 12th centuries were marked by further settlement expansion, with numerous communities establishing other villages in marginal situations, but in the following centuries most of them were abandoned as the settlement system contracted once more.

There was a second phase of village expansion in the 15th, 16th and 17th centuries. The contraction to the present villages took place in the 18th century (Lalli 1978). In the second half of the last century the rural population increased inexorably, and with it profound poverty for the peasant farmers. The inhabitants responded in two ways: in many parts of the valley marginal areas were taken into cultivation – there are huge areas of abandoned terraces around the Boiano basin, for example, that were intensively cultivated at this time – and the period also witnessed the beginnings of the process of massive emigration from the valley, as from so many other regions of the Italian Mezzogiorno, to northern Europe, America and Australia, a process that has only been halted in recent years (Clark 1984).

5 DISCUSSION

To what extent can the two sequences be correlated? The prehistoric settlement trends outlined above cannot be correlated easily with the geomorphological record. The 'Neolithic/Bronze Age?' fluvial sediments identified in the valley (Table 2) are only correlated very loosely by pollen in one case and by rather undiagnostic lithic artefacts and potsherds, though the latter can be assigned to the Bronze Age with a reasonable degree of confidence. The ensuing Samnite/Roman aggradation has generally overwhelmed the earlier Holocene evidence, making its survival extremely exiguous. However, soil development suggests that several centuries at least separate the formation of the earlier sediments from the Samnite/Roman phase, the beginning of which can probably be dated to the last two or three centuries BC. The earlier sediments are indeed more likely to be Neolithic and/or Bronze Age than Iron Age. The pollen and molluscan evidence in the 'Neolithic/Bronze Age?' sediments generally indicates small openings in a predominantly closed landscape.

Several palynologists have argued from pollen diagrams in central Italy that an increasingly open landscape can be discerned in the later prehistoric period, suggesting that this developed in response to increasing aridity, particularly in the second half of the 2nd millennium BC (Bonatti 1970; Follieri et al. 1988; Frank 1969; Kelly and Huntley 1991). There is some evidence from lake-level changes to support this hypothesis (Carancini et al. 1986; Fugazzola Delpini 1982; Hunt and Eisner 1991). Other palynologists recently have tended to interpret similar evidence in terms of forest clearance by farmers (Watts 1985). The debate is likely to continue until we have much better dated pollen sequences, better understanding of the pollen signatures created by natural clearings and agricultural clearances, and better understanding of local settlement trends and land use changes in the Italian context. It may well be that in the Biferno Valley, as elsewhere in central Italy, the early 'clearances' discerned reflect both the trend to aridity over this period and the increasing impact on this changing landscape of an agricultural system that was both expanding and intensifying.

The evidence of the archaeological survey and excavations for the dramatic expansion of rural settlement and intensification in land use between the 3rd century BC and 1st/2nd centuries AD correlates closely with geomorphological evidence for a phase of massive aggradation at this time. These sediments, identified in all three sections of the valley, are dated by the occurrence of black glazed and Italian *sigillata* pottery; the absence of sherds of later imperial fine wares of the 3rd to 5th centuries AD is striking. Given the preservation of fine pottery in the sediments, and the frequency of later imperial wares on settlement sites, the absence of later imperial wares from the alluvial deposits seems likely to be significant, rather than a function of survival and recovery. The lower sediments are much coarser than those of the preceding and ensuing episodes of aggradation. The thick colluvial deposits in the gorge of the middle Biferno Valley containing treestump casts and molluscan fauna of woodland species indicate the presence of woodland beside the river. Far more numerous in these beds, however, are the indicators of the open country beyond. The implication would seem to be that the dramatic expansion of population and the associated extension of cultivation and intensification in production systems (new crops, new husbandry systems, market production) had a dramatic effect on the landscape in terms of soil erosion.

For the post-Roman period, the level of correlation between geomorphological change and settlement change is remarkably good. Parallel with the establishment of the modern system of hilltop villages and satellite villages from the 10th century onwards is the evidence in all three sections of the valley for sediments with Medieval pottery, including sherds of the 13th century. What is also noticeable, however, is that these deposits resemble those of the pre-classical 'early clearance' episodes rather than those of the Samnite/early Roman cultivated landscape in scale and thus perhaps in the intensity of landscape impact. The same is true of the next group of fluvial sediments, assigned to the 'early postmedieval period', with pottery of the 16th, 17th, and 18th centuries, broadly contemporary with the second phase of village expansion. The expansion of settlement in the latter part of the last century has also left its imprint, the 'late 19th century' sediments in Table 2 generally containing pottery dated from the 1860s onwards. In short, the repeated correlation between periods of settlement expansion and episodes of alluviation in the Biferno Valley point to human land use practices as the major agent of landscape change during the later Holocene.

Of the three separate phases of fluvial activity identified for the 20th century, the last two, of the post-war decades, have been by far the most profound. In the 1970s it was still common to see hand-cultivation in the upper valley, and cultivation with a light plough drawn by a donkey, horse or pair of oxen in the upper and middle valley. In the lower valley, however, the 1960s and 1970s witnessed the change to mechanisation that began elsewhere in Italy in the pre-war years, that has fundamentally changed the landscape. This has since extended to much of the middle valley. Animal traction gave way to heavy tractors with caterpillar tracks capable of not bogging down on heavy soils. Cultivation technology changed from light wooden ploughs with metal shares, or all-metal ploughs, capable of cutting down to 15-20 cm, to single- or multi-shared steel ploughs which commonly reach down to 75-80 cm or more.

The change has been accompanied by a dramatic extension of the cultivated zone, because the new technology has enabled farmers to plough slopes that formerly could only be cultivated by hand. In the lower valley, moreover, the heavy alluvium of the floodplain of the Biferno below the dam formerly given over to scrub and rough grazing has been drained and also taken into cultivation. Field boundaries have been bulldozed to make larger and more efficient ploughing units. Polyculture has increasingly given way to monoculture, with separate vineyards, olive groves, orchards, and cereal fields. The

trend has been in response to both the new technologies and their efficiency demands, and market forces in the form of subsidies from the Italian state and the EU. The changing nature of land use in the valley during the 1970s was a major factor in the success of the archaeological survey, but the damage to the landscape in terms of dramatically accelerating erosion rates is frightening. The equivalence between the classical and recent sedimentation processes has already been noted, except that one is a record of three or four centuries of agricultural intensification, the other of scarcely a generation.

6 CONCLUSION

The Biferno Valley survey provides some of the most detailed archaeological and geomorphological sequences from the Mediterranean basin. The early history of the valley predominantly reflects tectonic activity in the Lower Pleistocene and a mixture of tectonics and climate change in the Middle and Late Pleistocene. In the Holocene, it seems clear that human activity plays a crucial role in shaping sedimentation patterns. It is surely more than coincidental that the archaeologically-defined major phases of land use expansion and/or agricultural intensification coincide with aggradation phases. The important late Samnite/early Roman phase does not compare well with the Vita-Finzi model of a climatically-controlled late Roman/early Medieval Younger Fill. Instead, where palaeoecological evidence is present, signs of cleared landscapes and soil erosion are evident. In the Biferno Valley at least, human activity is the dominant influence on fluvial activity in the Holocene.

REFERENCES

Alessio, M., L. Allegri, F. Bella, G. Calderoni, C. Cortesi, G. Dai Pra, D. De Rita, D. Esu, M. Follieri, S. Improta, D. Magri, B. Narcisi, V. Petrone & L. Sadori 1986. C–14 dating, geochemical features, faunistic and palynological analyses of the upper 10m core from Valle di Castiglione (Rome, Italy). *Geologica Romana* 26: 287-308.

Barker, G. 1974. A new neolithic site in Molise, southern Italy. *Origini* 8: 185-200.

Barker, G. 1981. *Landscape and society – prehistoric central Italy*. London: Academic Press.

Barker, G. 1991. Two Italys, one valley: an Annaliste perspective. In J. Bintliff (ed) *Archaeology and the Annales School*: 34-56. Leicester: Leicester University Press.

Barker, G. 1994. *Mediterranean landscape change: the Archaeology and History of the Biferno valley, Italy*. Leicester: Leicester University Press (in press).

Barker, G. & A. Grant (eds) 1991. Ancient and modern pastoralism in central Italy: an interdisciplinary study from the Cicolano mountains. *Papers of the British School at Rome* 59: 15-88.

Bell, M. 1982. The effects of land-use and climate on valley sedimentation. In A. F. Harding (ed) *Climatic change in later prehistory*: 127-142. Edinburgh: Edinburgh University Press.

Bintliff, J. 1976. Sediments and settlement in southern Greece.

In D. A. Davidson & M. Shackley (eds) *Geoarchaeology*: 267-275. London: Duckworth.

Bintliff, J. 1992. Erosion in the Mediterranean lands: a reconsideration of pattern, process and methodology. In M. Bell & J. Boardman (eds) *Past and present soil erosion: archaeological and geographical perspectives*: 125-131. Oxford: Oxbow Monographs 22.

Bonatti, E. 1970. Pollen sequence in the lake sediments. In G. E. Hutchinson (ed) *Ianula: an account of the history and development of the Lago di Monterosi, Latium, Italy*: 26-31. *Transactions of the American Philosophical Society* 64: 5-175.

Carancini, G.L., S. Massetti & F. Posi 1986. L'area tra Umbria meridionale e Sabina alla fine della protostoria. *Dialoghi di Archeologia* 3,2: 37-56.

Chester, D. K. & P.A. James 1991. Holocene alluviation in the Algarve, southern Portugal: the case for an anthropogenic cause. *Journal of Archaeological Science* 18: 73-87.

Clark, M. 1984. *Modern Italy 1871-1982*. London: Longman.

Davidson, D.A. 1980. Erosion in Greece during the first and second millennia B.C. In R. A. Cullingford, D. A. Davidson and J. Lewin (eds), *Timescales in geomorphology*: 143-158. New York, John Wiley.

Di Niro, A. 1981. *Necropoli arcaiche di Termoli e Larino: campagne di scavo 1977-78*. Campobasso: Soprintendenza Archeologica e per i Beni Architettonici Artistici e Storici del Molise.

Donahue, R. 1988. Microwear analysis and site function of Paglicci Cave level 4a. *World Archaeology* 19: 357-75.

Follieri, M., M. Magri & L. Sadori 1988. 250,000-year pollen record from the valle di Castiglione (Roma). *Pollen et Spores* 30 (3-4): 329-56.

Frank, A. H. E. 1969. Pollen stratigraphy from the Lake of Vico (central Italy). *Palaeogeography, Palaeoclimatology and Palaeoecology* 6: 67-85.

Fugazzola Delpino, M. A. 1982. Rapporto preliminare sulle ricerche condotte dalla Soprintendenza Archeologica dell'Etruria Meridionale nei bacini lacustri dell'Apparato vulcanico sabatino. *Bollettino d'Arte, Supplemento* 4: 123-149.

Gutierrez-Elorza, & Pena-Monne, 1990. Upper Holocene climatic change and geomorphological processes on slopes and infilled valleys from archaeological dating (northeast Spain). In M. Boer & R. De Groot (eds), *Precedings of a European Conference of Landscape Ecological Impact of Climatic Change, 3rd-7th Dec 1989*: 1-18. Amsterdam: IOS Press 1990.

Hodges, R., K. Wade & G. Barker 1980. Excavations at D85 (S. Maria in Civita): an early medieval hilltop settlement in Molise. *Papers of the British School at Rome* 48: 70-124.

Hunt, C.O. & W. R. Eisner 1991. Palynology of the Mezzaluna core. In A. Voorrips, S. Loving & H. Kamermans (eds), *The Agro Pontino survey project*: 49-59. Amsterdam, Studies in Prae- und Protohistorie 6.

Hunt, C.O., D. D. Gilbertson & R. E. Donahue 1992. Palaeoenvironmental evidence for agricultural soil erosion from late Holocene deposits in the Montagnola Senese, Italy. In M. Bell & J. Boardman (eds) *Past and present soil erosion: archaeological and geographical perspectives*: 163-174. Oxford: Oxbow Monographs 22.

Kelly, M. G. & B. Huntley 1991. An 11000-year record of vegetation and environment from Lago di Martignano, Latium, Italy. *Journal of Quaternary Science* 6 (3): 209-24.

Lalli, R. 1978. *Conoscere il Molise*. Campobasso: Edizioni Enne.

Lloyd, J. A. 1991. Farming the highlands: Samnium and Arcadia in the Hellenistic and early Roman Imperial periods. In G. Barker & J. A. Lloyd (eds) *Roman landscapes: archaeological survey in the Mediterranean region*: 180-93. London: British School at Rome, Archaeological Monographs 2.

Miall, A. D. 1977. A review of the braided-river depositional environment. *Earth Science Reviews* 13: 1-62.

Peretto, C., C. Terzani & M. Cremaschi (eds) 1983. *Isernia la Pineta: un accampamento più antico di 700.000 Anni*. Bologna: Calderini.

Pope, K. O., & Van Andel, Tj. H. 1984. Late Quaternary alluviation and soil formation in the southern Argolid: its history, causes and archaeological significance. *Journal of Archaeological Science* 11: 281-306.

Potter, T. W. 1979. *The changing landscape of south Etruria*. London: Elek.

Rust, B. R. & E. H. Koster 1984. Coarse alluvial deposits. In R. G. Walker (ed), *Facies models*: 53-69. Toronto: Geoscience Canada Reprint Series 1.

Salmon, E.T. 1967. *Samnium and the Samnites*. Cambridge: University Press.

Turq, A. 1978. Note preliminaire sur l'outillage en quartzite entre Dordogne et Lot. *Bullétin de la Société Préhistorique Française* 75: 136-9.

Van Andel, Tj. H. & Zangger, W. 1990. Landscape stability and destabilisation in the prehistory of Greece. In S. Bottema, Entjes-Nieborg, & W. Van Zeist (eds), *Man's role in the shaping of the eastern Mediterranean landscape*: 139-157. Rotterdam, Balkema.

Vita-Finzi, C. 1969. *The Mediterranean valleys*. Cambridge: Cambridge University Press.

Ward-Perkins, J. B., A. Kahane & L. Murray Threipland 1968. The Ager Veientanus survey north and east of Veii. *Papers of the British School at Rome* 36: 1-318.

Watts, W. A. 1985. A long pollen record from Laghi di Monticchio, southern Italy: a preliminary account. *Journal of the Geological Society*, London 142: 491-499.

CHAPTER 14

Holocene sedimentary sequences in the Arc River delta and the Etang de Berre in Provence, southern France

M. PROVANSAL
Institut de Geographie, Université d'Aix-Marseille, Aix en Provence, France

ABSTRACT: This paper describes the Holocene development of the Arc River delta in Provence, southern France. Three main sources of information have been used to reconstruct the history of this small Mediterranean delta. These are; (1) geomorphological fieldwork in the Arc River basin, (2) archaeological investigations of the Arc littoral plain and (3) an evaluation of the Holocene sedimentary sequence of the Etang de Berre. Anthropogenic-induced soil erosion appears to be the main source of fine sediment for delta growth, but this material only actually contributes to the growth of the delta if it is associated with an increase in stream discharge. Examination of the history of land occupation and agricultural activity and of the climate of Provence over the last 5000 years suggests that four main environmental controls promote delta sedimentation and enlargement. These are; (1) steepening of river gradients, (2) human disturbance of catchment soils, (3) increases in precipitation and stream discharge and (4) eustatic changes in sea level. The evolution of the Arc River delta is not the product of a single cause. Two main periods of deltaic sedimentation have taken place during the Holocene. The first of these commenced during prehistory up to the Versilian transgression when infilling of marshy littoral plains began. The second major phase began at the end of the Middle Ages when the combination of intensive agricultural activity with the net climatic deterioration of the Little Ice Age produced highly favourable conditions for increased sediment supply and delta growth.

KEYWORDS: Delta, Holocene, deltaic sedimentation, Provence, southern France, human activity, soil erosion, sediment supply, river discharge, geomorphology, palaeoclimate, clay minerals.

1 INTRODUCTION

Alluvial plains and deltas often form sedimentary sinks that may record the geomorphological evolution of the upstream fluvial system. The development of such depositional features reflects the interplay between river basin runoff and sediment supply, and littoral erosion/deposition dynamics. Deltas and alluvial plains may therefore provide valuable sources of evidence to elucidate the Holocene history of the landscapes of Provence, particularly the relative importance of catchment slope erosion and climatically-driven changes in flood regime and fluctuations of sea level. The research reported here is based on palynological (Laval et al. 1990, 1991), sedimentological (Roux 1991) and archaeological investigations (Leveau and Provansal 1991) conducted in the the Arc River delta and the Etang de Berre (a shallow lagoon) of southern Provence, in order to reconstruct the Holocene environmental history of the area.

2 BACKGROUND

The extent of deltaic deposition and progradation is largely controlled by rates and patterns of river sediment supply and the ability of the river to transport fine-grained alluvial sediment to the coastline. In southern Provence,

Holocene soil erosion over the last 6000 to 7000 years has been mainly the result of agricultural activities, although changes in hydroclimate have also been an important controlling factor. Sediment conveyance to the coastal zone also depends upon runoff magnitude and stream competence. Each of these variables can be increased by agricultural developments on catchment slopes. For example, land clearance may increase runoff volume (Clauzon et al. 1971), whilst river management practices (e.g. riparian forest clearance, bank protection and channel straightening) often produce increased stream flow velocities and enhanced suspended sediment conveyance downstream (see Jorda et al. 1990).

Holocene climatic variations have also played a significant role in changing hydrological regimes. For example, increases in annual rainfall totals and more particularly, fluctuations in the *seasonal* distribution of precipitation can enhance both catchment runoff volumes and sediment delivery. This can result in an increase in flood frequency in the delta region and greater transfers of sediment to the littoral zone. It has been shown by Provansal (1992) that such hydrological conditions marked the start of Sub-Boreal Period (5000 BP) and also characterised several parts of the Sub-Atlantic Period such as the end of the first Iron Age around 500 BC, the Early Middle Ages around 500 to 1000 AD and the beginning of the Modern Age at c. 1400 to 1800 AD.

In summary, delta development is controlled by both basin physiography and land use as well as fluvial dynamics and climate. Understanding the complex interaction of these factors over the last five to six thousand years in the Arc River delta requires detailed investigation of both the littoral plain and the upstream drainage basin. These areas must be considered together as a related unit and this is the nature of the approach adopted here.

3 THE ARC RIVER BASIN AND THE ETANG DE BERRE BASIN

The Arc River drains a catchment of c. 775 km² in Provence, southern France (Fig. 1). The river rises below the Sainte Victoire limestone massif and then passes through successive basins divided by narrow valleys. The geomorphology of the Arc River basin is characterised by extensive gravel rich terraces of periglacial origin. At the beginning of the Holocene the spread of woodland (mainly pines and deciduous oaks) induced an episode of river

incision and at this time the formation of red-brown soils and travertine deposits took place.

Two Holocene alluvial units are present along the Arc River Valley. The oldest and most extensive depositional phase began during the Boreal Period (about 8000 BP), before the first evidence of human soil disturbance. This alluvial fill is related to a relatively dry Mediterranean ecosystem when widespread natural erosion took place (cf. Triat-Laval 1983). The Roman settlements are built on the surface of this terraced alluvial unit. The more recent episode of alluviation is dated to Medieval or modern times. Between the formation of these two alluvial deposits the Roman Period is characterised by a more stable landscape with only limited fluvial activity when further soil development took place. The deltaic sedimentary sequences and the sediments in the Etang de Berre record the timing of these depositional phases and their impact on the littoral plain.

The archaeological record of the Arc basin reveals a complicated sequence of changing population and settlement distribution. Some Mesolithic sites (8000 to 7000

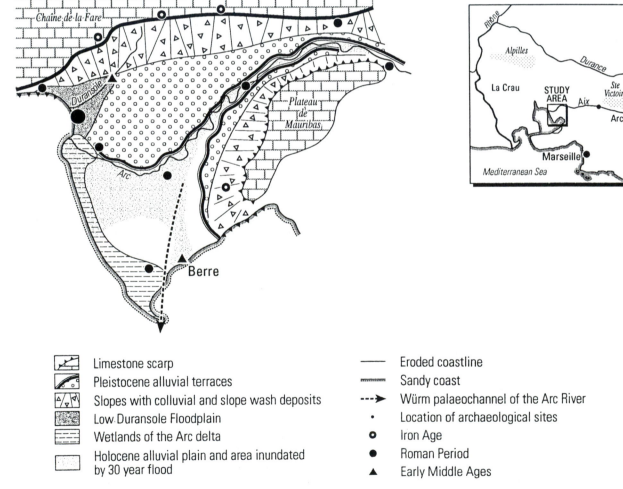

Figure 1. Geomorphological map of the Arc Delta showing the distribution of Pleistocene river terraces, Holocene alluvial sediments and colluvial sediments and the position of the Würm palaeochannel. The inset map shows the location of the Arc Delta downstream of Aix in southern France and the partially enclosed basin of the Etang de Berre.

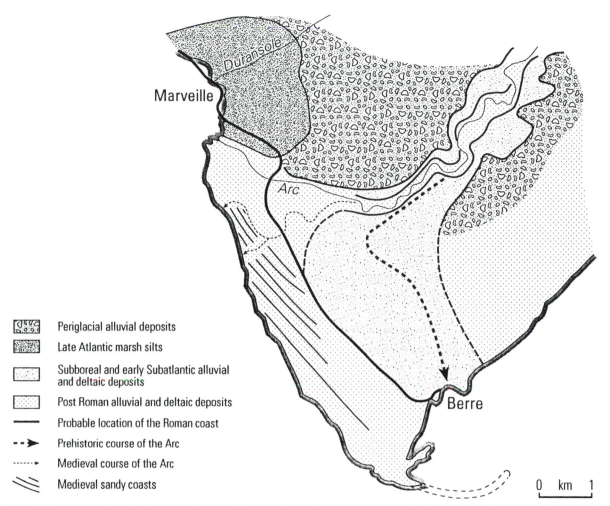

Figure 2. The prehistoric and historic evolution of the Arc Delta. The map shows the distribution of sediments relating to the late Atlantic, Sub-Boreal to Sub-Atlantic and Post Roman periods. The migration of the main channel of the Arc River during the Holocene is also indicated.

Legend:
- Periglacial alluvial deposits
- Late Atlantic marsh silts
- Subboreal and early Subatlantic alluvial and deltaic deposits
- Post Roman alluvial and deltaic deposits
- Probable location of the Roman coast
- Prehistoric course of the Arc
- Medieval course of the Arc
- Medieval sandy coasts

BP) have been recorded in the valley and the most important site in the region was located on the southern border of the Etang de Berre. However population growth only became important from Middle and Late Neolithic times, around 5000 to 4000 BP when agricultural settlement of the upper basin and on the hills surrounding the Etang took place. The Bronze Age and the start of Iron Age (4000 to 2500 BP) were characterised by a relatively small population. A population increase from the Second Iron Age onwards (Tene period, towards 400 BC) saw the building of several *oppida* on higher land and the development of a polyculture (Marinval 1988).

In the wet zones at low elevation, especially in the Arc littoral plain, settlement dates only from early Augustan Times (1st century AD) when several villas and one village were settled near the river and the Etang de Berre. At this time the Arc River basin was densely populated. By the late Roman Period (3rd-5th centuries AD), the lowland zones were abandoned again and remained marshy throughout the Middle Ages until they were drained and cultivated again during the Modern Period

(14th-16th centuries) (Leveau and Provansal 1991). Olive cultivation began at the end of Middle Ages (Leveau et al. 1991) and this land use activity seems to have played an important part in soil erosion.

The present littoral plain is set within periglacial silts which were deposited during the Late Pleistocene (Fig. 2). The centre of the delta complex covers a relatively small area (c. 190 hectares) and is connected towards the south with a more stable area of wetlands. Elevated parts of the delta are enclosed by stony banks and it is in these areas where settlement has generally been concentrated.

The Etang de Berre (a coastal lake basin) is fed by the Arc River and the smaller Touloubre and Vallat Neuf streams that drain similar catchments (Fig. 1). The Etang de Berre comprises two basins separated by an area of shallow water. In the underlying sediments of the southern basin, a palaeochannel was cut by the Arc River during a Pleistocene low sea level stand (see Leenhardt et al. 1967). This former channel controlled the course of the river until the end of the prehistoric period. The more recent shift of the mouth of the river towards the northern

162   M. Provansal

basin may have been induced by neotectonic deformations. At the beginning of the Iron Age (2700 BP), the silting of the Passe de Caronte transformed the Etang into a shallow (< 10 m deep) low energy lagoon.

Four sediment cores were extracted from the basins and were used for detailed sedimentological (Roux 1991) and palynological (Laval and Medus 1990; Laval et al. 1991) investigations. Two sediment cores (cores 6 and 10) were taken from the southern basin and two cores (cores 2 and 3) were taken from the northern basin (Fig. 3). The core sequences were dated by eleven radiocarbon dates. This research was conducted to establish the chronology, volume and nature of Holocene sedimentation in the littoral zone.

Evidence derived from the field examination of channel morphology, the deposits of the Arc River in its littoral plain, the sedimentology of cores from the Etang de Berre and also from research carried out by the Shell Oil Company at the Pointe de Berre, demonstrate that the deltaic plain has been built up during four main phases (Figs 2 and 3). These phases are discussed in turn below.

4 THE EARLY HOLOCENE TO ATLANTIC PERIOD (10,000 TO 4700 BP)

A marine transgression occurred in the Berre basin between 8000 BP (southern basin) and 7500 BP (northern basin) and is recorded in the malacofaunal, microfaunal and pollen records (see Laval et al. 1990, 1991). This transgression was slightly higher than the present level of the lagoon (see Provansal 1991, 1993a) and promoted the spread of littoral marshlands. At the northern part of the present mouth of the Arc River this expansion has been radiocarbon dated to 5430 ±190 years BP (Ly 5 423).

There is some evidence within the sedimentary sequence of fluvial activity which predates the early Holocene transgression. The Würm palaeochannel of the Arc River at the Pointe de Berre formed an important sediment sink, where 4 to 6 m of Holocene silts accumulated before the start of Atlantic Period (Fig. 3). In core 6, two layers containing weathered clay minerals confirm the incidence of limited top soil erosion prior to human interference as early as the start of the Post-glacial Period (Fig. 3).

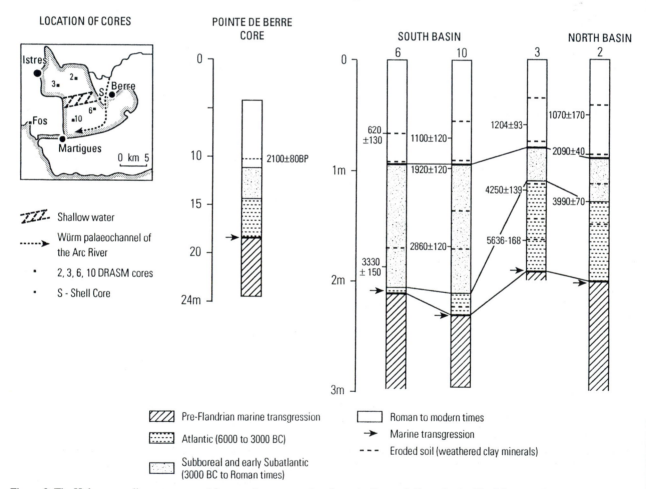

Figure 3. The Holocene sedimentary record from various cores taken from the Etang de Berre basin. The 24 m core from the Pointe de Berre was recovered by the Shell Oil company. Cores 6 and 10 (southern basin) and 2 and 3 (northern basin) were taken by D.R.A.S.M. (Direction des Recherches d'Archéologie Sous Marines du Sud Est de la France). The presence of weathered or degraded clay minerals is used to indicate an influx of eroded topsoil from the catchment slopes. Laboratory numbers for radiocarbon dates are available from the author.

The gradual emergence of the deltaic plain from Atlantic times was effective in trapping most of the incoming sediment and thus depriving the southern basin (then the Arc River outlet) of fine alluvial material. From around 7500 to 4000 BP, channels of relatively steep gradient flowed straight into the Etang de Berre and deposited 70 to 80 cm of silts in the northern basin. During this period, the sedimentary prism becomes thickest at the mouth of the Arc palaeochannel (2 to 4 m thickness of silt at the Pointe de Berre) whereas the equivalent layer in the southern basin only records a 5 to 20 cm thickness (Fig. 3). This difference in sedimentation rate was probably the result of sediment deposition in a marsh environment and/or filtration through riverside forest which trapped a large part of the sediment load of the river.

The sedimentary sequence characterised by unweathered coarse clay minerals (except two beds of probable pedogenic origin in core 3) shows that sediment was derived mainly from channel erosion in the bedrock rather than top soil sources (Fig. 3). Anthropogenic disturbance of the soil and vegetation cover appears also to have been modest during Middle and Late Neolithic times.

5 SUB-BOREAL TO EARLY SUB-ATLANTIC (4700 to 2500 BP)

This period is characterised by an increase in stream sediment loads and by alluvial deposition in the littoral plain. Along the Arc River an extensive silt- and gravel-rich alluvial unit several hundred metres wide suggests an increase in stream competence and conditions at this time were more favourable for riverside forest development (see Laval et al. 1991). The youngest flood silts were deposited before the 6th century BC.

The increase in sedimentation recorded at the delta front (3 m of silt deposited within 2000 years at the Pointe de Berre) and in the southern basin of the lake (1 m deposition in cores 6 and 10 in comparison to 30 to 40 cm in cores 2 and 3 from the northern basin) during this period suggests a shift in the areas of deposition towards the lower part of sedimentary prism (Fig. 3). This appears to be caused by the combined effects of the following influences:

(i) Alluvial deposition on the littoral plain increased river gradients and sediment delivery to the Etang de Berre.

(ii) The growing human pressure on catchment soils increased sediment supply. Sedimentary layers characterised by an increase in fine-grained material, degraded clay minerals and increases in the organic matter content indicate the transfer of weathered top soil derived from cultivated areas. In all four cores these features are associated with the recession of mixed oak forest and with the development of sclerophitic open vegetation following the expansion of cultivated areas (see Laval et al. 1990).

During the Iron Age (2800 to 2000 BP), the southern basin records an episode of soil erosion associated with increasing settlement density on the margins of the lower

Arc Valley (Provansal 1993a). In addition, riparian forests continued to trap sediment resulting in different sediment particle sizes being transported to the southern and to the northern basins (Roux 1991).

(iii) A wetter climate produced higher river discharges between the Early Bronze Age and the end of the First Age (Provansal 1992). The Arc River basin which drains the upland massifs of southern Provence appears to have been more sensitive to climatic variation than the smaller catchments on the northern edge of the Etang de Berre. The Final Bronze Age and/or First Iron Age humid phase was associated with particularly high rates of sedimentation (about 80 cm in core 6 during the last one thousand years BC – approximately 0.8 mm per year).

(iv) The eustatic recession of the Etang water level during the Sub-Boreal Period accelerated these processes – at the beginning of the Tene phase the shoreline was only about 10 cm higher than its present level (Provansal 1993a).

Rapid deltaic progradation during this two thousand-year period is therefore related to several contributing factors: human-induced soil erosion, climate-related increases in river discharge and marine regression.

6 SECOND IRON AGE TO LATE ROMAN TIMES (400 BC to 500 AD)

The lowering of the Etang de Berre water level continued until the start of the Roman Empire. It reached its present level at c. 2110 ±60 years BP (Ly 5 399) between 350 and 4 BC. This fall in lake level resulted in lower water tables and the drying out of wetland areas. This transition is clearly marked in the archaeological record with an increase in settlement in the downstream plain and on former marshlands (Fig. 2). Dense settlement of areas previously subjected to regular inundation suggests a reduction in flood magnitude.

The Roman Period was characterised by a general reversal of the sedimentary processes that had characterised the coastal plain for previous three to four thousand years. There was a general shift from alluvial deposition to river incision and this shift is also evident at many other sites in the Provence region (see Ballais et al. 1993). These changes in the Arc River system probably relate to a decrease of flows connected with a drier climate during the Roman Period (Provansal 1993a). It seems likely that human activity was also important through improved channel management and more effective soil conservation measures on catchment slopes.

In the Etang de Berre, sedimentation rates were also low during this period (0.25 to 0.35 mm per year in cores 3 and 10 for the Roman Period and Early Middle Ages, and a sedimentation rate of 0.21 mm a year in core 6 until the Late Middle Ages (620 ±130 years BP). A relative rise in sedimentation rate (0.4 mm per year between 2090 and 1070 BP) is only evident in core 2 and probably reflects the proximity of this site to the mouth of the Arc River at the end of the Roman Period.

In each of the four core sequences the Roman Period

corresponds with a layer containing strongly weathered clay minerals. This sedimentary feature reflects the enlargement of cultivated areas on catchment slopes and in the low alluvial plain.

7 EARLY MIDDLE AGES (c. 500 AD) TO THE PRESENT

During this period deltaic progradation once again increased in importance and rising water levels and greater risk of flooding resulted in the abandonment of low-lying land. Evidence for this phase of abandonment has only recently been revealed following drainage in modern times (cf. Leveau and Provansal 1991).

Small eustatic increases of the lagoon level (9-10th century at Berre, 15-16th century at Martigues-Provansal 1993a) are believed to have contributed to the enlargement of littoral marshlands. This process was also encouraged by an increase in the rate of lateral river channel migration. During this period the Arc River built a levée of 1 to 2 m in thickness (overlain by pottery kilns dated from the 1st century) which generally confines floodwaters in the upper part of the delta, which then spread out downstream (Fig. 2). The increase in flood frequency resulted in the formation of marshland and the displacement of the river mouth to its present site (Fig. 2). The significant increase in sedimentation rate in core 2 is a result of its proximity to the river mouth at this time.

Renewed catchment erosion is also discernible in the Pointe de Berre sequence where a 6 m thickness of silts dated from c. 2100 ±80 years BP indicate a sedimentation rate of c. 5 mm a year (Fig. 3). All the cores signal an increase in sedimentation rate – 0.6 to 1.6 mm a year from the end of Middle Ages). Core sedimentology is characterised by very fine grain sizes (silts and clays) and clear evidence of strongly degraded clay minerals (Fig. 3). This information, together with historical and palynological data, points to an increase in agricultural activity and the destruction of the remaining forests. An important period of erosion, between the 13th and 15th Centuries, is associated with an increase in olive tree pollen (Leveau et al. 1991).

In comparison with previous periods of deltaic growth, for example during the Sub-Boreal Period, the Middle Ages is characterised by more rapid progradation, higher suspended sediment loads and accelerated infilling of the Etang basin. This was the result of a combination of both socio-economic (and these appear to have become predominant) and of natural factors. It is also likely that the increased humidity of the Little Ice Age increased water and sediment yields from the Arc drainage basin. The systematic transformation of catchment slopes into cultivated terraced units and the decline of riparian forests are also likely to have accelerated these processes (cf. Jorda et al. 1990). At the end of the 19th century, renewed incision and delta and coastline recession took place. This can be explained by a reduction in agricultural activity, especially on catchment slopes, and by an increase in reafforestation.

8 CONCLUSIONS

Holocene deltaic sedimentary sequences in the Mediterranean region do not display a simple relationship with catchment land use change. This is well illustrated by the Holocene development of the Arc River delta. During the two thousands years before Christ, land reclamation had an important impact on the sedimentary processes operating on the Arc delta and the littoral plain. However, the presence of riverside forests trapped part of the sediment load upstream and restricted its transport to the delta head.

During the Roman era, the Arc River was embanked and only limited delta sedimentation took place. This was probably also the result of a drier climate, large-scale water and soil management, and the transfer of agricultural activities towards lowland plains. The modern period, up to the 14th century, was characterised by intensive soil management such as the erection of terrace walls and the expansion of olive cultivation. These practices accelerated the transfer of sediment towards the delta with rapid phases of delta progradation corresponding to periods of higher rainfall.

The period of agricultural decline and of careless land management since the beginning of the present century has been characterised by agricultural recession, afforestation and limited river channel maintenance. Overall, these trends have reduced downstream sediment transfer and have resulted in the recession of the delta.

The sedimentary record of the delta of the Arc River may provide a useful model of the interaction between the physical environment and human activity. Both its small size and relatively homogeneous basin character decrease the number of controlling variables and make it a useful case study. However, comparisons with larger delta systems must only be made with caution, as changes in scale can introduce a number of problems relating to the relative importance of natural process and land use in a drainage basin.

The Holocene development of other delta systems in the Mediterranean basin does, however, show a number of similarities with the delta of the Arc River. Delta progradation in the Mediterranean basin generally started with the spread of agriculture during the Neolithic Period (towards 5500 BP) and accelerated at the beginning of the Sub-Boreal Period (towards 5000 BP). The small Andalusian deltas in southern Spain and the main Rhône River delta provide useful examples. The main progradation stage, however, is always *after* the Roman Period (see for example the Pö River and Medjerda River deltas) and usually at the start of modern times (e.g. the Rhône and Ebro deltas). In the Rhône delta, the Roman Period is characterised by limited fluvial activity leading to the use of low-lying areas.

It has been demonstrated that the evolution of the Arc River delta is not the product of a single cause and two periods in particular during the Holocene have been identified as playing a major part in effecting deltaic sedimentation. The first of these commenced during prehistory up to the Versilian transgression when infilling of marshy littoral plains began. The second major phase began at the end of Middle Ages when the combination of intensive agricultural densities with the net climatic deterioration of the Little Ice Age produced highly favourable conditions for increased sediment supply and delta growth.

9 ACKNOWLEDGEMENTS

This research was carried out with the assistance of CNRS between 1986 and 1990. The cores from the Etang de Berre was taken by D.R.A.S.M. (Direction des Recherches d'Archéologie Sous Marines du Sud Est de la France) and we are grateful for this collaboration. Nine radiocarbon dates have been determined by the Laboratoire de Géologie du Quaternaire (LGQ), Faculté des Sciences de Luminy, and two dates have been determined by the Laboratoire du Radiocarbone (Ly), Université Claude Bernard, Lyon.

REFERENCES

Ballais, J-L., M. Jorda, M. Provansal & J. Covo 1993. Morphogénèse holocène sur le périmètre des Alpilles. Archéologie et environnement, de la Sainte Victoire aux Alpilles. *Publ. Université de Provence*: 515-545.

Clauzon, G. & J. Vaudour 1971. Ruissellement, transports solides et transports en solution aux environs d'Aix-en-Provence. *Rev. Géog. phys. et Géol. dyn. fasc.* 5: 489-504.

Jorda, M. & M. Provansal 1990. Terrasses de cultures et bilan érosif en région méditerranéenne. Le bassin-versant du Vallat de Monsieur, (Basse Provence), *Méditerranée* 3(4): 55-62.

Jorda, M., P. Parron, M. Provansal & M. Roux 1991. Erosion et détritisme Holocène en Basse Provence calcaire. L'impact de l'anthropisation. *Physio-Géo.* 22/23: 37-47.

Laval, H. & J. Medus 1990. Pollen analyse de sédiments du Quaternaire récent de l'Etang de Berre, Bouches-du-Rhône, France. *C.R. Acad. Sci., Paris* 309: 2135-2141.

Laval, H., J. Medus & M.R. Roux 1991. Palynological and sedimentological records of Holocene humans from the Etang de Berre, southeastern France. *Holocene* 1(3): 26-272.

Leenhardt, O. & M.R. Roux 1967. Morphologie du substratum de l'Etang de Berre. *Bull. Soc. Geol. France* 7: 88-92.

Leveau, Ph. & M. Provansal 1991. Construction deltaòque et histoire des systèmes agricoles, le cas d'un petit delta: l'Arc étang de Berre. *Revue archéologique de Narbonnaise* 23: 111-131.

Leveau, Ph., C. Heinz, H. Laval, Ph. Marinval & J. Medus 1991. Les origines de l'oléiculture en Gaule du Sud. Données historiques, archéologiques et botaniques. *Revue d'Archéométrie* 15: 83-94.

Marinval, Ph. 1988. Cueillette, agriculture et alimentation végétale, de l'Epipaléolithique jusqu'au 2ĭ Age du Fer en France Méridionale. Apports paléethnographiques de la Carpologie. Thèse EHESS, Paris.

Neboit, R. 1983. *L'homme et l'érosion.* Publ. Fac. Lettres. Clermont-Ferrand (France), 183 pp.

Provansal, M. 1991. Variations verticales du trait de côte en Provence depuis 5000 ans. Quelques données nouvelles. *Méditerranée* 4: 1-8.

Provansal, M. 1992. Le rôle du climat dans la morphogénèse à la fin de l'Age du Fer et dans l'Antiquité en Basse Provence. Colloque *Le climat entre − 500 et + 500, méthodes d'approche et résultats'in Les Nouvelles de l'Archéologie* 50:21-26.

Provansal, M. 1993a. *Les littoraux holocènes de l'Etang de Berre. Archéologie et environnement, de la Sainte Victoire aux Alpilles.* Publ. Université de Provence, 279-284.

Provansal, M. 1993b. *Les sédiments holocènes de l'Etang de Berre. Archéologie et environnement, de la Sainte Victoire aux Alpilles.* Publ. Université de Provence, 417-424.

Roux, M.R. 1991. Les sédiments de l'Etang de Berre, témoins de la pression anthropique holocène. *Méditerranée* 4: 3-14.

CHAPTER 15

Human activity, landscape change and valley alluviation in the Feccia Valley, Tuscany, Italy

C.O. HUNT
Department of Geographical and Environmental Sciences, University of Huddersfield, Queensgate, Huddersfield, UK

D.D. GILBERTSON
Institute of Earth Studies, UCW Aberystwyth, Aberystwyth, Dyfed, UK

ABSTRACT: In this paper we describe late Holocene coarse- and fine-grained alluvial deposits from the Feccia Valley, Tuscany, Italy. Two sets of palaeochannel fills and three sets of coarse alluvium are present. The palaeochannel fills contain pollen and other evidence for accumulation in a river draining relatively well-vegetated landscapes, while some of the coarse alluvium was laid down after clearance phases. Phases of gravel sedimentation may be related to historical and archaeological evidence for periods of intensification of human activity and the expansion of farming in the area. Changes in depositional mode were rapid: two sets of palaeochannel fill deposits and two sets of coarse sediments have accumulated in the Feccia Valley since approximately the fifteenth century AD.

KEYWORDS: Italy, valley alluviation, late Holocene, human activity, rapid environmental change.

1 INTRODUCTION

In an important review of later Holocene valley alluviation in the Mediterranean, Bell (1982) showed that valley fill events took place diachronously across the region, and thus in most cases climatic change was unlikely to be as important a causal agent for alluviation as human disturbance of landscapes. He further showed that very little direct evidence of environmental conditions during the valley fill episodes was available and that the link between human activity and alluviation rests largely upon circumstantial evidence and the use of modern analogue. Several notable regional studies have largely confirmed Bell's (1982) position (among others Davidson 1980; Pope and van Andel 1984; van Andel et al. 1986; Chester and James 1991).

In earlier papers (Hunt et al. 1992; Gilbertson et al. 1992) we put forward evidence for anthropogenic causes for late Holocene valley alluviation events in central Tuscany. This evidence came from stratigraphical, sedimentological, palaeobiological and archaeological studies of slope and alluvial sequences, which provided relatively unambiguous palaeobiological evidence for exposed agricultural landscapes and rapid soil erosion during the deposition of colluvial sediments and coarse fluvial deposits.

In this paper, we present detailed stratigraphical and some palynological evidence from one of these Tuscan alluvial sequences, in the Feccia Valley (Figs 1 and 2). The alluvial deposits of the Feccia Valley contain a range of palaeobiological indicators including pollen, algal microfossils, molluscs, plant macrofossils and artefacts which together with sedimentary evidence can be used to build a sequence of the environmental and land management changes that took place in the valley over the last thousand or so years. The palynological evidence points to woodland clearance as a causal factor in initiating phases of gravel alluviation.

2 THE FECCIA VALLEY

The Feccia Valley lies 20 km southwest of Siena between the ranges of the Montagnola Senese and the Colline Metallifere. The valley is a graben with phases of fill of Miocene to early Quaternary age (Gilbertson and Hunt 1987) and is flanked by horsts of metamorphosed Mesozoic mudrocks, conglomerates and limestones (Giannini and Lazzarotto 1970). The hills above the valley have become covered with dense evergreen *macchia* and deciduous oak woodland since abandonment for agricultural purposes during the last war. Recent agricultural development has been discouraged by the steep slopes, poor soils and the damage to crops caused by wild boar, but the woods are managed on a fifteen-year cycle for firewood and cork extraction. Agricultural terracing is extremely rare, even on the steepest slopes, except very close to Medieval settlement. The once-marshy valley floor was drained in the late nineteenth-century and sunflowers, wheat and sweetcorn are grown. In this part of Tuscany, rainfall events, particularly those associated with late summer thunderstorms, can be very severe and we measured ground lowering rates of around 30 mm per year at a number of localities where agricultural soils were exposed (Gilbertson et al. 1992). The last major fill of the Feccia Valley graben is over 100 m of lacustrine silts of Early Pleistocene age (Gilbertson and Hunt 1987). This fill has subsequently been incised and an inset terrace over 1 km wide and underlain by a complex gravel sequence up to 3 m thick now occupies most of the floor

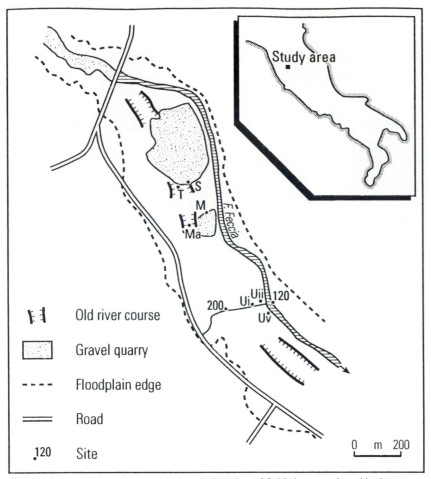

Figure 1. Map of the study area showing the location of field sites mentioned in the text.

Legend:
- Old river course
- Gravel quarry
- Floodplain edge
- Road
- .120 Site

of the valley. The surface of this feature shows an indistinct slightly meandering channel in some places (Fig. 1). The modern Fiume Feccia has incised through the gravel fill, but is considerably disrupted by gravel extraction. Where undisturbed, its morphology conforms to the 'wandering gravel' style of Blacknell (1981), in spite of severe agricultural soil erosion in the area today. The impact of modern soil erosion is moderated because much runoff and eroded sediment is absorbed by dense shrubby and woody streamside vegetation which seems to act as a filter.

3 FIELD AND LABORATORY METHODS

The Feccia gravels were observed over several kilometres of drainage ditch, river cutbank and gravel quarry sections. Representative sections (Appendix 1) were measured and drawn and samples taken for palaeobiological analysis. Pollen was extracted from sediments by boiling in 5% KOH for 5 minutes, sieving on nominal 7 micron mesh and swirling on a clock glass. A minimum of 250 land pollen grains per sample were counted except in the case of sample 6, where 100 land pollen grains were counted. Mollusc and plant macrofossil counts were pre-

viously described in Hunt et al. (1992). The age estimates were made by radiocarbon assay (uncalibrated here), youngest potsherd fabric seriation and by correlation with other sequences in the Montagnola Senese described by Hunt et al. (1992), but the number of radiocarbon dates is limited and there are large standard deviations, so we emphasise the importance of regarding our chronology as provisional.

4 ALLUVIAL STRATIGRAPHY IN THE FECCIA VALLEY

A complex sequence of fluvial deposits was exposed in drainage ditches and gravel pits in the Feccia Valley (Figs 1-3). A preliminary description of some of these deposits was given by Gilbertson et al. (1983; 1992) and Hunt et al. (1992). The sequence (Fig. 2) consists of plane and cross-bedded silts, sands and gravels (F8), overlying epsilon cross-bedded gravels, silty palaeochannel fills and overbank silts (F7) with a radiocarbon date of 170 ±40 BP (SRR-3253) on a fish-weir in one of the palaeochannel fills. Below this, plane and trough cross-bedded gravels (F6) overlie silty palaeochannel fills (F5) with a radiocarbon date of 400 ±50 BP (SRR-3252) on the outermost layers of a poplar stump that had grown in the top of a palaeochannel fill. Further plane-bedded and trough cross-bedded gravels (F3) overlying rather eroded remnants of early Holocene deposits (F2) and then Lower Pleistocene siltstones (F1) of the last graben fill. The ages of the lowest two units were determined by Gilbertson et al. (1983; 1992) on biostratigraphical grounds.

5 PALYNOLOGY OF THE RIVER SEDIMENTS

The pollen diagram was aggregated from samples from a variety of localities; it cannot therefore be 'read' in the normal stratigraphical sense, sample by sample, except where 'runs' of samples are present. Thus, sample 1 was a mud intraclast within unit F6, and thus can only loosely be associated with unit F5. Sample 2 came from a palaeochannel fill in a position stratigraphically equivalent to the palaeochannel fill from which samples 3-5 were taken and indeed is a relatively close palynological match for these samples. In the same way, stratigraphically equivalent palaeochannel fills in three localities, represented by sample 7, samples 8 and 9, and samples 10 and 11, were sampled in unit F7a. Again, there are points of similarity between the two pairs of samples (8, 9) and (10, 11), most notably a *Quercus* decline. The two samples from unit F5 are from the same locality. The lateral and vertical variability of the assemblages from the palaeochannel fills must reflect both small scale patterns of land use and vegetational colonisation of the floodplain, and rapid temporal changes in these patterns. These points are discussed in detail below.

A number of general points emerge. Most samples contained both Quaternary pollen and recycled pre-Quaternary palynomorphs. The Quaternary pollen and spores were generally well preserved. Most pollen grains

were preserved in three dimensions and took stain. Most of the pre-Quaternary palynomorphs showed a degree of thermal maturity, varying in colour from very pale yellow to dark brown. They did not take stain except in rare cases. Few were well preserved and many were unidentifiable.

The Quaternary pollen assemblages were variable in composition, with *Pinus* or *Quercus* usually the most important tree species. Graminese, Ericales, cereal pollen and herbs of grassland and open ground were usually present, with Compositae, both Tubuliflorae and Liguliflorae, often common. Spores were sometimes common; they were mostly psilate monolete (undifferentiated Filicales), *Polypodium* and *Pteridium*. Algal microfossils were common in most samples.

Some of the features of these assemblages were probably the result of taphonomic and pedogenic processes, or the re-sedimentation of palynomorph assemblages from eroded soils. Bisaccate pollen, including *Pinus*, is often 'over-represented' in waterlain sediments (Traverse and Ginsburg 1966) as are fern spores and Compositae pollen (Hunt 1987; Traverse 1988). The same taxa are also very resistant to decomposition in soil profiles (Havinga 1967, 1971, 1984; Dimbleby 1985).

Although the samples had many features in common, as described above, they did show some important differences. They are therefore described in more detail below, by lithological unit.

Unit F5

Samples from this unit were characterised by generally high values for *Quercus* (11.5-33.7%) and *Pinus* (4.1-41.2%), abundant Ericales (6.6-20.5%), Compositae (4.8-16.2%) and variable but often high values for Cyperaceae (1.0-28.7%). Gramineae (0-7.3%) and cereal (0-

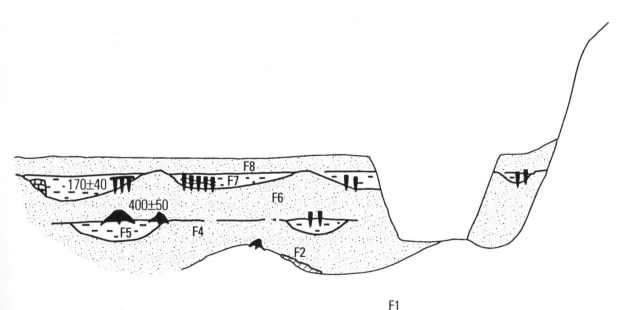

Figure 2. A schematic section illustrating the stratigraphical relationships between the alluvial units in the Feccia Valley. See Appendix 1 for individual unit descriptions.

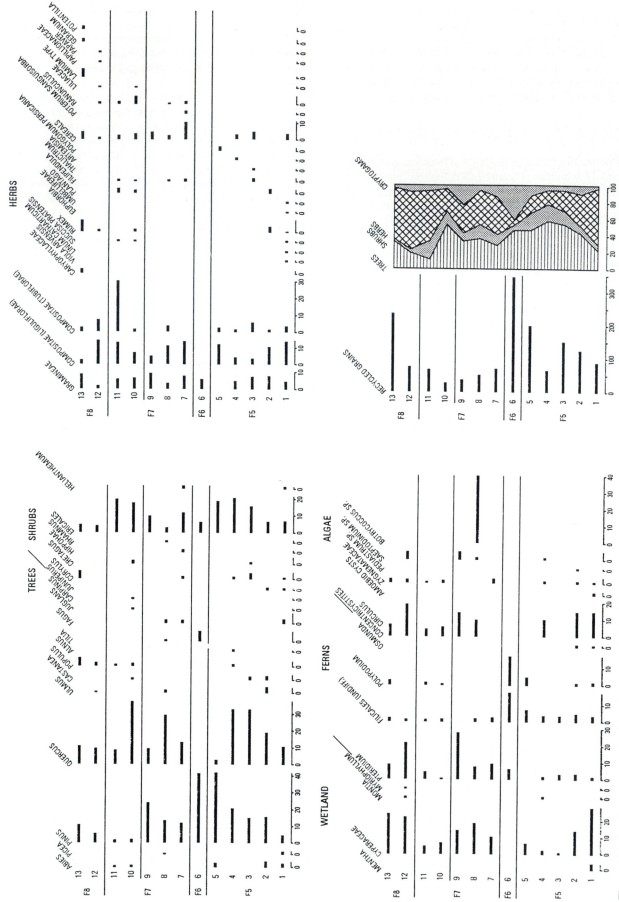

Figure 3. Pollen diagram from alluvial sediments in the Feccia Valley (for explanation see text). Summary pollen box diagram of vegetation types is in 0-100%.

4.2%) pollen, and spores of Filicales (3.1-7.0%) and *Pteridium* (0-3.3%) were also significant. Algal microfossils were present in some samples. Assemblages were dominated by *Concentricystes circulus* (up to 14.8% as a percentage of total Quaternary pollen), with low counts of Zygnemataceae (0.8-1.2%), and occasional Amoeboid cysts, *Pediastrum*, and *Saeptodinium*. Recycled palynomorphs were very common (65.0-200.0% as a percentage of total Quaternary pollen).

In unit F5 at site UII, *Pinus*, Compositae and fern spores become more common upwards, while *Quercus* and Gramineae decline in importance. These trends may reflect an episode of pedogenesis after the deposition of this unit, before the deposition of unit F6, although it is tempting to regard the fall in *Quercus* as reflecting a genuine decline in oak pollen (see also the arguments below, which might be taken to support this viewpoint). The presence of *in situ Populus* stumps rooted in the meander plugs of unit F5 is evidence for a break in deposition following the deposition of the unit, during which pedogenesis could have occurred.

Unit F6

The single sample from this unit gave a very sparse assemblage dominated by *Pinus* (41.2%), with some *Fagus*, Ericales and Gramineae and abundant Filicales (17.6%), *Polypodium* (17.6%) and *Pteridium* (5.9%). Recycled grains were very common (345.0% as a percentage of total Quaternary pollen).

Erosion of soils and bedrock in the Feccia catchment might be expected to liberate large numbers of 'resistant' Quaternary grains such as fern spores and *Pinus* together with thermally mature pre-Quaternary palynomorphs. The waters of the Feccia may therefore have contained a palynomorph assemblage rich in 'resistant' and recycled material during the deposition of unit F6. The high counts for recycled grains, fern spores and *Pinus* may also reflect taphonomic processes in the fluvial environment (Hunt 1987). Thermally mature palynomorphs are known to be denser than thermally immature specimens of the same species (K.J. Dorning, pers. comm., 1986), and therefore are more likely to be deposited in a turbulent fluvial environment than are most relatively light Quaternary grains. The fern spores, and the *Pinus* pollen (if waterlogged) would also be larger and heavier than most Quaternary pollen and thus also likely to be preferentially deposited. It is also possible that the sample has been significantly 'skewed' by post-depositional weathering.

Unit F7

Assemblages from this unit were characterised by rather variable but often high counts for *Quercus* (8.8-38.1%), Ericales (2.5-20.0%), *Pinus* (1.6-23.8%) Compositae (4.8-43.2%), Gramineae (3.3-9.5%), cereals (2.4-10.5%), Cyperaceae (0-18.9%) and *Pteridium* (0.7-23.8%). Algal microfossils were present in all samples and *Concentricystes circulus* was usually the most common (up to 14.3% as a percentage of total Quaternary pollen), with

some low counts for Zygnemataceae (0-2.6%) and *Saeptodinium* (0-4.8%). *Botryococcus* was very common in one sample (40.0% as a percentage of total Quaternary pollen). Recycled palynomorphs were common (25.9-70.0% as a percentage of total Quaternary pollen).

In both section 200 (samples 8 and 9) and MAO (samples 10 and 11) there was again an apparent indication of taphonomic effects in the relatively high counts for 'resistant' pollen and spores – *Pinus*, Compositae, *Pteridium* – in the upper sample from each palaeochannel fill. In both sections the rise in 'resistant' pollen and spores was accompanied by a decline in *Quercus* percentages from over 30% to under 10%. This may be the result of the preferential weathering of the oak pollen, but it is possible that it might genuinely reflect a decline in the importance of the tree in the Feccia Valley. The presence in sample 11 of the fragile pollen of *Populus* can be taken as evidence to support this latter viewpoint, since these grains are certainly very vulnerable to weathering (Havinga 1984). In this case, therefore, (and thus possibly also in the case of unit F5) the rise in 'resistant' pollen must reflect the recycling of this material from eroding soils in the catchment and the fall in *Quercus* must relate to forest clearance, which caused the soil erosion.

There were considerable differences between the three sites sampled, for instance Site UII (sample 6) had a very high count for cereal pollen (10.5%), while Site MAO had lower counts for *Pinus* and higher counts for Ericales than Site 200. These differences may reflect differences in pollen and spore taphonomy, but may also be the result of localised patterns of land use on the floodplain. The generally lower counts for recycled palynomorphs in Unit F7 (25.9-70.0%) than in Unit F5 (65.0-200.0%) might reflect slower fluvial flow during the deposition of F7 than during F5 and thus enhanced deposition of Quaternary pollen, or it might reflect less vigorous erosion of bedrock during the accumulation of F7.

Unit F8

The assemblages from this unit are characterised by moderate counts for *Pinus* (5.7-11.1%) and *Quercus* (9.4-11.1%), some *Populus* (1.9-4.4%) and Ericales (3.8-4.4%), variable Gramineae (1.9-8.8%) and Compositae (4.4-39.8%), some cereal pollen (0.9-4.4%), high Cyperaceae (22.6-24.4%) and *Pteridium* (8.8-9.4%). The algal microfossils are mostly *Concentricystes* (6.6-18.9% as a percentage of total Quaternary pollen) and Zygnemataceae (1.9-2.2% as a percentage of total Quaternary pollen). Recycled spores are very common (75.0-23% as a percentage of total Quaternary pollen).

The presence of the fragile grains of *Populus* may be taken as evidence that these samples have not been greatly affected by post-depositional weathering. The very high count for recycled grains, the high counts for *Pteridium* and especially the high count for Compositae (39.8%) in sample 12 may be consistent with the recycling of 'resistant' material from eroding soils and bedrock, or possibly with their concentration in the fluvial environment.

6 PLANT MACROFOSSILS AND MOLLUSCS

Plant macrofossil and mollusc evidence from the Feccia Valley was previously described and discussed by Gilbertson et al. (1983) and Hunt et al. (1992). We now regard many of the aquatic mollusc taxa listed by Gilbertson et al. (1983) as recycled from the Lower Pleistocene. Plant macrofossil and molluscan analyses of fluvial deposits essentially provides information concerning the environment in and close to the river, but does not provide an overview of the landscape in the same way as palynology. In the case of the Feccia Valley, the mollusc and plant macrofossil evidence essentially reinforces the palynological evidence for an alternating pattern of more and less well vegetated landscapes. The plant macrofossil assemblage from unit F5 contain macrofossils from trees, shrubs, herbaceous taxa and wetland taxa. The molluscs include woodland, catholic and aquatic taxa. Samples from unit F6 did not yield plant macrofossils or molluscs. The plant macrofossils from unit F7 include shrub, open ground and wetland taxa, but the uppermost sample in this unit from Site 200 contains only open ground and wetland taxa (Hunt et al. 1992). The mollusc assemblages include catholic, open ground and aquatic taxa. The plant macrofossil assemblages from unit F8 contained only open ground and wetland taxa, but no molluscs were recovered from this unit (Hunt et al. 1992).

7 INTERPRETATION: THE LATE HOLOCENE RIVER ENVIRONMENT

All of the palaeobiological evidence appears to reflect a mosaic of vegetation on and around the Feccia floodplain, most probably with considerable similarity to the modern vegetation pattern of the area. Coppiced and pollarded oak woodland and managed scrub woodland with large permanent oak stands and brushwood cut for firewood on a fifteen-year cycle is still common in the area today, mostly on interfluves and steep valley sides. Oak woodland is probably represented by the often high counts for *Quercus* in the pollen diagram. Pines are used locally today as roadside and shade trees, while poplars are the common waterside tree. Other trees are not common in the area today, and are not heavily represented in the pollen diagram, though the presence of a fairly diverse broadleaved woodland during F5 times is attested by the plant macrofossil evidence (Hunt et al. 1992). It is clear that during the deposition of the palaeochannel fills (F5, F7) oak woodland was an important component of the vegetation of the Feccia catchment. The decrease in oak pollen at the top of each palaeochannel fill points to a sharp decline of oaks before the end of each palaeochannel depositional phase. It is hypothesised that these oak declines reflect large scale woodland clearance for agriculture. Oak then remained of lesser importance and landscapes were generally open during the aggradation of the gravel units (F6, F8).

Heathland and shrubby *maccia* are found on the high ground around the Feccia Valley today and heathers also occur on local acid metamorphic outcrops as an understorey, either to undisturbed woodland, or to managed woodland regenerating after partial cutting for firewood or charcoal. Pollen of Ericales is most important during the deposition of the palaeochannel fills (F5, F7a) and becomes less important during the aggradation of the gravel units (F6, F7c). It declines, however, after the decline of oak in each case, presumably because with each clearance phase the tree cover was removed first but the heather understorey was at first relatively undisturbed. During the later stages of clearance the heathers were apparently removed to a significant extent. Taxa such as *Juniperus, Rhamnus* and *Cretaegus* have locally colonised agricultural land abandoned for more than about 10 years, but there is no particular pattern to their occurrence here.

The pollen of herbaceous plants includes a variety of taxa typical of grassy vegetation, including *Thalictrum, Plantago* and *Poterium* together with abundant cereal grains and pollen and macrofossils of taxa often found in cultivated ground, such as *Viola arvensis, Artemisia, Polygonum persicaria*, Chenopodiaceae and *Papaver*. Although most agriculture in the area today is arable as the result of EC policy, it is clear from this new palynological evidence that both pastoral and arable agriculture were practiced in the Feccia Valley during the period reflected by the aggradation of units F5-F8. There does not seem to be a recognisable shift in the counts of pollen of herbaceous taxa during the different fluvial phases, so it is possible that there was little change to the pattern of agriculture during the period represented by the pollen diagram, only in its extent.

Taxa such as Cyperaceae, *Scirpus, Mentha, Montia, Filipendula* and *Succisa pratensis* suggest areas of damp ground beside the stream, though it must be borne in mind that some sedges prefer dry land habitats. The only pollen found of a truly aquatic plant are the few grains of *Myriophyllum* from sample 12, but achenes of aquatic Ranunculaceae are fairly common in both palaeochannel fill units, and are also present in the gravels of F8 (Hunt et al. 1992). Aquatic molluscs are confined to the palaeochannel fill units, perhaps because the Feccia was too energetic or (more likely) largely ephemeral during the gravel deposition phases. Alternatively, the absence of molluscs from the gravels may reflect taphonomic factors. Briggs et al. (1990) have commented on the low probability of mollusc deposition and preservation in high energy gravel environments.

The rise of 'resistant' taxa – pine, Compositae, fern spores – and recycled grains at the top of each palaeochannel fill points either to an increase in soil erosion before the end of each palaeochannel depositional phase or to *in situ* weathering of fragile grains and relative concentration of 'resistant' types. In the case of F7, the presence of fragile grains of *Populus* in a sample from the top of this unit make the former hypothesis most likely. A stillstand of sedimentation occurred at the top of F5, during which time poplars grew on the alluvial deposits, so in this case the rise in 'resistant' types could reflect weathering and preferential preservation of grains. The importance of 'resistant' and recycled types in the gravel sedimentation phases suggests that soil erosion continued during the aggradation of the coarse sediment units.

The algal microfossils in the meander plug sediments include both planktonic types (*Pediastrum, Saeptodinium* and *Botryococcus*) and benthonic forms (*Concentricystes* and Zygnemataceae). *Concentricystes* is usually the commonest algal microfossil. This form is widespread in calcareous fluvial environments and in SW England can be extremely common in calcareous tufas which accumulated in pools among woodland and thus in the shade (M.G. Macklin 1983 pers. comm.). The Zygnemataceae require shallow, slow moving, sun-warmed water to produce their spores (Van Geel 1976). The other taxa are less demanding in their requirements. The impression is given of a shallow, sluggish, perhaps shaded stream, though since the algae would grow mostly in the spring and summer this might not be a complete picture.

8 LATE HOLOCENE RIVER BEHAVIOUR

Evidence from the fluvial deposits in the Feccia Valley points to three phases of late Holocene coarse sediment aggradation, separated by two phases when channels incised slightly into the previously aggraded sediment. The latest phase of gravel aggradation (F8) appears to match the latest phase of slope deposit accumulation identified elsewhere in the Montagnola Senese, of late nineteenth century date (Hunt et al. 1992; Gilbertson et al. 1992). Coarse alluvial deposition of unit F6 in the Feccia Valley appears to have started during or after the sixteenth century and ended before the late eighteenth century, reflecting a phase of slope erosion and colluvial sedimentation which probably took place in the sixteenth and earliest seventeenth century (Hunt et al. 1992). There is no dating evidence for the earliest fluvial aggradation phase, but an earlier phase of slope deposit accumulation appears to have occurred in the eleventh and twelfth centuries AD (Hunt et al. 1992). The sedimentation episodes found here are out of phase with climatic episodes seen in southern Europe over the last 1000 years (Hunt et al. 1992) and thus climatic change is not likely to have been causal.

The phasing of slope and fluvial sedimentation episodes seems to correspond with archaeological and historical evidence for periods of intensification and expansion of land use in the 11th-12th, 16th and late 19th centuries found by the Sheffield-Siena Monterrenti Project and the Michigan State University Palaeolithic Archaeology Program (Barker and Symonds 1984; Hunt et al. 1992; Gilbertson et al. 1992; G.W.W. Barker, pers. comm. 1987; R. Colten, pers. comm. 1988). The evidence for woodland clearance and enhanced soil erosion described here is consistent with the archaeological and historical evidence.

9 CONCLUSIONS

The palaeobiological information from the Feccia Valley points directly and unambiguously to changes in Medieval and post-Medieval landscape management as a causal factor in changing runoff and sediment supply characteristics and hence changes in alluviation style. In the Feccia Valley, the palaeochannel fills are characterised by high incidences of tree pollen, especially oak, and plant macrofossils and molluscs indicative of arboreal and/or shrubby vegetation close to the river. At the end of each channel fill phase the percentage of oak pollen decreases in a way that cannot always be explained by taphonomic or pedogenic processes and therefore most probably reflects clearance on a large scale – there is also some indication of clearance from the plant macrofossils. The large coarse sediment units are characterised by low arboreal pollen percentages and there is a lack of macrofossil evidence for local arboreal or shrubby vegetation, though there is at times evidence for local herbaceous vegetation. Even if the chronology provisionally adopted herein is 'out' by several hundred years, it is clear that changes in alluviation style occurred very rapidly in the Feccia Valley.

The aggradation of coarse fluvial deposits, particularly on braided floodplains, is conventionally linked with 'flashy' discharges and high sediment supply, both characteristic of relatively devegetated landscapes, whereas streams depositing predominantly fine grained alluvium are usually characteristic of well-vegetated landscapes, more even discharges and low levels of sediment supply. The data from the Feccia Valley conform to this model and point to the importance of management strategies in influencing process style in this part of the Mediterranean landscape.

10 ACKNOWLEDGEMENTS

This study owes much to the kindness of Dr H. Patterson (potsherd identification), Dr D. Harkness (radiocarbon assays), Prof. G. Barker and R. Colten, (discussion of the fieldwalking evidence), Dr S. Mitchell (historical discussion). The Provincia di Siena provided hospitality and numerous students from the universities of Sheffield and Siena took part in the surveys on which this account is based. Mr J. Jacyno drew the pollen diagram. The manuscript was substantially improved by constructive criticism from Prof. J. Lewin and an anonymous referee.

REFERENCES

Barker, G.W.W. & J. Symonds 1984. The Monterrenti Survey, 1982-1983. *Archeologia Medievale*, 278-289.
Bell, M. 1982. The effects of land-use and climate on valley sedimentation. In Harding, A.F. (ed) *Climatic Change in Later Prehistory*, Edinburgh University Press, Edinburgh, 127-142.
Blacknell, C. 1981. Morphology and sedimentary features of

point bars in Welsh gravel bed rivers. *Geological Magazine*, 118, 181-192.

Briggs, D.J. & D.D. Gilbertson 1980. Quaternary processes and environments in the Upper Thames Basin. *Transactions of the Institute of British Geographers*, NS 5, 53-65.

Briggs, D.J., D.D. Gilbertson & A.L. Harris 1990. Taphonomy in a braided river environment and its implications for studies of Quaternary cold-stage river deposits. *Journal of Biogeography*, 17, 623-637.

Chester, D.K. & P.A. James 1991. Holocene alluviation in the Algarve, Southern Portugal: the case for an anthropogenic cause. *Journal of Archaeological Science*, 18, 73-87.

Dimbleby, G.W. 1985. *The palynology of archaeological sites*. London: Academic Press.

Giannini, E. & A. Lazzarotto 1970. Studio geologico della Montagnola Senese. *Memorie della Societa' Geologica Italiana*, 9, 451-495.

Gilbertson, D.D., D.A. Holyoak, C.O. Hunt & F.N. Paget 1983. Palaeoecology of late Quaternary floodplain deposits in Tuscany: the Feccia Valley near Frosini. *Archeologia Medievale*, 10, 340-350.

Gilbertson, D.D. & C.O. Hunt 1987. An outline and synthesis of the geoarchaeological development of the southern Montagnola Senese, Tuscany. *Archeologia Medievale*, 14, 349-408.

Gilbertson, D.D., C.O. Hunt, R.E. Donahue, D.D. Harkness & C.M. Mills 1992. Towards a palaeoecology of medieval and post-medieval Tuscany, *in* Bernardi, M. (ed) *L'Archeologia del Paesaggio*. Firenze, Consiglio Nationale della Richereche, editzione All'Insegna del Giglio, 205-248.

Godwin, H. 1975. *The History of the British Flora: A Factual Basis for Phytogeography*. Cambridge: Cambridge University Press.

Havinga, A.J. 1967. Palynology and pollen preservation. *Review of Palaeobotany and Palynology*, 2, 81-98.

Havinga, A.J. 1971. An experimental investigation into the decay of pollen and spores in various soil types. In J. Brooks et al. (eds) *Sporopollenin*. London: Academic Press, 446-479.

Havinga, A.J. 1984. A 20-year investigation into the differential corrosion susceptibility of pollen and spores in different soil types. *Pollen et Spores* 26, 541-548.

Hunt, C.O. 1987. Comment: the palynology of fluvial sediments: with special reference to alluvium of historic age from the upper Axe Valley, Mendip Hills, Somerset. *Transactions of the Institute of British Geographers* N.S. 12, 364-367.

Hunt, C.O., D.D. Gilbertson & R.E. Donohue 1992. Palaeobiological evidence for late Holocene soil erosion in the Montagnola Senese, Tuscany, Italy. In M. Bell & J. Boardman (eds) *Past and present soil erosion: archaeological and geographical perspectives*. Oxford, Oxbow Monograph 22, 163-174.

Pope, K.O. & T.N. van Andel 1984. Late Quaternary alluviation and soil formation in the Southern Argolid: its history, causes and archaeological significance. *Journal of Archaeological Science*, 11, 281-306.

Traverse, A. 1988. *Palaeopalynology*. Boston: Unwin Hyman.

Traverse, A. & R.N. Ginsburg 1966. Palynology of the surface sediments of Grand Bahama Bank, as related to water movement and sedimentation. *Marine Geology*, 4, 417-459.

Van Andel, T.J., C.N. Runnels & K.O. Pope 1986. Five thousand years of land use and abuse in the Southern Argolid, Greece. *Hesperia*, 55, 1, 103-128.

Van Andel, T.J. & C. Runnels 1987. *Beyond the Acropolis: A rural Greek Past*. Stanford University Press, Stanford.

Van Andel, T.J., E. Zangger & A. Demitrack 1990. Land use and soil erosion in prehistoric and historical Greece. *Journal of Field Archaeology*, 17, 379-396.

Van Geel, B. 1976. Fossil spores of Zygnemataceae in ditches of a prehistoric settlement at Hoogkarspel (The Netherlands). *Review of Palaeobotany and Palynology*, 22, 337-344.

Appendix 1. Representative section logs from the Feccia Valley. (Sample depths are in each case measures from the top of the unit concerned).

Unit	Thickness	Description
Site 200		
F8	0.7-1.7 m	Silts and sands, plane laminated, epsilon cross-bedded and trough cross-bedded and sandy cobbly gravels, imbricated and plane and trough cross-bedded, with a silty soil profile developed at the top. There is a strongly erosive contact with underlying units. Samples 12 (a gravel horizon) and l3 (a silt) were taken respectively 1.6 m and 1.0 m from the top of this unit.
F7	0.0-1.0 m	Clayey silts, plane laminated, containing fish wiers and stakes. Twigs from the hurdles of a fish wier gave a radiocarbon date of l70 ±40 BP (SRR 3252). A worked flint of upper palaeolithic style and two sherds of Maiolica (16-17th century) were found. The unit is incised into the underlying unit. Samples 8 and 9 were taken from respectively 0.8 m and 0.1 m from the top of this unit.
F6	> 2.0 m	Gravels and coarse cobbly gravels, imbricated, plane and occasionally trough cross-bedded. Base unseen at this site but this unit can be traced laterally into neighbouring exposures.
Site 120		
F6	> 1.0 m	Coarse cobbly gravels, imbricated, crudely plane bedded, containing towards the base cobbles of mollusc-rich calcareous tufa. The upper surface of this unit was removed during gravel extraction, the lower surface is erosive.
F2	0.2-0.4 m	Fine sands, silts and peats, plane-bedded and occasionally trough cross-bedded. The deposits are draped over the eroded surface of the underlying unit.
F1	> 1.0 m	Silts, plane-laminated, strongly weathered at the surface. The upper surface is irregular, penetrated by scour holes and with tree stumps rooted into it.
Site M, MA		
F8	0.9-1.1 m	Sandy silts, massive, top surface bulldozed. Sharp junction with underlying units.
F7	0.0-2.6 m	Palaeochannel fill: grey sands and silts containing occasional twigs, small-scale trough cross bedded, overlying silts and sands, epsilon cross-bedded, incised into underlying unit. Samples l0 and ll (respectively 0.3 m and 0.1 m from the top) came from this unit.
F6	2.4-2.5 m	Sandy silty gravels, trough cross-bedded, containing rolled Roman brick and silty intraclasts derived from the underlying unit. Erosive lower contact with underlying unit. Sample l came from a silty intraclast.
F5	0.0-0.2 m	Palaeochannel fill: grey structureless silts with wood fragments, incised into the underlying unit. Sample 2 came from this unit.
F3	> 0.3 m	Sandy silty gravels, trough cross bedded. Base unseen.
Site S		
	0.3 m	Made ground – randomly orientated stones in a sandy matrix overlying a bulldozed surface.
F8	0.4 m	Sandy silts and thin gravels, plane and trough cross bedded. Sharp junction with
F6	0.5 m	Sandy gravels, trough cross-bedded. Erosive base.
F5	0.0-0.4 m	Palaeochannel fill trending 260° N, blue-grey plane laminated silts with a drifted oak trunk. Incised into
F3	> 0.8 m	Sandy coarse gravels, trough cross-bedded, base unseen.
Site T		
F7	0.0-0.6 m	Palaeochannel fill trending 275° N: grey silts, laminated and with root casts, upper surface bulldozed, incised into
F6	1.8 m	Sandy gravels, trough cross bedded, with a large silt intraclast containing a poplar stump at 1.5 m. Erosive lower contact with
F5	0.0-0.5 m	Palaeochannel fill trending 290° N: blue-grey plane-laminated silts with wood fragments and with tree stumps in growth position rooted in the highest horizon (the outermost layers of one of these stumps gave a radiocarbon date of 400 ±50 BP SRR3252). Incised into
F3	> 1.0 m	Sandy gravels, trough cross-bedded, base unseen.
Site Ui		
F8	1.75 m	Silty sands, sandy silts and gravels, generally fining up, trough cross-bedded and plane-laminated and showing weak pedogenic features at the top. Sharp lower boundary with
F7	0.0-0.8 m	Palaeochannel fills: a complex of silt channel-fills passing down into epsilon cross-bedded sandy gravels. Near the top of one channel fill, the remains of a wooden ?platform anchored by large split oak stakes. Incised into
F6	> 0.75 m	Sandy silty gravels, trough cross-bedded, base unseen.
Site Uii		
F8	0.7-0.9 m	Sandy silts and silts, generally fining upward, with weak pedogenic features at several levels, draped over
F7	1.0-1.3 m	Palaeochannel fills: blue grey silts, passing downward and laterally into epsilon cross-bedded gravels and laterally into overbank silts. Two silty channel fills were visible in the exposure. One contained two parallel lines of stakes, driven in from near the top of the unit and Roman brick was found in this unit. Incised into underlying units. Sample 7 came from the lower of the two channel fills, 1.2 m below the top of the unit.
F6	0.3-1.6 m	Gravels, trough cross-bedded, erosive base.
F7	0.0-0.9 m	Palaeochannel fills with many drifted twigs and branches, passing downward and laterally into epsilon cross-bedded gravels. Two silty channel fills were visible in the exposure. A fragment of possibly worked wood and two

post-medieval (16th-17th century) postherds were found in this unit. Incised into underlying unit. Samples 3, 4 and 5 (respectively 0.8, 0.4 and 0.1 m from the top came from this unit.

Site Uv

F8	0.3-0.6 m	Sands and silts, with weak pedogenic features, draped over
F7	0.0-1.0 m	Palaeochannel fill: epsilon cross-bedded silts passing down into fine gravels, passes laterally into 0.3 m of overbank silty sand with weak planar bedding. Two stakes were driven into the underlying deposits from horizons in the overbank sediments. The unit is incised into the underlying deposits.
F6	1.5-2.0 m	Gravels, trough cross bedded and with occasional silt lenses. Drapes over topography of underlying deposits. Sample 6 comes from a silt lens 1.45 m below the top of this unit.
F2	0.0-0.3 m	Palaeochannel fill, silty gravel rich diamicton, incised into
F1	>0.5 m	Shaly mudstones, dipping a c. 5°, with a weathered, eroded surface.

PART 3

Geochronology, correlation and controls of Quaternary river erosion and sedimentation

Geochronology, correlation and controls of Quaternary river erosion and sedimentation

MARK G. MACKLIN
School of Geography, University of Leeds, Leeds, UK

It is probably true to say that nowhere outside the Mediterranean basin has the debate concerning the underlying controls of Quaternary river erosion and sedimentation been so contentious or protracted. Much of this discussion was prompted, and subsequently structured, by Claudio Vita-Finzi's remarkably influential work *The Mediterranean Valleys*, and opinion since his book was published in 1969 has remained strongly divided as to the primary causes of river instability in the Mediterranean. During the Pleistocene it is generally accepted that tectonics and climate were the principal controls of river evolution through their influence on local and regional base-level, vegetation, sediment supply and runoff. In the Holocene, however, although these factors have continued to influence river systems, human disturbance of stream channels and catchments has become of equal and in many cases of primary importance in river development. Although the major extrinsic factors (i.e. tectonics, climate, human activity) that influence alluvial system dynamics have been well documented, interpretation of causality for particular episodes of river sedimentation, or erosion, is frequently more problematic. Within tectonically active parts of the Mediterranean, for example, it is often difficult to distinguish between Pleistocene aggradation and dissection cycles related to climate, from those initiated by uplift or subsidence. Similarly, limited high resolution palaeoclimatic and cultural data, and generally poor dating provision for many Quaternary alluvial sequences in the region does not normally allow cause and river response to be related with any degree of confidence. Indeed, one of the main reasons that the relationship between Holocene fluvial activity and climate- and/or human-induced environmental change has remained so controversial in the Mediterranean basin is the difficulty of distinguishing between climatic and anthropogenic signals in the various forms of proxy data. Interpretation is further confounded by complex response (Schumm 1973) within drainage basins whereby externally imposed change may diffuse through the system in a complex manner, with many lags, and where various reaches of a river system respond to a given initial stimulus at different times and in varying ways. This means that river behaviour is indeterminate in so far as it is not always possible to predict a unique, immediate pattern of outcomes resulting from a given stimulus for change (Chorley et al. 1984). Complexity of response is also

compounded by the existence of geomorphic thresholds, which may be due to either extrinsic or intrinsic causes. The threshold for sedimentation or erosion might be crossed at different times in different reaches, or river basins, during a phase of climate change, resulting in an uneven pattern of alluviation or river incision.

Elucidating behaviour-forcing variables of Quaternary river erosion and sedimentation is the principal, and linking, theme of all seven papers in this part of the volume. Critical in this respect for establishing cause and effect in a river system is independent and accurate dating of fluvial surfaces and sedimentary sequences. A wide range of dating techniques has been employed in this series of studies including isotopic (^{14}C), geomorphic (e.g. soil profile development) and correlated age methods (artefacts, palaeosols, calcretes). In a number of papers sediment provenance has also been studied in order to elucidate the nature of catchment erosion processes and also to determine the location of eroding areas within a drainage basin and the type of material (e.g. topsoil as opposed to subsoil sources) being eroded.

In Chapter 17 Ballais reports a detailed regional study of Holocene alluvial chronologies in the eastern Maghreb, North Africa. The principal aim was to distinguish between the role of natural factors (mainly climate) and anthropogenic effects on the Holocene development of river systems in the region. The investigation provides a transect from the subhumid, northern part of Tunisia at the Mediterranean coast to the arid Saharan zone in the south. There is evidence of at least three Holocene alluvial terraces in the eastern Maghreb. The oldest is dated to between 8300-5000 years BP and formed during a period of wetter climate. Pollen and lake level data in the region have shown this resulted from an intensified and northward displacement of the monsoonal system (COHMAP 1988). A second younger historical terrace, which is not found in the Saharan part of the eastern Maghreb, began to accumulate in the 2nd century AD and continued to about the 6th century (1470-1350 years BP). Sedimentation rates for this unit averaged 7.4 mm year^{-1}, five times higher than the early to middle Holocene alluvial fill. Evidence suggests that, although landscape disturbance associated with the development of large scale cultivation (beginning around the 5th century BC) was the precursor to alluviation during this period, accelerated soil erosion and floodplain sedimentation in the 1st and 2nd centuries AD was initiated by a small change in rainfall characteristics. A third, less extensive, terrace developed locally after 600 years BP and is similarly attributed to climate change. Over the last few centuries high rates of incision have resulted in the formation of a lower bench 1.0-1.5 m above mean river flow levels. It comprises very coarse gravel, in contrast to the finer-grained material of the earlier fills, indicating that in very recent times there has been an overall increase in the competence of fluvial erosion and considerable bedrock erosion. Ballais suggests that this is in part due to overgrazing and destruction of vegetation but also may relate to recent climatic variation. He concludes that natural bioclimatic variations remain the driving force of morphogenesis of river systems in the eastern Maghreb, although they are only effective in river basins disturbed by human activity.

In the next chapter geomorphic and archaeological investigations conducted by Abbott and Valastro in the territories surrounding the ancient Greek cities of Croton and Metapontum, southern Italy, reveal a complex set of relationships between regional tectonics, human activity and episodes of valley aggradation and incision. Dating control is provided by a suite of ^{14}C dates plus a wealth of archaeological material. Differences in the degree of Holocene valley floor incision between these areas is controlled by regional variations in rates of uplift and subsidence, which have produced contrasting alluvial sequences. Holocene alluvial deposits in the chorai of Metapontum consist of vertically stacked sediment packages that accumulated primarily by vertical accretion during valley-wide flooding. Each unit comprises a fining-upward cycle, between 2-10 m thick, capped by a palaeosol. Channels associated with each unit are relatively deep and a few hundred metres wide, suggesting downcutting and subsequent aggradation was relatively rapid. Truncation of soils, or other evidence of increased erosion on the floodplain, is extremely rare suggesting that the primary systems change was an increase in the rate of sediment influx rather that a shift in rainfall amount and river discharge. Episodes of instability and floodplain aggradation are dated to the late Neolithic, the Greek occupation and the Medieval period. In marked contrast, the early to middle Neolithic, middle Bronze Age and Roman periods were characterized by stability and soil formation on the valley floor. Unlike the stacked alluvial stratigraphies of the Metapontum streams, the middle Cacchiavia Valley exhibits a cut-and-fill sequence of inset alluvial units. Holocene river development was characterized by long periods of slow fine-grained sedimentation separated by brief incisional episodes. Subsidence is probably the dominant reason that streams of the choras of Croton are not incised but also severe agricultural erosion has introduced very large quantities of fine-grained sediment. The timing of alluvial aggradation and incision in the Croton chorai does not appear to match the sequence at Metapontum, indicating that regional climatic trends are in themselves incapable of accounting for changes in fluvial activity. Abbott and Valastro conclude that both choras exhibit alluvial sequences that reflect

fluctuating sediment yields through the Holocene. In the Metapontum area episodes of rapid catchment erosion seem to coincide with increasing population levels in the later Neolithic, and during the Iron Age and Classical periods. At least one phase of soil formation appears to coincide with a period of depopulation at the end of the Punic War. At Croton, relatively slow aggradation of silts and clays seems to have occurred almost continuously through the middle to late Holocene with very brief incisional episodes that are more likely to be related to tectonic uplift than human activity or climatic perturbations. It is interesting to note the strong similarities between the Croton and Metapontum chorai alluvial sequences and arroyo stratigraphies in the semi-arid western USA (e.g. Balling and Wells 1990), although in the United States climate, as opposed to anthropogenic activity, is generally considered to be the primary stimulus for river instability in the pre-Colombian period. Evidence of equifinality in alluvial systems with markedly different cultural histories once again highlights the many pitfalls that exist when trying to establish causality.

A study of late Quaternary alluvial fan sedimentary sequences around the Konya basin of south central Anatolia, Turkey, is reported by Roberts in Chapter 19. Now largely dry, the basin was occupied by an extensive lake between c. 23,000-17,000 years BP. Absence of tectonic deformation of palaeoshorelines developed during this period, across which a series of alluvial fans have been built, suggests that fan regimes have been controlled primarily by shifts in hydro-climatic parameters. The development of one of the larger fans in the region, that at Ibrala, has been investigated in association with geo-archaeological work on Neolithic settlement mounds. Three principal lithostratigraphic units have been identified. A lower fan member comprising fluviatile gravels, debris flow deposits and eroded *terra rossa* soil material. These are overlain by calcareous lacustrine marls and fine sands; the latter unit was laid down at the distal margin of the alluvial fan at the edge of the former lake. They correlate with the last phase of high lake levels radiocarbon-dated to between 23,000-17,000 years BP and were deposited under cold, semi-arid conditions. Between c. 17,000 and 9000 years BP lake levels were low as a result of a shift to more arid conditions and no sediment deposition occurred on the edge of the fan. At the end of the Pleistocene alluvial sedimentation recommenced on the fan and the third, and most recent, unit consists of 3 to 5 m of silts and clays. These fine-grained sediments indicate a steady rather than flashy river regime with a high suspended sediment load from the catchment. The onset of alluviation was partly due to the re-activation of springs, which feed the Ibrala River. No Holocene lacustrine sediments are recorded in the fan sequence implying lake levels have remained low during the last 9000 years. The evidence that *terra rossa* soils within the Konya basin (formed under relatively warm, subhumid climatic conditions prior to 25,000 years BP) were eroded before the Holocene is of more general importance as it cautions against attributing episodes of accelerated soil erosion solely to human disturbance.

Middle and Late Pleistocene fan, piedmont and coastal plain alluvial sequences in southeast Cyprus are examined by Stevens and Wedel in Chapter 20. Detailed sedimentological and facies analyses have been used to isolate, and evaluate, extrinsic and intrinsic controls of alluvial system behaviour. Their investigations have focused on a terraced fill which interdigitates coastwards with raised marine deposits of Tyrrhenian I age (c. 200,000 years BP) and younger. The sequence is characterised by bounding erosional surfaces, palaeosols and calcrete development and an upward trend from axial stream to alluvial fan facies. Lithological contrasts (differences between mafic and limestone components) are used to differentiate between axial river and tributary stream sediments sources. Allo- and autocyclic trends revealed by Markov-chain analysis show genetic relationships and the influence of tectonics and intrinsic factors on river development.

Anketell et al. have carried out similarly detailed sedimentological studies of Quaternary alluvial deposits in the Tripolitania region of northwest Libya in Chapter 21. Their investigations have been concentrated in two regions: the Jifarah Plain, including the foothills of the Jabal Nafusah, and the major valleys of the Beni Walid region on the gently dipping south-facing slope of the Jabal. Fluvio-aeolian deposits on the Jifarah Plain have been subdivided into six aggradational units (Q1-Q6) using erosion surfaces, palaeosols and calcretes. The age of the oldest unit is not yet clear but there is evidence to assign unit Q1 to the Early Pleistocene, Q2 to the Middle Pleistocene, and Q3 and Q4 to the Late Pleistocene. A major phase of downcutting and terrace formation occurred in the early Holocene beginning with the erosional surface which defines the base of Q5. Radiometric dates on valley basalts indicate that the Tripolitanian valleys are of considerable antiquity and were formed in the Early Eocene following uplift in the Late Palaeocene. The valleys continued to form conduits for lava flows from the Oligocene through to the Early Pleistocene. The absence of reliable dating of Tripolitanian Quaternary deposits precludes accurate correlation with sequences in neighbouring areas of North Africa or Southern Europe, however, some broad comparisons can be made on the basis of lithostratigraphy and stratigraphic sequence.

In the final two chapters in this section by James and Chester (Chapter 22) and by Harvey et al. (Chapter 23) the correlation of fluvial units, and their relative ages, have been established on the basis of different degrees of soil development on river terraces in two basins located respectively in the southwest and southeast of the Iberian Peninsula. Both studies employ mineral magnetic and chemical analyses of alluvial soils to identify sediment sources and elucidate catchment erosion histories. In the semi-arid Mediterranean environment, progressive development and increasing rubification of argillic B

horizons, and subjacent accumulation of pedogenic carbonate, are the primary morphological features of soil development over time. These properties have been used to develop a soil chronosequence and chronological framework for river landform development. James and Chester (Chapter 22) have studied Quaternary river sediments and soils in the Algarve, southern Portugal, and identified a clear distinction between the brown-coloured, charcoal-rich late Holocene alluvium and the redder, older Pleistocene fill. Contrasts between the Holocene and Pleistocene river sediments are shown in some part to relate to age-dependent transformation of iron oxi-hydroxides (rubification) but primarily reflect different sediment sources. It appears that the younger fill largely comprises the former topsoil of the catchment which, following widespread human disturbance of the vegetation, has been almost entirely stripped during the last three thousand years and now lies, in part, in the valley bottoms. Lack of subsoil indicates that erosion over this period has been more superficial than deep-seated, more consistent with processes of overland flow or rilling, than gullying. In Chapter 23 Harvey et al. use soil chronosequences to provide a relative chronology of long-term Pleistocene drainage development, stream capture and aggradation/incision episodes in the tectonically active Sorbas basin of southeast Spain. One particularly important finding in this study was that soil characteristics are not sensitive enough to identify any time transgressive behaviour of individual terraces. This implies that during the progressive incision of the drainage basin over the Quaternary as a whole, switches between dissection and aggradation took place relatively rapidly, more rapidly than can be detected from variations in soil characteristics. These dissectional and aggradational changes occurred throughout the basin, almost irrespective of the local tectonic context. This suggests response to basin-wide controls of sediment supply (probably climate in the case of the older terraces and human-induced change more recently), rather than response to major tectonically- or capture-induced changes in base-level.

REFERENCES

Balling, R.C. and S.G. Wells 1990. Historical rainfall patterns and arroyo activity within the Zuni River drainage basin, New Mexico. *Annals of the Association of American Geographers* 80: 603-617.

Chorley, R.J., S.A. Schumm & D.E. Sugden 1984. *Geomorphology*. London, Methuen.

COHMAP members, 1988. Climatic changes of the last 18,000 years: observations and model simulations. *Science* 241: 1043-1052.

Vita-Finzi, C. 1969. *The Mediterranean Valleys: Geological Changes in Historical Times*. Cambridge, Cambridge University Press.

CHAPTER 17

Alluvial Holocene terraces in eastern Maghreb: climate and anthropogenica controls

JEAN-LOUIS BALLAIS
Université d'Aix-Marseille I, France

ABSTRACT: There are usually at least two Holocene terraces along the watercourses of the eastern Maghreb. The lower Prehistoric Holocene terrace is present from the humid part of the Mediterranean climate zone to the lower Saharan zone. This feature is equivalent to the 'remblaiement Holocène principal' as found in the southern French Alps. This terrace formed during a climatic period of greater humidity which lasted from the beginning of the lower Holocene (Boreal – 8300 BP) to the end of the Climatic Optimum (Atlantic – 5000 BP), or even perhaps the beginning of the Subboreal (3700 BP). In the Maghreb, the Pleniglacial period was generally arid and was followed by an increase in humidity during the Late glacial . The lower Prehistoric Holocene terrace developed during more humid conditions following the Late glacial period.

The very low Historical Holocene terrace is not found in the Saharan part of the eastern Maghreb. Its sediments began to accumulate from the 2nd century AD and commonly ceased formation in about the 6th century (1470 to 1350 BP). The sedimentological characteristics of these terraced sediments indicate deposition under similar conditions to the modern fluvial environment. As a result of the nature of human settlement and land use, the rate at which aggradation and incision took place was generally greater at this time than during the development of the lower terrace. The evidence suggests that only slight climatic variation resulted in alluviation because of the slope instability created by human activity. A second very low terrace developed locally after the High Islamic era (600 BP). River terraces of similar age have been noted from several locations around the Mediterranean basin, in particular in the Lower Provence region of France.

KEYWORDS: Eastern Maghreb, river terraces, palaeoclimate, anthropisation, alluviation, aggradation rate, morphogenesis.

1 INTRODUCTION

The synthesis presented here is the result of a study of almost seventy sections; half of which were examined in the field by the author and half were considered using existing documents (Ballais 1991c). Detailed section descriptions and the main results of sedimentological and mineralogical analyses have been processed and published elsewhere (Ballais 1976, 1991c, 1992; Ballais et al. 1979, 1988, 1989, 1990, 1993; Lubell et al. 1976). The area under consideration is formed for the most part by Tunisia, where more than 75% of the findings are grouped, together with the eastern border of Algeria and most of the area around Tripoli in Libya. Over this vast region, the prevailing climate is at present a Mediterranean one, from the Algerian-Tunisian Tell (humid zone of the Mediterranean climate according to Emberger's terminology (1955)) to the Sahara (Saharan zone). The average annual rainfall thus ranges from more than 1000 mm to approximately 50 mm. Fluvial activity, involving perennial rivers in the north and wadis elsewhere, has produced two groups of alluvial terraces: a 'lower Prehistoric Holocene terrace' and a 'very low Historic Holocene terrace'.

The general aim of this study is to distinguish between the role of natural factors (mainly climate) and anthropogenic effects in the Holocene fluvial morphogenesis of the eastern Mahgreb. In effect, this distinction is only useful for the Neolithic and later periods since it is generally assumed that human activity had only a minimal impact upon the landscape of this region prior to cultivation and livestock breeding.

2 THE LOWER PREHISTORIC HOLOCENE TERRACE

2.1 *Spatial distribution*

The map showing the distribution of this terrace in Tunisia (Fig. 1) is incomplete. For reasons of consistency, the map shows only those locations identified by the author. These locations were noted within an area whose southern limits lie at approximately 32°N and northern limits at around 37°N. When studied in greater detail, this distribution does not appear to be homogeneous. In fact, it is only particularly well-defined in the far south and in the centre and north. On the other hand, on both sides of the Grands Chotts, this terrace is either not easily distinguishable from a glacis (el Hallouf-Oum ez Zessar wadi, el

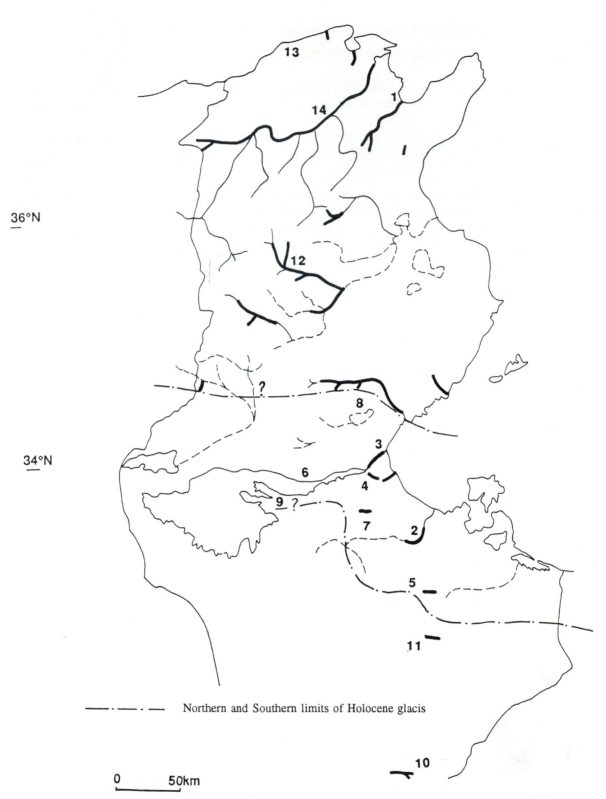

Figure 1. Distribution of the lower Prehistoric Holocene terrace in Tunisia: (1) Tunis; (2) wadi el Hallouf-Oum ez Zessar; (3) wadi el Akarit; (4) wadi Seradou; (5) Smila; (6) Bir Oum Ali; (7) Melrhir Toujane; (8) Bled Bedour; (9) wadi Limaguess; (10) wadi Jenain; (11) wadi Bir el Amir; (12) wadi es Sgniffa; (13) Kef Abed; (14) wadi Medjerda.

Akarit wadi, Seradou wadi), or is replaced by a glacis (Smila, Bir Oum Ali, Melrhir Toujane, Bled Bedour, Limaguess wadi).

2.2 General characteristics

Generally speaking, the lower Prehistoric Holocene terrace is elevated above the major contemporary stream bed by several metres (3 to 5 m rising to 10 to 15 m in exceptional cases). It is also inset below the Late Pleistocene glacis or the Late Pleistocene terrace (Coque 1962; Ballais 1984). From the south to the north, the facies associations are similar to those being formed today. Thus, in the far south, on the Jenain wadi at approximately 32°N, it reaches a thickness of 6 m and is composed largely of sand and of grey and black silt (Petit-Maire et al. 1991). The upper gypsum-rich sand is covered by a gypsum crust. At approximately 2.6 m above the base, a freshwater malacofaunae level has radiocarbon been dated to 6580 ±350 years BP (Gif 8474).

On the Bir el Amir wadi (32°30' N), the facies are clearly finer-grained, silt- and gypsum-rich, generally whitish on blackish in colour becoming beige towards the summit and also covered by a gypsum crust. The middle section contains an association of land gasteropoda which have been dated to 7026 ±175 years BP (C 3541).

In central Tunisia, at approximately 36°N along the es Sgniffa wadi, the gypsum is no longer apparent and instead calcium carbonate is concentrated in the form of pseudo-mycelium or farinaceous nodules. The facies are still fine grained (clay/silt) but the colour tends towards a reddish beige. Gravel is rare and is only found in small localised beds. The terrace surface possesses a well-developed fossil soil exceeding 1 m in thickness which is brown in colour with a fine prismatic structure rich in land molluscan fauna. This fauna has been dated by radiocarbon to 7516 ±114 years BP (C 3732). This terminal fossil soil is a common feature, especially at the summit of the lower Medjerda terrace.

In northern Tunisia, on the flysch substratum, the carbonate concretions are replaced by ferruginous compounds indurating the pebbly beds. A good example of this is found at Kef Abed at more than 37°N. These sedimentary associations are very similar to modern conditions. For example, in the lower Saharan zone (Jenain wadi) aeolian morphogenesis is predominant at present, in the upper Saharan zone gypsum-rich soils are currently forming (e.g. Bir el Amir wadi), in the arid and semi-arid zones calcareous soils are under development (e.g. es Sgniffa wadi, Medjerda wadi) and in the humid zone (Kef Abed) weathering and leaching processes are important under the forest cover.

2.3 Chronology and morphogenesis

Each time this lower terrace has been dated, the dates obtained usually indicate a lower to middle Holocene age, particularly in the south (Ballais 1991a) towards the end of the Versilian transgression (Ballais et al. 1988).

These terrace sediments therefore accumulated at the same time as the deposition of the main Holocene lacustrine sediments throughout the Sahara (Petit-Maire et al. 1991) and at the same time as the Holocene Climatic Optimum in temperate Europe. During this period, therefore, it appears that the climate of tropical and temperate zones evolved in a consistent manner – at least in the Euro-African belt. In twelve cases (Fig. 2) it was possible to calculate the rate at which these terrace (or glacis) sediments accumulated. It has been assumed that this process began around 10,000 BP (Table 1), which seems highly probable, at least in the south (Ballais 1991a). The age of the upper part of the accumulation has been estimated, either by using the presence of Neolithic sites, or it has been assumed that the incision began around 3500 BP (Table 1). The resulting average rate of 1.4 mm yr⁻¹ (a minimum value) remains relatively modest.

In the north it is possible that the situation was more complex, as suggested by evidence in the Cheria-Mezeraa wadi in the Nemencha (Ballais and Heddouche 1991). Indeed, without a sufficiently accurate means of dating these sediments it is difficult to differentiate between a terrace which probably dated from the Late Pleistocene and contains Ibero-Maurusian industries (Mathlouthi 1988), and a Holocene terrace.

However, the prehistoric Holocene terrace sediments accumulated in the south during a period of rising humidity and of a significant increase in the total annual rainfall – following the maximum aridity of the Late Pleistocene

Table 1. Morphoclimatic evolution (mainly in arid and saharan zones) during the Holocene period.

Isotopic chronology (¹⁴C BP)	Morphogenesis	Molluscs
30 ?	Shifting sands Incision	*Leucochroa candidissima*
610 ±110	Terrace	
1 470 ±190	Aggradation	
1 850 ?	Aggradation	
2 380 ±155	Flood deposits, Deflation Incision	*Helicella*
3 680 ±160	Terrace	
4 220 ±50	Aggradation	
5 195 ±105	Aggradation	
5 640 ±100	Flood deposits	*Helicella*
5 858 ±66	Pedogenesis	*Helix melanostoma*
5 930 ±87	Pedogenesis Gypcrete	
5 995 ±50		*Melania*
6 750 ±130		
		Eobania vermiculata
7 026 ±175	Terrace aggradation Shifting sands	
7 890 ±90	Terrace aggradation	*Biomphalaria pfeifferi*
8 010 ±160	Terrace aggradation	*Lymnaea natalensis*
8 230 ±70	Lake	*Cerastoderma glaucum*
8 260 ±180		*Leucochroa candidissima* Melanoïdes
10 530 ±349	Fixed dunes	

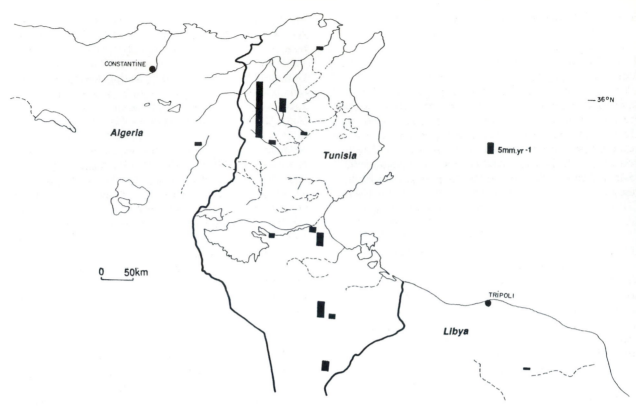

Figure 2. Aggradation rate of the lower Prehistoric Holocene terrace in the eastern Maghreb: (1) wadi Chéria-Mezeraa; (2) wadi es Sgniffa; (3) wadi Medjerda; (4) wadi Seradou; (5) wadi Jenain.

(Ballais and Ben Ouezdou 1992). More particularly, in places where there is now an annual rainfall of approximately 50 mm, aquatic Gasteropoda lived in fresh water pools, notably *Lymnaea natalensis* and *Biomphalaria pfeifferi* (Petit-Maire et al. 1991). In general, therefore, water discharges increased and became more effective once again, sometimes even before 10,000 BP (Ballais and Heddouche 1991). If, when compared to the high aridity of the Pleniglacial, this can be interpreted as being due to an increase in local rainfall, this rainfall must occasionally have been sufficiently intense to produce streams on the recently formed sand dunes in the south. The overall rise in rainfall has been confirmed by palynological analyses (Brun 1989).

However, the fine facies, which predominate in this lower terrace, indicate that in the main, stream competence remained generally low and that the underlying sediments and rocks did not make a significant contribution to the stream sediment loads. It could be assumed that during this increase in humidity, vegetation returned to the slopes, so that, at the Climatic Optimum, fluvial erosion was limited. It is at this time that the soils developed; gypsum soils in the far south, brown carbonate-rich soils in the centre and leached ferruginous soils in the north. In contrast to the present day, this pedogenic activity indicates that the climatic zones shifted significantly, some tens of kilometres, towards the south. For the Jenain region, this resulted in a two-fold increase in the total annual rainfall (approximately 100 mm as opposed to

approximately 50 mm at the present time). The dry phase identified at around 7500-7000 BP in the south (Ballais 1991a) does not seem to have permanently interrupted the aggradation of the terrace sediments. Moreover, the el Akarit wadi terrace was still being formed around 3910 BP (Zouari 1988).

Furthermore, the area in which the lower terrace is either unclearly defined, or replaced by a glacis (Fig. 1), has two distinctive features. Firstly, if relief is excluded, it corresponds in general to the upper Saharan zone (100 mm yr^{-1} < P < 150 mm yr^{-1}), or to almost pre-Saharan Tunisia (Coque 1962). The second is that it can be identified with the area in which Late Pleistocene loess has been deposited, including those of the Monts de Matmata (Coudé-Gaussen and Rognon 1988).

2.4 *The role of tectonics*

The role of tectonics in the development of this terrace is of minor significance and only occurs locally. The most obvious cases are those of the es Sgniffa wadi and Medjerda at the apex of its delta where the considerable thickness of sediment (up to 15 m on the es Sgniffa wadi) must be attributed to subsidence. Another example is the Seradou wadi where the alluvial beds on the left bank, downstream from the Roman dam (see below), have been distorted by the reactivation of a fault – although the surface of the terrace has not been disturbed.

3 THE VERY LOW HISTORIC HOLOCENE TERRACES

Following the formation of the lower prehistoric Holocene terrace, a very widespread phase of incision took place (Fig. 3) at a medium rate (average rate: 1.2 mm yr[1]). This resulted in the subsequent terrace being stepped or inset into the previous terrace, sometimes directly into the Late Pleistocene formations, or even bedrock. However, in a number of cases (e.g. the Cheria-Mezeraa wadi, the es Sgniffa wadi and the Medjerda) the historic formation was located on the prehistoric alluvial layer.

3.1 *The very low main Historic Holocene terrace.*

3.1.1 *Characteristics of the very low main Historic Holocene terrace*

The historic aggradation is widespread along most of the watercourses, from the north to the south, from the humid zone to the upper Saharan zone, but with the exception of the lower Saharan zone (Fig. 4). To date, this terrace, unlike the former terrace, has not been identified in the far south of Tunisia.

As with the former terrace, this feature developed for the most part in those catchments formed in loose, erodible rocks and the sediments are generally fine in texture, beige in colour to the south and greyer to the north. The sediments nearly always contain fragments of Roman or even earlier pottery (more than twenty recorded examples) and often charcoal, hearths or other artefacts. This terrace also has an extensive surface area, particularly in the north, and forms a sediment unit which varies in thickness from 1 m (along watercourses which are of the order of 2 to 3 m deep) to 5 to 6 m (along major rivers).

On the largest rivers detailed study of sedimentation patterns shows variation with latitude. Thus, in the arid zone, in the Leben wadi terrace (Ballais 1991b) towards 35°N, desiccation cracks appear in thin beds formed by washed clay and have been filled by sand during later flooding. These patterns, which are characteristic of a wadi, disappear in the more humid zones, with the exception of the Kébir-Miliane wadi, which is today perennial. Conversely, further to the north, the terraced sediments of many watercourses contain dark, hydromorphic silt-like facies, rich in organic matter or in manganese oxide. In eighteen different cases, it is possible to correlate the presence of these facies with the perennial nature of the watercourse, or inversely, their absence with the intermittent nature of the watercourse. In three other cases, this correlation is not apparent. In most cases (21 recorded examples), there is only one terrace whose sediments contain Roman pottery. The characteristics of this terrace

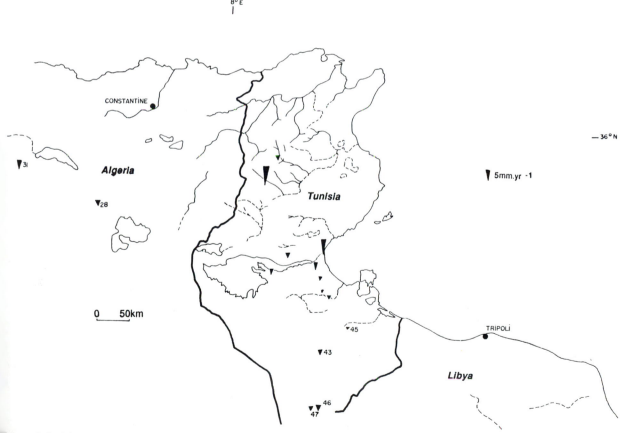

Figure 3. Incision rate of the lower Prehistoric Holocene terrace in the eastern Maghreb.

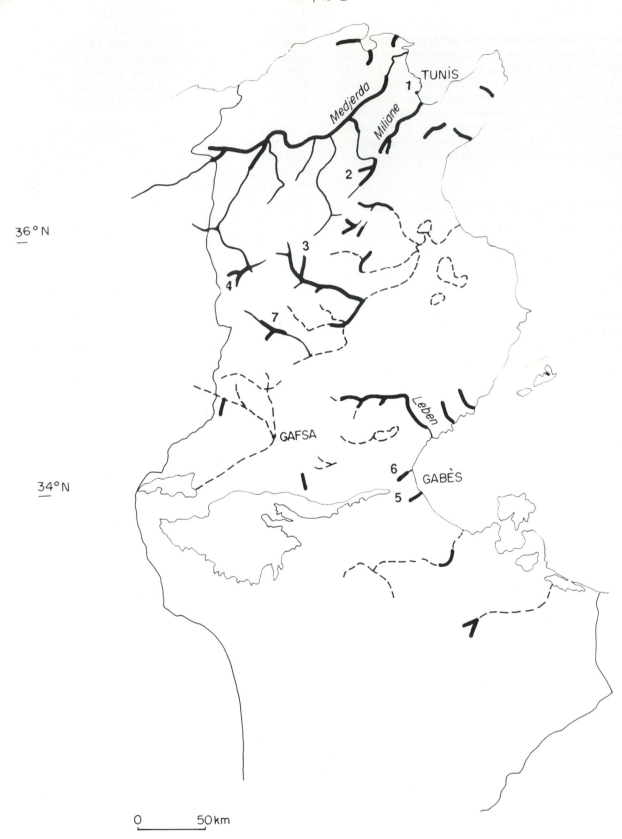

Figure 4. Distribution of the main very low historic Holocene terrace in Tunisia: (1) Carthage; (2) Henchir Rayada; (3) wadi es Sgniffa; (4) Haïdra; (5) wadi Seradou; (6) wadi el Akarit; (7) Kasserine.

are described above. However, sediment texture becomes coarser in the south and in the north when approaching the slopes. In some cases, which at present have been identified mostly in the south, a very low terrace of coarse texture could be Historic as it is set into the low and dated Prehistoric Holocene terrace. More generally, this terrace is less developed in the south of Tunisia and in Libya, than further to the north. In many cases the subsequent down cutting within the lower terrace continued, without interruption or aggradation, particularly along the minor wadis. Lastly, there are also three cases of pseudo terraces formed by an alluvial formation upstream from the artificial Roman dams.

3.2   The age of the main very low Historic terrace

3.2.1   Archaeological evidence

By identifying this very low terrace, it is possible to compare the present water level of the channel at mean flow conditions with that of the Roman constructions (bridges, aqueducts, and particularly dams). In ten cases it can be demonstrated that the watercourses have reverted to their Roman levels (sometimes even, to be more precise, to 2nd-3rd centuries AD) and that, in four cases, the present incision is 0.50 m to 2 m deeper than at Roman times. On the Seradou wadi, the pseudo-terrace developed behind a dam dates from the 2nd-3rd century BC (Ballais 1990). It is therefore highly probable that this aggradation started in the 2nd-3rd centuries AD at the earliest. Should this be the case, an accurate age for the end of this aggradational phase at Haïdra (Ammaedara) can be established, as the foundations of a Byzantine bridge, which was constructed at the same time as the 6th century fort, were dug into the alluvial deposits. A specific example is provided by the Kébir-Miliane wadi (Bourgou and Oueslati 1987) where the terrace sediments appeared to start accumulating from the 4th century BC.

3.2.2   Isotopic dating

There are still very few isotopic dates for the very low Historic terrace. It is important to mention, however, the Chéria-Mezeraa wadi, radiocarbon dated using samples of land molluscs to 1350 ±70 years BP (Farrand et al. 1982) and the el Akarit wadi dated to 1470 ±190 years BP (Page 1972). Both dates appear to be reliable. On the other hand, two other dates, also obtained using land malacofaunae, give ages which are too early: 2043 ±48 years BP (C 3733) on the el Hattab wadi terrace at Kasserine and 2050 ±58 years BP (C 3734) on the es Sgniffa wadi terrace. These dates must be too early, as the alluvial sediments which either carried the Roman pottery (Kasserine) or covered a Roman town (es Sgniffa wadi) could only have occurred much later than the creation of Africa Novus (46 BC).

3.2.3   The genesis of the very low main Historic
         Holocene terrace

The role played by neotectonics in the build-up of this terrace can be disregarded despite the distortions from the Historic period to be found at Monastir, on the eastern Tunisian coast and at Ras Angela on the northern shore (Ballais 1991b). Similarly, the variations in sea level, which was first higher than at present at around 2700 BP, but then several decimetres lower during the Carthaginian then the Roman period, at least in Carthage (Ballais et al. 1988), could not have influenced the intermittent or endoreic watercourses. Furthermore, the rate at which the main Historic terrace sediments were accumulating became considerable. Of the eighteen instances examined, the average reached 7.4 mm yr^{-1} (Fig. 5) – in other words – five times greater than the lower Prehistoric Holocene terrace, which suggests that the geosystems at this time differed significantly from the earlier ones.

3.2.4   Anthropisation and climate

It is of interest to note that aggradation only started when the area in question, with the exception of the Sahara, was occupied by sedentary populations – the result of a lengthy evolution. Generally speaking, the northeast of Tunisia, near Carthage, was cultivated from perhaps the 5th century BC onwards and the cereal plains in the northwest of Tunisia and some parts of eastern Algeria from the 3rd-2nd century BC. The steppes of Algeria, Tunisia and Libya were cultivated from the 1st-2nd century AD and the borders of the Sahara during the 2nd century, with the building of Severian limes (Trousset 1986). If the observed variation in alluvial deposits was linked to the development of this vast area, this would suggest that the imbalance of the geosystems resulting from this development would be noted at the same time in all places. In other words, the threshold which produced the change from incision to aggradation took place at the same time in the north (after 8 centuries) and in the south (after several decades). This coincidence is not impossible, but is highly unlikely, particularly if the tremendous differences in mean annual rainfall between the north and the south are taken into account. In addition, it has now been shown that a Historic terrace was deposited around the Mediterranean, and what is more, this occurred towards the end of Antiquity and the early Middle-Ages (Vita-Finzi 1969; Brückner 1986; Ballais and Crambes 1992). The extensive occurrence of this feature can best be explained by climatic factors. In view of the characteristics of the alluvial deposits, it seems likely that they originated from erosion of soils, particularly those which developed during the Holocene Climatic Optimum and at the beginning of the Roman era. It can therefore be assumed that the spreading of cultivation and ploughing destroyed a large part of the vegetation in the basins, reducing the cohesion of soils and superficial formations. It then required only a small change in rainfall characteristics, perhaps in the annual total, or at least in intensity, to produce considerable soil erosion and the start, if not the return, of water in the stream channels and increased discharges. This increase was nevertheless not enough to carry the large sediment load from the slopes to the base levels. The Kébir-Miliane wadi, a specific case in point, may therefore be explained by the morphogenetic effectiveness of the more abundant and intense rainfall found

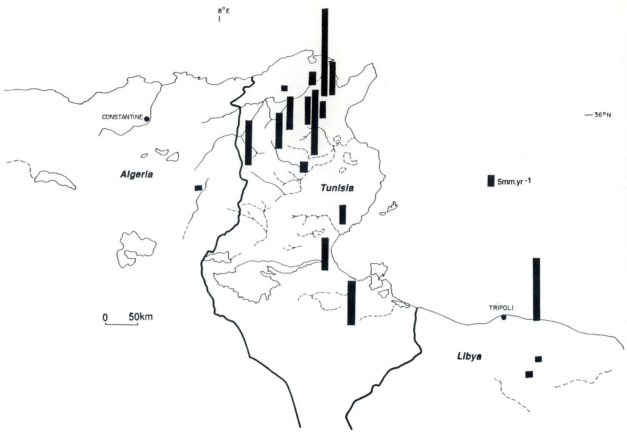

Figure 5. Aggradation rate of the main very low historic Holocene terrace in Tunisia.

in southern Tunisia around 2400 BP (Ballais 1991a, 1992).

For these very recent periods it is difficult to compare the climatic situation with that of the Sahara and temperate Europe. However, the fact that Acacia and Tamarix could be found in central Serir Tibesti at around 1700 BP and 1400 BP, indicates that, at least in tropical Sahara, the mean annual rainfall was more than the present day rainfall of 5 mm (Pachur 1974).

3.3 The very low Post-Islamic Holocene terrace

3.3.1 General characteristics
Following the aggradation of the main very low Historic terrace fill, the general trend for watercourses in the study area was vertical incision, with two or three interruptions. The main interruption can be seen in the very low Post-Islamic Holocene terrace, which is, as far as we know today, little represented (Fig. 6). One of the main reasons for this is its small size. With only one exception (Kébir-Miliane wadi), it covers very small areas, in particular in convex meander lobes, and its height above the high-water level rarely exceeds 2 m. This is occasionally a rocky terrace which was breached in the previous build-up. Elsewhere, the facies can sometimes be compared to that of the previous terrace, and are sometimes considerably coarser, at least at the base. The age of this terrace is still rather uncertain, as appropriate means of dating are

not available. However, at Henchir Rayada, it contains Islamic pottery from the 10th-11th century and on the el Akarit wadi, it was dated by radiocarbon using collagen to 610 ± 110 years BP (79/29).

3.3.2 Genesis
As for the previous terrace, the widespread presence of a terrace of the same age can be seen throughout the Mediterranean basin (Vita-Finzi 1969; Ballais and Crambes 1992). Moreover, in contrast to the final years of Antiquity (3rd to 5th centuries AD), population was probably low at the beginning of the Hafside period (12th century AD). Finally, from a climatic standpoint, this was the end of the Medieval Optimum and the beginning of the Little Ice Age (Le Roy-Ladurie 1967), which has also been indicated in Morocco (Lamb et al. 1989). Once again it seems most likely that this resulted from climatic causes, but further studies will be required to clarify this point.

3.3.3 Benches
Following the build-up of the very low Post-Islamic Holocene terrace, down cutting restarted at a greater rate of about 3.2 mm yr^{-1} (average over 38 examples: Fig. 7) with increases of approximately 10 mm yr^{-1} starting from the building of the railways less then one century ago e.g. the Bou Jebib wadi. Two benches were then formed locally.

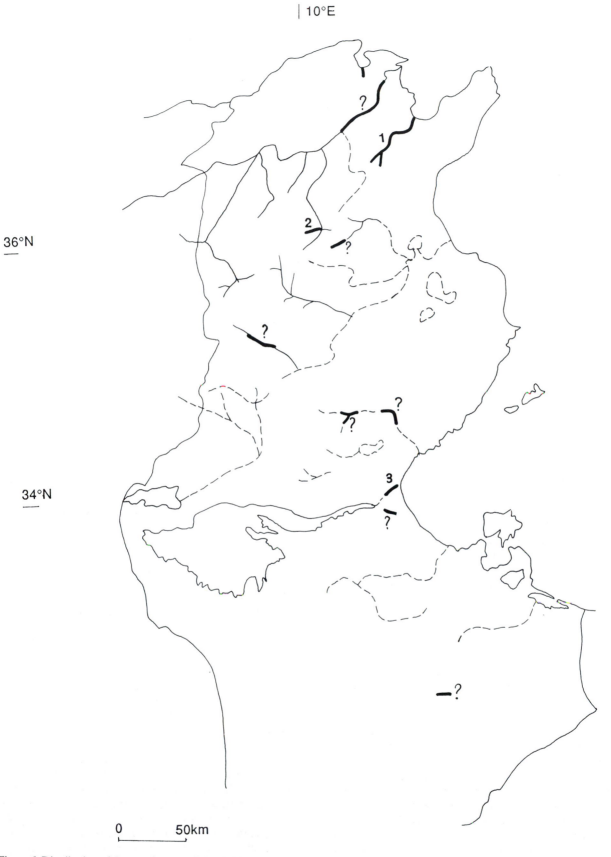

Figure 6. Distribution of the very low post-Islamic Holocene terrace in Tunisia: (1) wadi Kébir-Miliane; (2) Henchir Rayada; (3) wadi el Akarit.

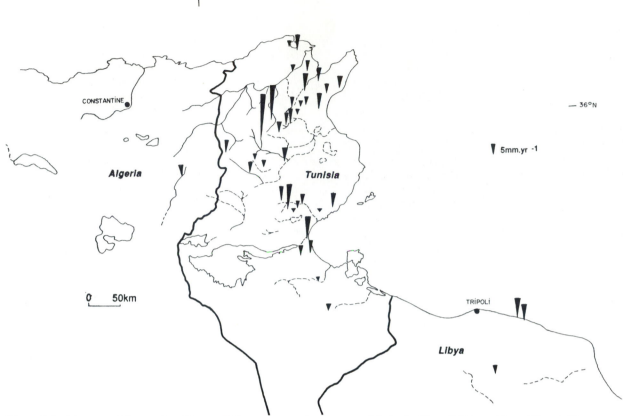

Figure 7. Incision rate of the very low Historic Holocene terrace in the eastern Maghreb: (1) wadi Krima; (2) wadi Bou jebib.

The upper bench

The upper bench is often narrow and elevated above the mean flow channel by 2 to 4 m. This feature can appear as a rocky terrace formed at the expense of the material found in the previous terrace, or as a coarse- or fine-grained aggradation several decimetres in thickness. In the absence of the necessary dating resources or detailed sedimentary data, the independence of this bench from the very low Post-Islamic Holocene terrace has not been demonstrated.

The lower bench

The lower bench may form a narrow terrace, particularly in convex meander lobes, nearly 1 to 1.5 m above the mean flow water channel. Often coarse-grained, with crude bedding this feature is occassionally inundated during exceptional flood events. The lower bench may be of very recent (20th century) origin as in the case of the Krima wadi which contains bolts and military shell fragments.

The contrast between the fine-grained material of the very low Historic terrace and the very coarse-grained bench, in the case of both the Krima wadi (subhumid zone) and the Seradou wadi (Saharan zone) is problematic. It suggests that during very recent times, there was both an overall increase in the competence of fluvial erosion throughout Tunisia and that the streams no longer simply transported eroded soil materials, but had effected considerable bedrock erosion. The main reason for this must be the significant increase in the area and depth the tilled surface, exacerbated by intensive farming techniques (tractors, multi-disc ploughs) and the general increase in over-grazing and the uprooting of ligneous plants.

The occurrence of both the terraces and the two benches remains exceptional. It is even rare to see the complete sequence of the very low Historic Holocene terrace and the two benches – only six such examples have been recorded from the north to the south of the study region.

3.3.4 Present day incision

Throughout the area the present day watercourses continue to incise, despite a tendency for the sea level to rise. The effectiveness of this process can only be seen, as is the case for all the Historic period, in unconsolidated materials. Here incision can reach a depth of 1 m in 100 years such as in the Bou jebib wadi.

The high-water channel can be identified throughout the region by the presence of grains which are significantly coarser than those found in the Holocene aggradations, and even the Late Pleistocene. Soil erosion was so extensive that the bedrock is now exposed in many areas thus providing material for erosion. As a result of this, rainfall interception by vegetation and its subsequent percolation into the soil profile is reduced, thus encouraging surface runoff.

A small change in climate cannot be ruled out, even if the representative nature of the increase in total annual rainfall in Tunisia (Bousnina 1986), which coincided with the end of the Little Ice Age is still being questioned.

4 CONCLUSIONS

An exceptional number of Holocene terraces can be found in the eastern Maghreb, and in particular in Tunisia. All the types of Holocene terraces previously identified around the Mediterranean are represented. The lower Prehistoric Holocene terrace resembles, for example, the 'remblaiement holocène principal' (main Holocene fill) in the southern Alps (Jorda 1985) or the Sainte-Victoire Mountain in Lower Provence (Ballais and Crambes 1992), both in France, or Bradano in Basilicate in Italy (Brückner 1986). This terrace is to be found from the humid zone to the lower Saharan zone. It formed during a climatic phase which was more humid than both the Pleniglacial period and the present time. Following this aggradation more arid conditions returned.

The two very low Historic Holocene terraces are similar to the several aggradations recorded from around the Mediterranean by Claudio Vita-Finzi (1969). They are particularly noteworthy because they are quite large and often complex landforms and are not usually found elsewhere. They are not found in the Saharan part of the eastern Maghreb, which was not cultivated during Antiquity, and their sedimentological characteristics indicate stratification which is comparable to that forming under present conditions. These findings confirm that natural bioclimatic variations remain the driving force of morphogenesis, and this was the case during the Historic period. However, these variations are only effective if the geosystems have been considerably anthropised and made significantly more fragile by human activity, which generally shows little concern for environmental stability. They form only short term variations (several tens of years to several centuries at most) in comparison to the large-scale variation (several millennia) which characterised the Holocene Climatic Optimum.

REFERENCES

Ballais, J.-L. 1976. Morphogenèse holocène dans la région de Chéria (Nementchas-Algérie). In *Actes Symp. versants en pays méditer*:127-131. Aix-en-Provence: C.E.G.R.M.

Ballais, J.-L. 1984. *Recherches géomorphologiques dans les Aurès (Algérie)*. Lille: A.N.R.T.

Ballais, J.-L. 1990. Terrasses de culture et jessours du Maghreb oriental. *Méditerranée* 3.4: 51-53.

Ballais, J.-L. 1991a. Evolution holocène de la Tunisie saharienne et présaharienne. *Méditerranée* 4: 31-38.

Ballais, J.-L. 1991b. Les terrasses historiques de Tunisie. *Z Geomorph. Suppl. Bd* 83: 221-226.

Ballais, J.-L. 1991c. Vitesses d'accumulation et d'entaille des terrasses alluviales holocènes et historiques au Maghreb oriental. *Physio-Géo* 22-23: 89-94.

Ballais, J.-L. 1992. Le climat au Maghreb oriental: Apports de la géomorphologie et de la géochimie. In: *Le climat à la fin de l'Age du Fer et dans l'Antiquité (500 BC-500 AD), Les Nouvelles de l'Archéologie* 50: 27-31.

Ballais, J.-L. 1993. Morphogénèse fluviatile holocène en Tunisie. *Travaux de l'URA 903 CNRS*. XXII: 63-78.

Ballais, J.-L., A. Marre & P. Rognon 1979. Périodes arides du Quaternaire récent et déplacement des sables éoliens dans les Zibans (Algérie). *Rev. Géol. dyn. et Géogr. phys.* 21, 2: 97-108.

Ballais, J.-L., M. Bourgou, R. Karray, J.-P. Lautridou, M. Levant, A. Oueslati, H. Ben Ouezdou, T. Bouhafa & A. Gragueb 1988. Premiers résultats du programme d'étude de l'Holocène de Tunisie. *Méditerranée* 2: 64-67.

Ballais, J.-L., J.-L. Dumont, M.-N. Le Coustumer & M. Levant 1989. Sédimentation éolienne, pédogénèse et ruissellement au Pléistocène supérieur-Holocène dans les Ziban (Algérie). *Rev. géom. dynam.* 2: 49-58.

Ballais, J.-L. & A. Heddouche 1991. Bas Sahara septentrional et Grand Erg Oriental. In: *Paléomilieux et peuplements préhistoriques sahariens au Pléistocène supérieur*:1-21. Solignac.

Ballais, J.-L. & H. Ben Ouezdou 1992. Sables éoliens quaternaires entre les chaînes de Gafsa et du Cherb (Sud Tunisien). *Z Geomorph. Suppl. Bd.* 84:89-99.

Ballais, J.-L. & A. Crambes 1992. Morphogenèse holocène, géosystèmes et anthropisation sur la Montagne Sainte-Victoire. *Méditerranée* 1.2:29-41.

Bourgou, M. & A. Oueslati 1987. Les dépôts historiques de la vallée du Kébir-Miliane (Nord-Est de la Tunisie). *Méditerranée* 1:43-49.

Bousnina, A. 1986. La variabilité des pluies en Tunisie. *Tunis: Fac. Sc. Hum. et Soc.*

Brückner, H. 1986. Man's Impact on the Evolution of the Physical Environment in the Mediterranean Region in Historical Times. *GeoJournal* 13.1:7-17.

Brun, A. 1989. Microflores et paléovégétations en Afrique du Nord depuis 30,000 ans. *Bull. Soc. géol. Fr.* (8).V.1:25-33.

Coudé-Gaussen, G. & P. Rognon 1988. Caractérisation sédimentologique et conditions paléoclimatiques de la mise en place de loess au nord du Sahara à partir de l'exemple du Sud-tunisien. *Bull. Soc. géol. Fr.* (8).IV.6:1081-1090.

Coque, R. 1962. *La Tunisie présaharienne. Etude géomorphologique*. Paris: A. Colin.

Emberger, L. (1955). Une classification biogéographique des climats. *Natur. Monsp., série Bot.* 7:3-42.

Farrand, W.R., C.H. Stearns & H.E. Jackson 1982. Environmental setting of Capsian and related occupations in the high plains of eastern Algeria. *Geol. Soc. Am.* 14.7:487.

Jorda, M. 1985. La torrentialité holocène des Alpes françaises du sud. Facteurs anthropiques et paramètres naturels de son évolution. *Cahiers lig. préh. et protoh.* 2:49-70.

Lamb, H.F., U. Eicher & V.R. Switsur 1989. An 18,000-year record of vegetation, lake-level and climatic change from Tigalmamine, Middle Atlas, Morocco. *Journal of Biogeography* 16:65-74.

Le Roy Ladurie, E. 1967. *Histoire du climat depuis l'An Mil*. Paris: Flammarion.

Lubell D., F.A. Hassan, A. Gautier & J.-L. Ballais 1976. The Capsian Escargotières. *Science* 191:910-920.

Mathlouthi, S. 1988. Les héritages continentaux et marins d'âge quaternaire dans les environs du système lacustre de Bizerte (extrême nord-est tunisien). *Méditerranée* 2:42-51.

Pachur, H.J. 1974. Geomorphologische Untersuchungen im Raum der Serir Tibesti (Zentralsahara). *Berliner Geogr. Abhandl.* 17:6-58.

Page, W.D. 1972. The geological setting of the archaeological site at oued el Akarit and the palaeoclimatic significance of gypsum soils (S.-Tunisian). Unpublished Thesis. University of Colorado.

Petit-Maire, N., P.F. Burollet, J.-L. Ballais, M. Fontugne, J.-C. Rosso & A. Lazaar 1991. Paléoclimats holocènes du Sahara septentrional. Dépôts lacustres et terrasses alluviales en bordure du Grand Erg Oriental à l'extrême-sud de la Tunisie. *C.R. Acad. Sci. Paris* 312. II:1661-1666.

Trousset, P. 1986. Limes et 'frontière climatique'. *C.T.H.S.*:55-84.

Vita-Finzi, C. 1969. *The Mediterranean Valleys. Geological Changes in Historical Times.* Cambridge: Cambridge University Press.

Zouari, K. 1988. Géochimie et sédimentologie des dépôts continentaux d'origine aquatique du Quaternaire supérieur du Sud tunisien: Interprétations paléohydrologiques et paléoclimatiques. Thèse ès-Sciences. Univ. Paris-Sud.

CHAPTER 18

The Holocene alluvial records of the chorai of Metapontum, Basilicata, and Croton, Calabria, Italy

JAMES T. ABBOTT
Department of Geography, University of Texas at Austin, Austin, Texas, USA

SALVATORE VALASTRO, JR.
Radiocarbon Laboratory – Balcones Research Center, University of Texas at Austin, Austin, Texas, USA

ABSTRACT: Ongoing geomorphic investigations being conducted in conjunction with archaeological reconnaissance in the territories surrounding the ancient Greek cities of Croton and Metapontum in southern Italy reveal a complex set of relationships between regional tectonics, human activity, and episodes of valley aggradation and incision. In the Metaponto area, incision by three relatively large rivers during the past several centuries has exposed thick suites of late Holocene deposits and intercalated palaeosols that record episodes of regional landscape stability and instability spanning the past 6000 years. Although regional subsidence has limited fluvial incision on the Crotone peninsula, entrenched streams on the margin of the peninsula record a sequence of alluvial cut-and-fill episodes dating back some 12,000 years. This paper details the alluvial sequences preserved in the two areas, and relates the geomorphic record to regional cultural and climatic sequences in an attempt to explain the controls on fluvial behaviour operating throughout the late Holocene.

KEYWORDS: Southern Italy, landscape evolution, Holocene alluvial stratigraphy.

1 INTRODUCTION

This paper describes preliminary results of an ongoing investigation of the preserved Holocene alluvial record in the chorai, or rural territories, of the ancient Greek cities of Metapontum and Croton in southern Italy (Fig. 1) being conducted in conjunction with archaeological investigations by the Institute for Classical Archaeology, University of Texas at Austin. Geomorphic research conducted to this point reveals that thick, complex sequences of Holocene alluvial deposits are present in the valleys of each chora. Although much work remains to be done before the loci of and impetus for the erosional episodes responsible for this suite of deposits can be satisfactorily detailed, the stratigraphic architecture, sedimentology, and temporal context of the valley fills are fairly clear at this time. This paper describes the character and timing of late Quaternary alluvial deposition in each area, and then briefly addresses the significance of the deposits in terms of the palaeoenvironmental and cultural record.

2 THE CHORA OF METAPONTUM

The Metapontum chora consists of the lower courses of the Bradano, Basento, and Cavone rivers and the intervening marine terraces extending some 20 km inland from the Ionian coast. The study area can be divided into four primary landform assemblages: (1) the successive, coast-parallel marine terrace treads, which are mantled with thin littoral accumulations of relatively coarse clastics and characterized by thick, well-developed soils;

(2) the modern coastal zone, which exhibits evidence of more than 2 km progradation into the Mediterranean in the last 2000 years (Boenzi et al. 1987); (3) the steep, erosive slopes of the lower valleys and the Lucanian hill country, which generally exhibit thin, poorly developed soils and are prone to extensive mass wasting and badland development; and (4) thick Holocene alluvial deposits in the valley bottoms (Fig. 2) and older, poorly preserved Pleistocene terrace remnants inset into the valley walls.

The study area is underlain by a wedge of clastic muds up to 500 m thick deposited during the Pliocene and Early Pleistocene in a large graben situated between the intensely folded Apennine Mountains to the west and a flat-lying carbonate platform to the east (Dainelli 1975; Servizio Geologico D'Italia 1978) (Fig. 3). Over the last million years, the region has been gradually uplifted several hundred metres, resulting in the sculpting of a series of step-wise marine terraces into the underlying muds and the deposition of a relatively thin to very thick cap of littoral sands and gravels (Boenzi et al. 1976; Servizio Geologico D'Italia 1978). Brückner (1982) identifies eleven marine terraces, ranging in age between 70,000 and 636,000 years BP, that rise to an elevation of approximately 450 m some 25 km inland from the coast. As regional uplift progressed through the Pleistocene, the ancestral Bradano, Basento, and Cavone gradually incised deep valleys perpendicular to the coast, and lateral erosion attacked the marine terraces, resulting in increasing dissection and exposure of the highly erodible clays with distance from the modern shoreline.

In their lower 25 km, each of the valleys is infilled with

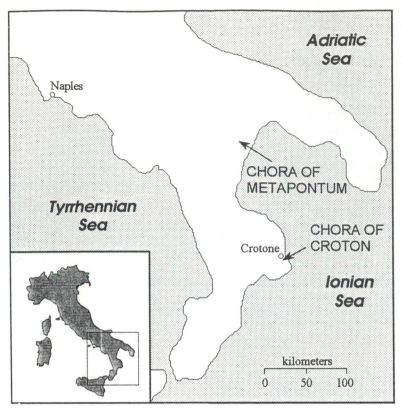

Figure 1. Location of the study areas.

Figure 2. View looking across the recent (T1) alluvial surface (foreground) at a cutbank formed in late Holocene alluvium (T2) at the Ferroleto section, Cavone Valley, illustrating the character of exposures in the Metaponto chora. The dissected T3 terrace and the margin of the marine terrace are visible in the background.

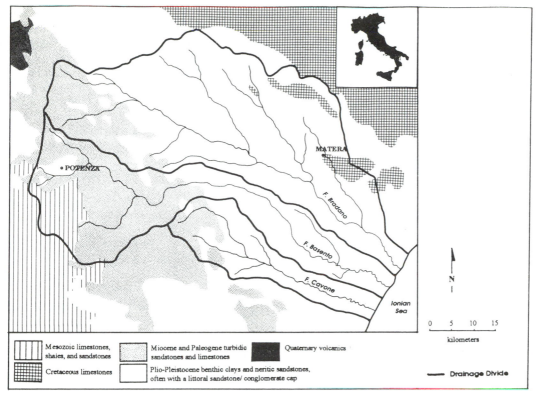

Figure 3. Simplified geologic map of the Bradano, Basento, and Cavone watersheds. The chora of Metaponto proper is limited to the lower drainages extending back approximately 20 km from the coast.

thick, fine-grained accumulations of alluvium containing intercalated palaeosols and stratified archaeological materials exposed by a low-gradient, incised meandering stream. Although some differences are apparent in the character, thickness, and geometry of the respective valley fills, broad similarities exist in the stratigraphy and alluvial architecture of the lower reaches of the three streams, suggesting that similar depositional controls affected each valley. In the following discussion, a uniform system of informal stratigraphic nomenclature has been applied to equivalent units and palaeosols in all three valleys to facilitate comparison; as a result, not all units are necessarily represented in each valley.

In addition to a relatively narrow modern floodplain (Unit I; surface T0), two alluvial surfaces of Holocene age are present in the lower reach of each valley. The lower of these (T1) consists of a series of alluvial strath and strath/fill benches mantled with relatively thin (1-3 m thick) sandy and silty alluvium (Unit H) at elevations of 5-12 m above the modern channel. The upper Holocene surface (T2) consists of a sequence of stacked, predominantly silty deposits with a total thickness in excess of 18 m. A minimum of four alluvial units (Units D-G), separated by weak to moderately developed palaeosols, underlie the T2 surface in the Bradano and Basento valleys. At present, the soil demarcating the top of the Unit E fill has not been identified in the Cavone Valley; although it is possible that a time-equivalent unit was not deposited, it is considered more likely that the soil marking the top of the fill was truncated prior to the

aggradation of Unit F. In the other two valleys, the Unit E palaeosol is also occasionally absent in individual sections, but occurs with enough regularity to indicate that it was a valley-wide feature and not a localized muddy deposit that accumulated in topographic swales.

Overlooking the broad T2 surface, several sets of relatively thin, strongly dissected gravelly terrace remnants (T3-T5) are inset into the valley walls at elevations up to 50 m above the modern channel. These remnants (Units A-C) represent dissection of Pleistocene-age fills, and are composed of planar and trough cross-bedded gravels and sands, indicating deposition under relatively high-energy conditions by a braided channel. The stratigraphic relationships in the downstream reaches of the three valleys are illustrated in Figure 4, and the planform relationships between different geomorphic surfaces in the coastal reach of the Basento River are illustrated in Figure 5. The other two rivers differ principally in the degree of development of the strath surfaces developed during the last phase of incision; in general, the extent of T1 surface is even greater in the Cavone Valley and much smaller in the Bradano.

Architecturally, the Holocene-age deposits represent a vertically stacked sequence of thick sediment packages that accumulated primarily by vertical accretion during valley-wide flooding. Channels associated with each unit are relatively deep and only a few hundred metres wide, indicating that little lateral migration occurred while the streams were incised and suggesting that downcutting and subsequent aggradation of the channels occurred relatively rapidly. The sedimentary character of the T2 deposits varies both spatially (between temporally equivalent strata in different parts of the valley) and temporally (between different strata at individual localities), but is broadly similar in each valley. Channel and channel margin deposits commonly consist of fine sands with small-scale trough crossbedding and ripple lamination. Interbedded silts and organic muds are common in the channels, while large-scale scour surfaces are relatively rare. Flood basin deposits are predominantly silty and vary from massive to laminated. Although there are numerous localized exceptions, Units E and F typically fine upward from interbedded laminae of fine sands and silts to massive silt loams and silty clays. In the lower valleys,

only a few metres of Unit D are typically exposed. The top of the unit commonly consists of thick accumulations of laminated or massive organic silty clays, often with abundant interbedded organic matter and very strong redox mottling, that probably represent lacustrine or palustrine deposition. The uppermost T2 fill (Unit G) is frequently strongly laminated to thinly bedded throughout the section. Some of the more distinctly laminated beds in this unit are thickest at the valley margin and thin towards the valley axis, indicting that they represent sheetwash and colluvial deposition from the slopes. Others thin away from the valley axis, indicating delivery by the trunk stream. Gravels are rare in the coastal reaches except where associated with degradation of an

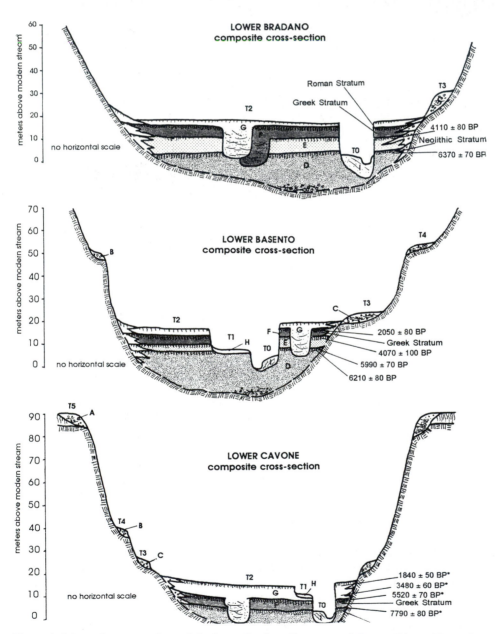

Figure 4. Schematic cross-sections of the lower Bradano, Basento, and Cavone Rivers, illustrating stratigraphic relationships and chronologic data.

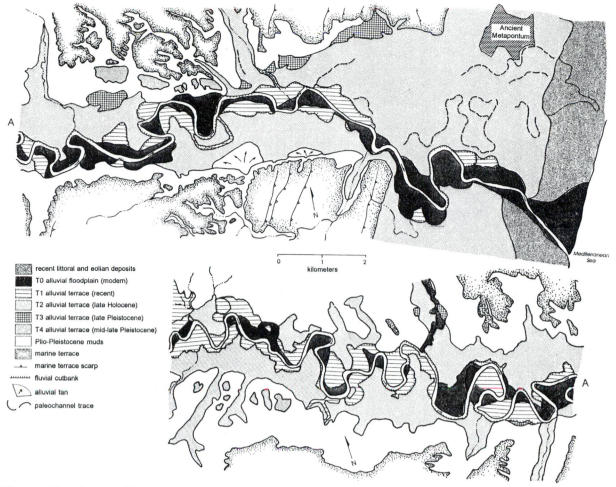

Figure 5. Map of geomorphic surfaces, lower Basento Valley.

adjacent gravelly high terrace; no gravelly trunk-channel deposits have been detected associated with units D through I within 25 km of the coast.

Soils interbedded in the T2 fill are almost exclusively cumulic and show little or no horizon differentiation; however, weak structural development and slight rubification is often apparent, as are krotovina, root casts, and redox mottling. In a few locations, weak to moderate carbonate filaments have developed in the Unit F palaeosol, and fine carbonate masses are common in some exposures of the Unit D palaeosol. In most locations, the lower 5 to 10 m of the T2 fill units are mottled and iron-stained, indicating periodic saturation by an elevated water table.

Farther from the coast, the similarity between the valleys largely disappears due to differences in lithology, relief, rate of uplift, and valley morphology. The Bradano system, which is largely underlain by erodible Plio-Pleistocene clays, maintains a suite of fine-grained late Quaternary deposits well into the interior. In contrast, the Basento and Cavone both shift abruptly to a gravel bed braided depositional pattern some 25-30 km from the coast and have largely or completely cannibalized the late Holocene valley fill. Artefacts and radiocarbon ages from

a preserved terrace remnant located approximately 50 km up the Basento suggest that as much as 35 m of late Quaternary deposits have been flushed completely from the kilometre-wide valley in the past few thousand years.

Several investigators have previously examined aspects of the alluvial sequence in the Metapontum chora (Brückner 1982, 1986, 1990; Neboit 1977, 1983; Vita-Finzi 1969, 1975). While Vita-Finzi argued that the Basilicata valley fills represent Pleistocene aggradation (the 'older fill', which he terms the San Leo alluvium) and a late Roman to high Medieval episode of aggradation (the 'younger fill' or Valchetta alluvium), both Neboit and Brückner recognized evidence for extremely rapid sedimentation – resulting in as much as 10 m of aggradation over the span of a few centuries – during the Greek occupation (roughly 700 BC-200 BC) and again during the Middle Ages. In addition to confirming these conclusions, the present study indicates that another, similar episode of rapid valley alluviation occurred several millenia earlier during the Late Neolithic occupation.

Chronologic control in the current study is provided by a series of 12 preliminary radiocarbon ages obtained from the sequence (Table 1) and the wealth of archaeological materials found in the sediments. All of the radiocarbon

Table 1. Radiocarbon ages on sediment and soils, Metaponto chora.

Sample number	Provenience	Material	$\Delta^{13}C$	Age (corrected years BP)
TX-7085	Lower Basento Valley Duck's Head Section Locality A, 9.1 m bgs top of Unit D	Sediment	−26.8%	6210 ±80
TX-7086	Lower Basento Valley Duck's Head Section Locality A, 6.1 m bgs Unit E paleosol	Cumulic Soil	−26.7%	4070 ±100
TX-7087	Basento/ La Canala Conflence Locality B, 4.0-4.2 m bgs Unit F paleosol	Cumulic Soil	−25.9%	2050 ±80
TX-7088	Basento/ La Canala Confluence Locality D, 8.0-8.2 m bgs top of Unit D	Sediment	−27.1%	5990 ±70
TX-7089	Lower Bradano Valley Cermignana Villa Locality, 10.5 m bgs top of Unit D	Sediment	−25.9%	6370 ±70
TX-7090	Lower Bradano Valley Cermignana Villa Locality, 5.5 m bgs Unit E paleosol	Cumulic Soil	−26.8%	4110 ±80
TX-7396	Lower Cavone Valley Ferroleto Locality, 6.5 m bgs base of Unit G	Sediment	−25.9%	3480 ±60**
TX-7397	Lower Cavone Valley Ferroleto Locality, 4.0 m bgs Unit G	Sediment	−26.7%	1840 ±70**
TX-7398	Lower Cavone Valley Ferroleto Locality, 7.0 m bgs Unit F paleosol	Cumulic Soil	−26.4%	5220 ±70**
TX-7399	Lower Cavone Valley Ferroleto Locality, 11.0 m bgs Unit D? paleosol	Sediment	−27.3%	7790 ±60 BP**
TX-7634	Middle Basento Valley Gróttole Locality, 0.3 m bgs Unit G	Modern Soil	−26.3%	280 ±50*
TX-7635	Middle Basento Valley Gróttole Locality, 2.2 m bgs Unit F? paleosol	Cumulic Soil	−26.8%	3130 ±60**

*Probable underestimate due to illuviation of organics in the soil zone.
**Probable overestimate due to incorporation of old allogenic organic matter.

ages obtained to this point are based on bulk samples of soil or sediment, which can sometimes provide biased ages if older allogenic carbon or illuvial soil organic matter is incorporated. However, the presence of artefacts stratified in the fills typically makes it possible to recognize these biases when they occur, and systematic deviations in radiocarbon ages on humates can be informative concerning characteristics of erosional processes in a basin. A good example is provided by the radiocarbon ages from the Cavone Valley, which are demonstrably several thousand years too old based on the presence of stratified Greek and Neolithic cultural material contained in the fill. In contrast, the radiocarbon ages from the Bradano and Basento dovetail perfectly with the ages indicated by archaeological inclusions (see Fig. 4). This suggests that areal erosion resulting in the introduction of large quantities of soil organic matter was more prevalent in the Cavone watershed than in the other two systems throughout the Holocene. Although the reason for this apparent widespread, long-term stripping of the upper solum in the Cavone drainage is unknown at present, it possibly may be related to more active tectonic uplift adjacent to the Apennine arc, differences in lithology, relief, vegetative cover, or a combination of two or more of these factors.

Collectively, the chronologic data indicate that periods of stability and soil formation occurred in the Metaponto chora during the Early to Middle Neolithic, middle Bronze age, and Roman occupation, while episodes of instability and floodplain aggradation are apparent in the Late Neolithic, during the Greek occupation, and in the Medieval period. Given the proximity of the coast, it is likely that deposits of the Late Pleistocene and early Holocene lie buried below the level of the modern stream in deep valleys cut in response to the Late Pleistocene low stand in sea level. Each episode of sedimentation is represented by a fining-upward cycle between 2 and 10 m thick, typically characterized by laminated sands and silts at the base, grading up into massive silty clays and silty loams, and finally into a relatively thin, clay-rich cumulic palaeosol.

3 THE CHORA OF CROTON

The general geological and geomorphic character of the Croton chora is similar to the Metapontum chora (Roda 1964; Selli 1977), with three major exceptions. First, while the scarps of marine terraces at Metapontum are arranged roughly parallel to each other and the modern coastline, the terrace sequence at Croton is much more complex, with cross-cutting scarps indicating encroachment of the sea from the east and the south at various times. Second, two sets of alluvial valleys are present, an older group oriented roughly east-west, and a more recent, north-south oriented group that has intersected and beheaded several of the east-west trending drainages through rapid headward cutting. Finally, the modern streams at Croton are poorly incised, in marked contrast to the deep channels of the Metaponto drainages.

Collectively, these three factors are indicative of the pervasive influence of tectonism on geomorphic processes in the Croton chora. Evidence of strong tectonic modification of the Croton landscape through the late Quaternary is provided by amino acid and $^{230}Th/^{234}U$ ages on a variety of marine terraces in the chora (Belluomini et al. 1988), which indicate equivalent ages for surfaces currently differing in elevation by up to 100 m. The erosional coastline, and a number of submerged Greek and Roman stone quarries and jetties on the margins of the peninsula (Cantafora n.d.), suggest that the peninsula has subsided as much as 8-10 m in the past few thousand years. Although this subsidence is probably the dominant reason for the fact that the modern streams are not incised, field observations indicate that agricultural practices are also resulting in the introduction of more fine-grained sediment into the fluvial systems than they are competent to remove. In fact, agricultural erosion is so severe in many locations that the boundaries between modern agricultural land and adjacent pastures on the valley slopes are marked by scarps up to 2 m in height.

Because of the poorly incised nature of the streams on the peninsula, investigation focused on streams on the margins of the peninsula, including the stream occupying the valle di Cacchiavia and the Neto, Vitruvo, and Tacina

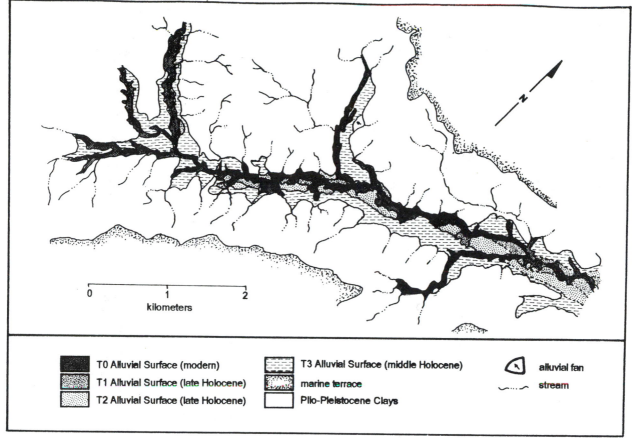

Figure 6. Map of geomorphic surfaces, middle valle di Cacchiavia.

Legend:
- T0 Alluvial Surface (modern)
- T1 Alluvial Surface (late Holocene)
- T2 Alluvial Surface (late Holocene)
- T3 Alluvial Surface (middle Holocene)
- marine terrace
- Plio-Pleistocene Clays
- alluvial fan
- stream

Rivers. Of these, only the Cacchiavia Valley is in the chora proper, and is therefore the primary focus of investigation. Unlike the larger Metapontine streams, which penetrate well into the interior and drain several diverse lithologic units, the valle di Cacchiavia is limited to the Plio-Pleistocene clays and silts of the peninsula. In addition to the larger streams, several small ephemeral upland valleys preserve small-scale alluvial sequences. Although no radiometric ages have been obtained to date from these small tributaries, cultural inclusions suggest that the deposits are limited to approximately the last 2000 years. In contrast, radiocarbon ages demonstrate that the deposits preserved in the Cacchiavia Valley span the past 12,000 years, while faulted, gravelly high terraces lying up to 60 m above the modern stream in the lower Neto Valley, although undated, are clearly of considerably greater antiquity.

Unlike the stacked stratigraphy exhibited by the Metapontine streams, the middle Cacchiavia valley exhibits a cut-and-fill sequence of inset alluvial units with interfingering aprons of slopewash alluvium and colluvium derived from the valley walls. Four distinct, discontinuous alluvial surfaces, designated T0 through T3, are present in the middle reach of the valley (Fig. 6). As the stream flows towards the coast, these surfaces gradually merge into the type of broad alluvial flat characteristic of streams on the peninsula. Figure 7

illustrates stratigraphic relationships in the middle Cacchiavia Valley. Although a few cultural inclusions were discovered in the fills, they were extremely scarce in comparison to the Metaponto chora, and no *in situ* features were discovered. Therefore, chronologic control is provided primarily by a series of nine radiocarbon ages obtained from bulk sediments and soils (Table 2; see also Fig. 7).

The radiocarbon suite suggests that Holocene development of the Cacchiavia Valley was characterized by long periods of slow, fine-grained sedimentation separated by brief incisional episodes. The oldest fill detected in the valley (Unit A) consists of massive to laminated silts and silty clays that were aggrading by the end of the Pleistocene. This unit contains a prominent, fluvially-reworked volcanic ash deposited approximately 12,000 years ago. It is not clear how long this unit continued to accumulate because sometime in the early to middle Holocene an episode of valley-wide erosive stripping truncated the upper deposits and completely removed any soil that may have developed. However, by roughly 5500 BP, a second unit (Unit B) composed primarily of massive silty clays had began rapidly to aggrade vertically on top of the older fill.

Sometime around 5000 BP, the stream incised a narrow casement valley into the older deposits, isolating the T3 terrace, and began to aggrade again. These deposits (Unit C) fine upward from massive silty loam to gleyed, mottled clays and contain two thin, unhorizonated cumulic palaeosols at depths of 1.3 m and 0.4 m bgs. Radiocarbon ages on these two soils suggest that the majority of the fill had aggraded by approximately 4700 BP, while slow, intermittent sedimentation continued for at least another millenium.

By approximately 2800 BP, the stream had again incised and then begun to fill a narrow valley in the older deposits, abandoning the T2 terrace surface. These deposits (Unit D) consist of fine sands and sandy loams that are thinly cross-bedded at the base and laminated farther up in the section. In contrast to the older fills, primary sedimentary structures are prominent in the unit. This surface was subsequently abandoned by approximately 1800 BP as the stream incised again, forming the T1 terrace. Currently, mixed sandy and muddy sediments are aggrading in the incised channel (Unit E).

In addition to the seven radiocarbon ages and cultural inclusions that define the timing of the cut-and-fill episodes, two anomalous radiocarbon ages were obtained in the suite of samples collected. The first of these is an age of 2820 ±40 BP obtained from a depth of 1.25 m in the Unit B fill. This age is several thousand years too young to reflect the time of deposition, and it is believed to represent illuviation of organic matter in the soil formed on the T3 surface. This soil consists of a granular structured A horizon approximately 40 cm thick underlain by a subangular blocky structured Btk horizon containing secondary filamental and fine nodular carbonate accumulations that extend to a depth of ap-

proximately 150 cm below the ground surface. The degree of horizonation and carbonate reprecipitation shown by the soil suggests a fairly long period of pedogenesis, and is inconsistent with the obtained age. Moreover, three ages ranging from 4700 BP to 3700 BP were obtained from the Unit C inset, clearly indicating that the T3 terrace was abandoned well before 2800 BP. Secondly, an anomalously old age of 10,120 ±220 was obtained from crossbedded channel sands at the base of Unit D, 0.5 m above where an age of 4140 ±90 BP was obtained from Unit C sediments. This sample was taken from an exposure butted up against the Unit A fill immediately above a lens of volcanic ash reworked from the thick ash lens in the older deposits. The result indicates that the basal deposits contain a considerable amount of locally-derived organic carbon reworked from the older deposits, in addition to the reworked ash.

4 DISCUSSION AND CONCLUSION

The results of field investigations completed thus far indicate that a number of phases of active areal erosion and valley aggradation affected both study areas through the late Holocene. In the Metaponto chora, these episodes have resulted in the accumulation of up to 25 m of fine-grained sediment during the last 6000 years. Aggradation in the downstream reaches appears to have resulted from the sudden introduction of large quantities of sediment into the alluvial valleys. Truncation of soils or other evidence of increased erosion on

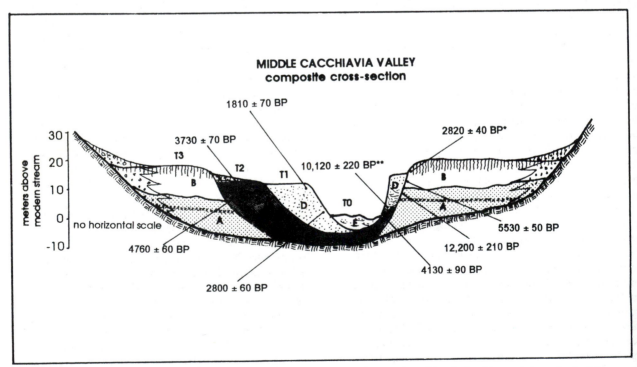

Figure 7. Schematic cross-section of the middle valle di Cacchiavia, illustrating stratigraphic relationships and chronologic data.

Table 2. Radiocarbon ages on sediment and soils, Croton chora.

Sample Number	Provenience	Material	Δ ^{13}C	Age (Corrected years BP)
TX-7400	Middle Cacchiavia Valley Papanice Section 2 (T3) 5.5-5.7 m bgs	Sediment	−27.4%	12,200 ±210
TX-7401	Middle Cacchiavia Valley Papanice Section 2 (T3) 1.25m bgs	Soil	−25.7%	2820 ±40*
TX-7627	Middle Cacchiavia Valley Cutro Dump Section (T2) 0.4-0.5 m bgs	Cumulic Soil	−26.7%	3730 ±70
TX-7628	Middle Cacchiavia Valley Papanice Ceramic Locality (T1) 2.4 m bgs	Sediment	−25.0%	10,120 ±220**
TX-7629	Middle Cacchiavia Valley Papanice Valley Section 2 (T3) 3.0-3.25 m bgs	Sediment	−25.6%	5530 ±80
TX-7630	Middle Cacchiavia Valley Papanice Ceramic Locality (T1) 3.0 m bgs	Sediment	−25.3%	4130 ±90
TX-7631	Middle Cacchiavia Valley Cutro Dump Section (T2) 1.3-1.35 m bgs	Cumulic Soil	−26.6%	4760 ±60
TX-7632	Middle Cacchiavia Valley Cutro Dump Annex (T1) 0.4 m bgs	Soil	−25.6%	1810 ±70
TX-7633	Middle Cacchiavia Valley Papanice Section 3 (T1) 4.6-4.8 m bgs	Sediment	−24.1%	2800 ±60

* Probable underestimate due to illuviation of organics in the soil zone.
** Probable overestimate due to incorporation of old allogenic organic matter.

the floodplains prior to renewed sedimentation is extremely rare, suggesting that the primary system change was an increase in the rate of sediment influx rather than a strong shift in rainfall and discharge. Aggradation in the valleys appears to have been accompanied by rapid progradation of the coastline and obstruction of the lower valleys, leading to poor drainage and the formation of marshes and swamps (Boenzi et al. 1987; Carter 1990). This trend probably had several significant impacts on the population, including a decrease in the amount of available arable land and an increase in the incidence of malaria (Delano Smith 1979).

If the character of sedimentation in the coastal reaches was controlled by changes in water table elevation, this control should be reflected in the sequence. At this stage of investigation, there is little data to contradict the hypothesis. The typical, subtle fining-upward sequence of the Holocene fill units at Metaponto (laminated sands and silts grading up into massive silts and silty clays and finally into a cumulic clay loam palaeosol) is consistent with deposition under the influence of a rising water table. All of the buried soils show signs of saturation (gley, redox mottling, etc.) and there is some archaeological evidence – principally an increase in the depth of burial of Greek materials beneath the Unit F palaeosol – that the formation of the palaeosol may have been time-transgressive moving up the valley. However, all support for this idea at this time is based on field observations, and additional information (e.g. micromorphological examination of the palaeosols, additional radiocarbon ages) is necessary before the depositional context can be fully understood.

In the vicinity of Croton, episodes of fluvial aggradation and incision resulted in a complex cut-and-fill sequence with a radically different architecture than the larger streams in the Metapontine chora. Five distinct inset alluvial fills, representing the terminal Pleistocene and most of the middle to late Holocene, are preserved in the valle di Cacchiavia. All of the units are lacking in gravels, reflecting the absence of hard bedrock in the drainage. Interestingly, the timing of alluvial aggradation and incision in the Croton chora does not match the sequence at Metaponto, suggesting that regional climatic trends are in themselves incapable of accounting for changes in fluvial activity.

Scholars working in the Mediterranean basin have long recognized that the region has undergone intense geomorphic and environmental change during the Holocene. In addition to historical evidence of deforestation and environmental degradation during the past several millenia (Semple 1931; Whatmough 1937; Hutchinson 1969; Hughes 1983), a number of studies have documented geomorphic and stratigraphical evidence of pervasive episodes of severe slope erosion and valley aggradation throughout the Mediterranean region (e.g. Kayser 1958; Eisma 1964; Vita-Finzi 1969, 1975; Bintliff 1975; Butzer 1974, 1990; Neboit 1977; Davidson 1980; Brückner 1982, 1986; Delano-Smith 1979; Pope and van Andel 1984; van Andel et al. 1990; Chester and James 1991; Coltorti and Da Ri 1985). Although progressive, this deterioration frequently appears to have occurred in a punctuated manner, with relatively short periods of pronounced slope instability and valley alluviation alternating with periods of relative stability and soil formation (van Andel et al. 1990; Pope and van Andel 1984; Butzer 1980). Opinion remains divided on the primary stimulus for these periods of instability, with some investigators favoring climatic mechanisms (e.g. Vita-Finzi 1969; Bintliff 1975; Devereaux 1982) and others preferring an anthropogenic explanation (e.g. van Andel et al. 1990; Brückner 1986; Butzer 1980). One reason that this issue remains controversial is that interpretation of causality is extremely difficult, given that both the paleoclimatic and cultural records from the region remain relatively poorly understood. Despite years of effort, reliable paleoclimatic data from the western Mediterranean remains scarce due to a combination of generally poor pollen preservation and the difficulty of distinguishing between climatic and anthropogenic signals in the various forms of proxy data (Pons 1991). A similar problem is presented by archaeological data from southern Italy, where research has focused on individual sites at the expense of regional inventories. As a result of this tendency, reliable estimates of population and settlement distribution are unavailable for most areas and time periods, while those attempts that have been made (e.g. Whitehouse 1984) have paid scant attention to the possibility of biases imposed through the comparison of regions with drastically different levels of geomorphic stability and potential for site visibility and preservation.

Although some information does exist suggesting that the climate fluctuated during the late Holocene (e.g. Giraudi 1989; Serre-Bachet 1991), the magnitude of this variation was probably not sufficient to account for rapid, drastic shifts in sediment yield. However, the pollen data for southern Italy (e.g. Sullivan 1983; Drescher-Schnieder 1985) and for the rest of the Mediterranean (e.g. Bottema and Woldring 1990; Jahns 1990) clearly indicate that humans had a powerful impact on the character of vegetative cover, and thus on hydrologic characteristics and rates of sediment yield, through forest clearing and agriculture. A third possibility that could explain some of the aggradational and incisional episodes is a complex response of the fluvial systems to system perturbations (Schumm 1973). This process, where various reaches of a stream respond to a given initial stimulus at different times and in varying ways, is one possible explanation for the episode of aggradation in the lower valleys and concomitant incision inland apparent during the Medieval period. It may be that this coastward flushing of the alluvial fills occurred not as a direct response to Medieval land use, but rather as a response to disequilibrium in the longitudinal profile created by the introduction of extremely high volumes of sediment during the preceding Classical period. Finally, the role of tectonic uplift in the interior remains an intangible factor that possibly had profound influence on fluvial behaviour at various times.

In summary, both chorai exhibit alluvial sequences that reflect fluctuating sediment yields through the Holocene. In the Metaponto area, identified episodes of rapid erosion seem to coincide with increasing population levels in the later Neolithic (Whitehouse 1984) and during the Iron Age and subsequent Classical period (de La Geniére 1979; Carter 1980; 1990), while at least one phase of soil formation in the valleys appears to coincide with a period of apparent relative depopulation at the end of the Punic Wars (Toynbee 1965; Carter 1980). At Croton, relatively slow aggradation of silts and clays seems to have occurred almost continuously through the middle to late Holocene, with very brief incisional episodes that are more likely related to tectonic uplift than human activity or climate perturbations. However, the Greek occupation witnessed the aggradation of a relatively coarse-grained sandy fill at Croton that almost certainly has its origins in the increased stress placed on the land to supply food for the large classical population.

5 ACKNOWLEDGEMENTS

Major funding for this project was provided by the Institute for Classical Archaeology, University of Texas at Austin and National Science Foundation Dissertation Improvement Grant SES-9206033, with additional support provided by Grants from Phi Kappa Phi and the Department of Geography, University of Texas at Austin. This research benefitted from fruitful discussions with many individuals, including Joe Carter, Charles Frederick, Karl Butzer, Jon Morter, and Steve Hall. Special thanks are due to the Metaponto and Croton crews, 1990-1992, and in particular to my capable part-time field assistants: Jean Alvarez, Gianna Ayala, Charles Frederick, Alexandra Layman, Yin Lam, J.B. Summitt, Susanna Van Sant, and Don Wade.

REFERENCES

Belluomini, G., E. Gliozzi, G. Ruggieri, M. Branca & L. Delitala 1988. First dates on the terraces of the Crotone peninsula. *Bolletin Societa Geologica Italia* 107: 249-254.

Bintliff, J.L. 1975. Mediterranean alluviation: New evidence from archaeology. *Proceedings of the Prehistoric Society* 41:78-84.

Boenzi, F., C. Cherubini & C. Giasi 1987. Dati e considerazione sull'evoluzione recente e sui caratteri idrogeologici della piana costiera Metapontina compresa tra il F. Bradano ed il F. Basento (Basilicata). *Geografica Fisica Dinamica Quaternaria* 10:34-46.

Boenzi, F., G. Palmentola & A. Valduga 1976. Caratteri geomorphologici dell'area del foglio 'Matera'. *Bolletin Societá Geologica Italiana* 95:527-566.

Bottema, S. & H. Woldring 1990. Anthropogenic indicators in the pollen record of the eastern Mediterranean. In S. Bottema, G. Entjes-Nieborg & W. van Zeist (eds.), *Man's Role in the Shaping of the Eastern Mediterranean Landscape*. Rotterdam: Balkema.

Brückner, H. 1982. Holozäne Bodenbildungen in den Alluvionen süditalienischer Flüsse. *Zeitschrift fur Geomorphologie N. F. Supplement* 48:99-116.

Brückner, H. 1986. Man's impact on the evolution of the physical environment in the Mediterranean region in historical times. *GeoJournal* 13:7-17.

Brückner, H. 1990. Changes in the Mediterranean ecosystem during antiquity: a geomorphological approach as seen in two examples. In S. Bottema, G. Entjes-Nieborg & W. van Zeist (eds.), *Man's Role in the Shaping of the Eastern Mediterranean Landscape*:127-137. Rotterdam: Balkema.

Butzer, K.W. 1980. Holocene alluvial sequences: Problems of dating and correlation. In R. Cullingford, D. Davidson, and J. Lewin (eds), *Timescales in Geomorphology*:131-141. New York: Wiley.

Cantafora, G., n.d. Personal communication, 1991.

Carter, J.C. (ed) 1980. *Excavations in the Territory, Metaponto, 1980*. Austin: Institute for Classical Archaeology, University of Texas at Austin.

Carter, J.C. (ed) 1990. *The Pantanello Necropolis 1982-1989: An Interim Report*. Austin: Institute for Classical Archaeology, University of Texas at Austin.

Chester, D.K. & P.A. James 1991. Holocene alluviation in the Algarve, Southern Portugal: The case for an anthropogenic cause. *Journal of Archaeological Science* 18:73-87.

Coltorti, M. & L. Dal Ri 1985. The human impact on the landscape: some examples from the Adige valley. In C. Malone & S. Stoddart (eds), *Papers in Italian Archaeology IV, The Cambridge Conference, part i: The Human Landscape*:105-134. BAR International Series 243.

Dainelli, P. 1975. An analysis of Italy's geomorphology. In C.H. Squyres (ed), *Geology of Italy*: 341-354. Tripoli: The Earth Sciences Society of the Libyan Arab Republic.

Davidson, D.A. 1980. Erosion in Greece during the first and second millenia BC. In R. Cullingford, D. Davidson & J. Lewin (eds), *Timescales in Geomorphology*:143-158. New York: Wiley.

De La Geniére, Juliette 1979. The Iron Age in southern Italy. In D. Ridgway & F.R. Ridgway (eds.), *Italy Before the Romans: The Iron Age, Orientalizing, and Etruscan Periods*: 59-93. New York: Academic Press.

Delano-Smith, C. 1979. *Western Mediterranean Europe: A historical geography of Italy, Spain and southern France since the Neolithic*. London: Academic Press.

Devereaux, C.M. 1982. Climate speeds erosion of the Algarve's valleys. *The Geographical Magazine* 54:10-17.

Drescher-Schnieder, R. 1985. Analyse palynologique dans l'Aspromonte en Calabre (Italie Meridionale). *Cahiers Ligures de Préhistoire et de Protohistoire*, N.S., n. 2.

Eisma, D. 1964. Stream deposition in the Mediterranean area in historical times. *Nature* 203:1061.

Giraudi, C. 1989. Lake levels and climate for the last 30,000 years in the Fucino area (Abruzzo – central Italy) – A review. *Paleogeography, Paleoclimatology, Paleoecology* 70:249-260.

Hughes, J.D. 1983. How the Ancients Viewed Deforestation. Contribution 1 to the symposium 'Deforestation, Erosion, and Ecology in the Ancient Mediterranean and Middle East'. *Journal of Archaeological Science* 10(4):437-445.

Hutchinson, J. 1969. Erosion and land use: the influence of agriculture on the Epirus region of Greece. *The Agricultural History Review* 17:85-90.

Jahns, S. 1990. Preliminary note on human influence and the history of vegetation in southern Dalmatia and southern Greece. In S. Bottema, G. Entjes-Nieborg & W. van Zeist (eds.), *Man's Role in the Shaping of the Eastern Mediterranean Landscape*:333-340. Rotterdam: Balkema.

Kayser, B. 1958. Recherches Sur les Sols et l'Erosion en Italie Meridionale: Lucanie. *Soc. d'Edition d'Enseignement Supérieur* (5 Place de la Sorbonne, Paris).

Neboit, R. 1977. Un exemple de morphogenese accélérée dans l'antiquite: les vallées du Basento et du Cavone en Lucanie (Italie). *Méditerranée: Revue geographique des pays méditerranées* 31(4):39-50.

Neboit, R. 1983. L'Homme et L'Erosion. Faculté des Lettres et Sciences Humaines de l'Univerité de Clermond-Ferrand II, *Nouv. Sér.* 17.

Pons, A. 1991. Pollen proxy data from western Mediterranean Europe. In B. Frenzel (ed.), *Evaluation of Climate Proxy Data in Relation to the European Holocene*:133-147.

Stuttgart: Gustav Fischer Verlag.

Pope, K.O. & T.H. van Andel 1984. Late Quaternary alluviation and soil formation in the southern Argolid: Its history, causes and archaeological implications. *The Journal of Archaeological Science* 11:281-306.

Roda, C. 1964. Distribuzione e facies dei sedimenti Neogenci nel bacino Crotonese. *Geologica Romana* 3:319-366.

Schumm, S. 1973. Geomorphic thresholds and complex response of drainage systems. In M. Morisawa (ed), *Fluvial Geomorphology*:299-310. SUNY Binghampton, Publications in Geomorphology, 4th Annual Meeting.

Selli, R. 1977. Excursion in Calabria. General geologic setting of the Crotone-Catanzaro Area. *Giornale Geologica* 41:410-458.

Semple, E. C. 1931. Ancient Mediterranean forests and the lumber trade. In: *The Geography of the Mediterranean Region*:261-296. New York: Holt.

Serre-Bachet, F. 1991. Tree rings in the Mediterranean area. In B. Frenzel (ed), *Evaluation of Climate Proxy Data in Relation to the European Holocene*:133-147. Stuttgart: Gustav Fischer Verlag.

Servizio Geologico d'Italia 1978. *Carta Geologica d'Italia, Foglio 3*. Rome: Ministero Dell'Industria del Commercio e dell'Artigianato.

Sullivan, D. G. 1983. Preliminary report on the Pizzica pollen samples. In J.C. Carter (ed.), *The Territory of Metaponto 1981-1982*. Austin: Institute for Classical Archaeology, University of Texas at Austin.

Toynbee, A. 1965. *Hannibal's Legacy: the Hannibalic War's Effect on Roman Life*. London: Oxford University Press.

van Andel, T.H., E. Zangger & A. Demitack 1990. Land Use and Soil Erosion in Prehistoric and Historical Greece. *Journal of Field Archaeology* 17(4):379-396.

Vita-Finzi, C. 1969. *The Mediterranean Valleys: Geological Changes in Historical Times*. Cambridge: Cambridge University Press.

Vita-Finzi, C. 1975. Late Quaternary alluvial deposits in Italy. In D.J. Squyres (ed), *Geology of Italy*:329-340. Tripoli: The Earth Sciences Society of the Libyan Arab Republic.

Whatmough, J. 1937. *Foundations of Roman Italy*. London: Methuen.

Whitehouse, R.D. 1984. Social organization in the Neolithic of southeast Italy. In W.H. Waldren, R. Chapman, J. Lewthwaite & R.-C. Kennard, (eds.), *The Deya Conference of Prehistory: Early Settlement in the Western Mediterranean Islands and Their Peripheral Areas*, part iv. BAR International Series 229(iv).

CHAPTER 19

Climatic forcing of alluvial fan regimes during the late Quaternary in the Konya basin, south central Turkey

NEIL ROBERTS

Department of Geography, Loughborough University, Leicestershire, UK

ABSTRACT: This study investigates alluvial sedimentary sequences around the Konya basin of south central Anatolia. This basin, now largely dry, was occupied by an extensive lake between 23,000 and 17,000 years ago. Absence of tectonic deformation in palaeo-shorelines suggests that alluvial fan regimes have been controlled primarily by shifts in hydro-climatic parameters. Sediments have been studied by a programme of coring, supplemented by analysis of sections, where available. In addition, geoarchaeological study of the basin margin shows that many hüyüks (settlement mounds) have been partially buried by subsequent deposition of alluvium. On the lower segment of the Ibrala fan, a veneer of fine-grained alluvium (< 3.5 m thick) overlies lake marl. In the upper segment this Upper Fan Member overlies fluviatile gravels along with debris flow deposits; these appear to pre-date the last phase of high lake levels. The Late Pleistocene fan was therefore much smaller than the modern one. The climatically-induced shift from a true alluvial fan regime to modern conditions of low-energy alluvial sedimentation occurred just prior to the appearance of Neolithic settlements on the fans c. 8500 years BP. However, active alluvial fan sedimentation has continued during the Holocene on smaller, higher-slope angle fans elsewhere in the Konya basin.

KEYWORDS: Alluvial fan, Turkey, climatic change, Neolithic, geoarchaeology, late Quaternary.

1 INTRODUCTION

The experimental method has great explanatory power in the natural sciences, but its application to problems of landform evolution is hindered by the short timescale of observation and difficulties of scaling-up forces from laboratory or 'miniaturized' field studies (e.g. Rachocki 1981, Schumm et al. 1987). The difficulty of holding constant, and hence isolating, the controlling variables in long-term geomorphological studies has tended to make such studies descriptive and prone to circular reasoning. An undated gravel terrace deposit above a modern meandering river is likely to be attributed to a past (?Pleistocene) phase when the climate was different (?colder, ?drier) compared to that of today. Alternative explanations, such as base-level change, tectonics, human impact or high magnitude events, are difficult to test under these conditions, because all these factors are likely to have varied at the same time. But history can be coaxed into conducting experiments (cf. Deevey 1969). In this effort, long-term studies at a single site potentially have an in-built advantage over ergodic or other experimental approaches, in that many factors (e.g. topography) will have effectively been unchanged through time (Bull 1991).

The problem considered in this paper is the relationship between climate change and the geomorphic processes operating on alluvial fans. An alluvial fan is a depositional landform whose surface morphology resembles a segment of a cone radiating downslope from the point where a stream or river leaves an upland area (Bull 1977; Rachocki 1981). Climatic conditions affect the hydrological regime on the fan surface, and these in turn modify the sedimentary environment, for example determining whether water-laid or mass movement deposition predominates. Climatic changes may also lead to an adjustment in fan slope, with results such as fan-head entrenchment and the creation of a new lobe at the margin. Similar responses can be induced by other, non-climatic agencies, however.

This study utilises alluvial fans in south central Turkey, where factors other than climate can be excluded from consideration in explaining changes in late Quaternary sediment stratigraphy, and where chronological control is relatively good. Additionally, lake levels and pollen data provide a record of climate change independent of the evidence for shifting geomorphological regimes on the fans themselves.

2 FIELD STUDY AREA

The Konya plain lies on the southern edge of the Anatolian plateau at an elevation of c. 1000 masl. The climate of the plain is semi-arid, with precipitation falling below 300 mm p.a., and it experiences a substantial seasonal temperature range, winter temperatures being around freezing and mean summer temperatures > 20°C. The plain and its immediate surroundings are treeless, except along river courses. By contrast, the catchment area for

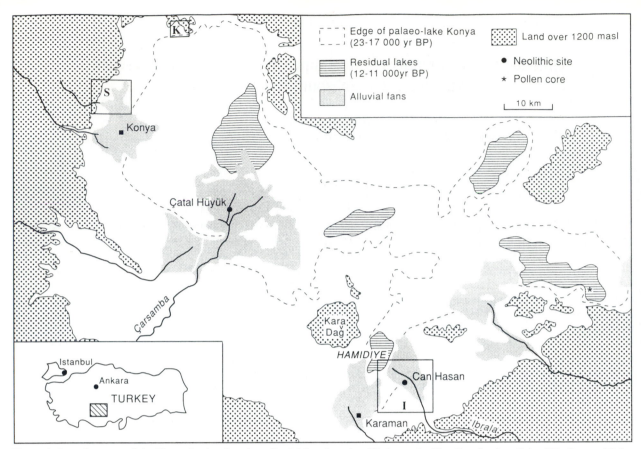

Figure 1. Location map of the Konya basin, showing alluvial fan deposits. (K shows the Karaömerler fan, S the Sille fan, and I the Ibrala fan).

Figure 2. Karaömerler fan and its catchment viewed from the marl plain. The fan has been quarried for aggregate, exposing a full section through the fan gravels.

Table 1. Characteristics of selected alluvial fans in the Konya basin.

Fan (and size class)	Fan surface area km^2	Catchment area km^2	Catchment: fan ratio	Mean fan slope	Discharge m$^3 \times 10^6$ pa	Dominant sediment type
Karaömerler (small)	0.063	0.68	10.8:1	6.8°	n.d.	Sieve and debris flow deposits
Sille (intermediate)	8	38	4.8:1	1.5°	2.4	Fluviatile sands and gravels
Çarşamba (large)	474	1720[1]	3.6:1	0.04°	94[1]	Flood-basin clays and backswamps

[1]Excluding overflows from the Beyşehir-Suğla basin.
Sources: de Meester (1970), D.S.I. (1975a).

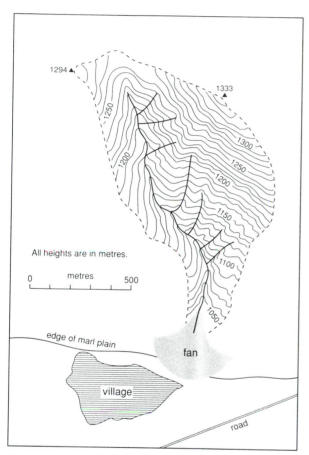

Figure 3. Topographic map of the Karaömerler fan and its catchment. The contour interval is 10 m.

inflowing rivers receives up to 1000 mm p.a. precipitation and is, or was until recent deforestation, well-wooded. Most of this is in the Taurus Mountains (Toros dağları) south and west of Konya, where peaks reach 3000 masl.

Hydrologically, Konya is a closed (i.e. non-outlet) basin, except for some karstic sink-holes (düdens). Most of the plain is today dry, although soils were waterlogged for much of the year prior to twentieth-century river regulation and irrigation. Shoreline depositional landforms and wave-cut cliffs are evidence that a shallow but extensive lake formerly occupied the floor of the plain (Erol 1978). Importantly, and in contrast to other Anatolian lake basins such as Burdur, none of these late Quaternary lake beds contains tectonic deformation structures,

and the elevation of lake marginal deposits stays constant over the full length of the Konya plain. This confirms historical seismicity data in showing that the basin is not, and has not been, tectonically-active during the radiocarbon time range.

As rivers and wadis enter the Konya plain, most of their sediment load is deposited close to the basin edge as fan-shaped masses of alluvium (Fig. 1). Alluvial fans range in size from those fed by short, ephemeral streams (e.g. at Karaömerler, Figs 2 and 3) to deposits covering, in the largest cases, several hundred km^2 (Table 1). The Karaömerler fan, an example of a small fan system, has been partly excavated for building aggregate, and the resulting sections reveal up to 9 m of coarse 'sieve' gravels and debris flow deposits overlying bedrock. There is no sign of an overall change in depositional environment through these sections, such as might have been caused by deforestation or climate change, although there are patches of uneroded soil in the otherwise denuded catchment. On the basis of this, it can be assumed that this fan, and others like it, have been accumulating under the presently-prevailing climatic regime.

The Sille fan, intermediate in size, has a more complex morphology, with two high-level terraces and a third, lowest terrace, currently being dissected by the main river channel. All of these comprise braided fluviatile sands and gravels, and include palaeochannel features; the lowest contains pottery sherds, indicating that it has been deposited during the later Holocene. As at Karaömerler, therefore, the fan is active geomorphologically, although its sedimentary environment has been dominated by fluviatile rather than mass movement processes.

The largest alluvial deposits, associated with perennial rivers such as the Çarşamba, are also broadly fan-shaped, but in other respects their hydro-geomorphological characteristics are today more akin to an alluvial floodplain than an alluvial fan environment. Overbank deposition of silts and clays over very low slopes has pushed these alluvial features towards the centre of the plain, on top of the lacustrine beds of palaeo-lake Konya. These shallow alluvial 'fans' also contain great numbers of ancient settlement mounds (hüyük in Turkish). The most famous of them, Çatal hüyük, is also one of the oldest, dating to 8200-7500 years BP, and this provides a minimum age for the last lacustral phase in the Konya basin. Çatal hüyük, like other archaeological sites, has been subject to burial by Holocene alluviation (Roberts 1983).

The lithostratigraphic sequence from one of the larger 'fans', that at Ibrala near Karaman (Fig. 1), has been investigated in association with geoarchaeological work at another prehistoric mound, at Can Hasan. This provides evidence of changing late Quaternary sedimentary regimes on the fan tied to a relatively well-established chronology.

2.1 *The Ibrala fan*

The Ibrala fan is bounded to the south by a broad erosional platform of Neogene limestone while to the north is the flat marl plain of the former lake bed (Fig. 4). (The new name Ibrala is used here, rather than the old name Selereki employed by de Meester (1970)). The Ibrala River enters the Konya basin from the south, and has built up a shallow alluvial fan north of Selereki. The river originally drained into an open-water marsh north of Hamidiye, but it has been diverted since the 1960s into a drainage canal which leads to a sinkhole, Düden gölü. The Hamidiye marsh is the final remnant of an embayment of the Konya lake which occupied all the low-lying central parts of the basin during the last lake high stand (Fig. 1). The Ibrala River rises in the central Taurus Mountains and has a catchment area of c. 350 km². A significant proportion of the annual discharge of 55 million m³ derives from springs that emerge from the Miocene limestone (D.S.I. 1975b, 29). The river is entrenched into this limestone with no clear traces of river terraces, and on the fan it occupies one main and several secondary distributary channels. Here, channel morphology is modified by the use of river water for irrigation, and there is evidence (e.g. from broken weirs) that fanhead trenching might be occurring if the river were not regulated.

The Ibrala fan today covers an area of 101.6 km² (de

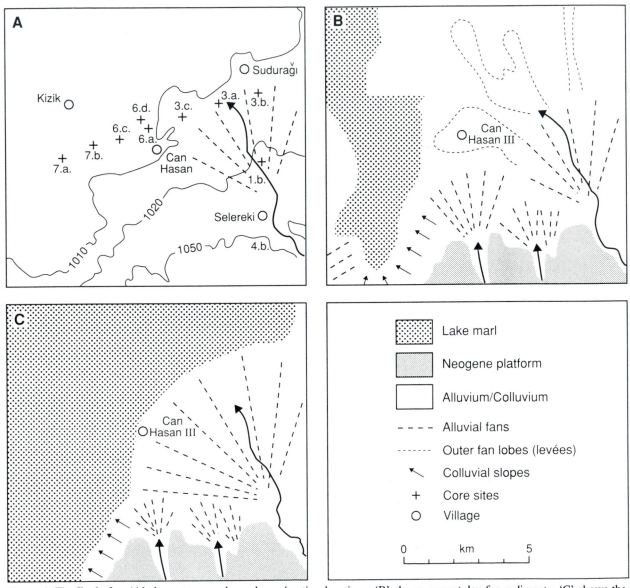

Figure 4. The Ibrala fan. 'A' shows topography and core/section locations; 'B' shows present-day fan sediments; 'C' shows the reconstructed early Holocene fan.

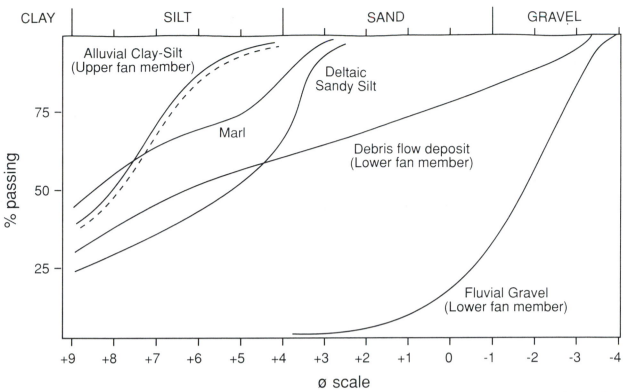

Figure 5. Selected grain-size curves for major lithofacies at the Ibrala fan edge. The dashed line shows a mud-brick sample from Can Hasan III.

Meester 1970). However, its 69.2 km² outer segment comprises a veneer of alluvial silt and clay less than 3.5 m thick overlying lake marl. Away from the outer fan lobes, which consist of low ridges representing former levées, the soils are poorly drained and the water table is at less than 2 m. The outer fan zone has an average slope of only 0.05° over 10 km and is much more like a floodplain than an alluvial fan environment. The same is also partly true of the inner fan segment, where the ground slope increases to 0.3°, but surficial sediments remain typical of levée and flood basin depositional environments.

2.2 Sedimentary analyses

The absence of sections except at the fan apex has meant that sedimentary sequences have been studied mainly by a programme of coring. The hand-auger used in most of this work penetrated up to 510 cm and had exchangeable screw and sampling heads. The cores form transects across the fan, running north-south and west to east, the latter passing next to the prehistoric settlement mound of Can Hasan III (Fig. 4).

Sediments were characterised initially in the field on the basis of colour and texture, and then analysed in the laboratory for carbonate content and granulometry. Grain-size analysis was carried out by dry sieving for particles larger than +4 φ and by pipette analysis for particles between +4 φ and +9 φ. Carbonate content was measured by the gas volumetric method. On the basis of this analysis, late Quaternary sediments of the Ibrala fan

have been classified into the following major lithofacies:

1. Lake marl. Light grey to white in colour and highly calcareous (CaCO$_3$ eq. > 70%). This was formed by palaeo-lake Konya during previous highstands.

2. Deltaic silty sand, with a distinctive positively skewed grain-size distribution which has a modal grain-size class between +3 and +4 φ.

3. Flood-basin clay and clay-silt, deposited from suspension in standing water away from the main river channels.

4. Fluviatile silts, sands and gravels. This includes a range of granulometric types based on facies changes from (generally coarser) inner to (finer) outer fan segments. Sediment sorting is usually moderately good, pointing to deposition by moving water in braided channels, on point bars, etc.

5. Colluvium. This has some variation in particle size range, but in all cases the sediments are extremely poorly sorted and include angular gravels. This suggests deposition by small-scale mass movement processes such as debris flows. All known examples on the fan margin come from more than 4 m below the present surface of the fan.

Grain-size curves of selected sediment samples are shown in Figure 5 and mean values for measured sedimentary parameters are listed in Table 2. Two different sorting indices, the Trask sorting coefficient, So, and the phi quartile deviation QDφ, have been applied to the samples. Mudflow deposits have sorting values and clay contents similar to those reported by Bull (1972, 71), with

Table 2. Sedimentary characteristics of major lithofacies on, or adjacent to, the Ibrala Fan.

	CaCO₃ eq.	S₀	QD∅	% clay	Md (μm)
Lake marl	78%	–	–	44	2.8
Deltaic silty sand	65%	5.8	2.5	23	34
Flood-basin clay silts	59%	4.1	2.0	34	14
Fluviatile gravel	–	3.6	1.7	–	3500
'Mudflow' colluvium	–	13.5	3.7	30	12

Mean values of at least 3 samples.

the mean for So being 13.5 (cf. 9.7 in Fresno County, California), that for QD∅ 3.7 (cf. 3.1) and with clay content averaging 30% (cf. 31%).

The ratio between median and maximum grain size (CM pattern) was used by Bull (1972) to differentiate between different types of alluvial fan deposits. Experience in coring unconsolidated sediments has shown the danger of contamination from overlying layers. A single pebble falling into a silt or clay deposit would not affect the overall grain-size distribution of the sample, but it would form the coarsest percentile and hence give an entirely spurious CM ratio. For this reason the C values have been replaced in this analysis by the 91st percentile (N), and the resulting NM pattern (Fig. 6) clearly distinguishes between stream channel gravel, debris flow/mudflow deposits and fine grained alluvium.

3 FIELD STRATIGRAPHY

3.1 *Distal fan sequence*

The series of sediment cores has been used to build up a lithostratigraphic sequence at the fan margin in order to examine the relationships between alluvial, lacustrine and archaeological sediments, and to define the changing position of the fan margin during the late Quaternary. The most informative sequence comes from the 8 km long east-west transect which extends from inner to outer fan and onto the marl plain through Can Hasan. This contains a wedge of lacustrine sediments representing a phase of lake transgression and regression up to an elevation of c. 1007 m (Fig. 7). The sequence of alluvial, lacustrine and deltaic sediments is well displayed in core 6d, where 310 cm of fine-grained alluvial clay-silt overlie lake marl and subaqueous silty sand (310-405 cm), below which is 80 cm of colluvium (Fig. 8).

Core 6d was taken next to the aceramic Neolithic mound of Can Hasan III, which was excavated in 1969-70 by the British Institute of Archaeology at Ankara (French et al. 1972). The site has ¹⁴C dates ranging from c. 8600 to 7900 years BP (Ergin 1979, 148; Burleigh et al. 1982; Tite et al. 1987). Today it stands only 2.5 m above the general fan surface, but excavation uncovered a greater depth of archaeological deposits buried by Holocene alluvium. Excavation uncovered 4 m of deposit, and archaeological levels below this were continuously

sampled by a Dachnowsky corer at the time of excavation (French et al. 1972, 184). The archaeological fill ends at a depth of 255 cm in this core (6j) and below is 45 cm of white clay-marl and 14 cm of subaqueous silty sand, a sequence of lacustrine sediments identical to that found nearby in core 6d.

Another sequence of particular interest is that from core 3c, where angular gravel lay beneath 490 cm of silt-clay. A sample taken at 410 cm was granulometrically almost identical to two samples of mudbrick from Can Hasan III (Fig. 5). It is possible that a similar silty alluvium lay at the surface of the fan during the early Holocene and was used for mudbrick making by the inhabitants of Can Hasan III. Chronologically, this suggests that the onset of fine-grained alluviation at the distal part of the fan occurred only a short time before the Neolithic period (Roberts 1991).

The sedimentary sequence at the Ibrala fan margin can be organised into three principal lithostratigraphic units. The oldest of these, the Lower Fan Member, comprises coarse-grained and/or poorly sorted sediments which were either water-lain or deposited by debris flows. The Lower Fan Member lies stratigraphically beneath the second unit, which comprises calcareous marl and fine sands, the latter laid down at the distal margin of a Gilbert-type fan-delta at the edge of the former lake (Nemec and Steet 1988). These lacustrine and fan-delta sediments have not been dated directly on the Ibrala fan but they correlate with the last phase of high lake levels, ¹⁴C dated from marginal beach facies elsewhere in the Konya basin to between 23,000 and 17,000 years BP (Roberts 1980, 1983). The third and most recent unit, the Upper Fan Member, comprises fine-grained alluvium 3.5-5.0 m thick. Deposition of this appears to have begun just prior to the establishment of Can Hasan III, that is, around the beginning of the Holocene, and has continued up to the present-day. There consequently appears to have been a phase of non-deposition at the fan margin during the Late Pleistocene between units 2 and 3. Elsewhere in the Konya basin, de Meester (1971, 16) has shown that the upper 50-150 cm of some marl deposits were laid down under sub-aerial conditions, as marls exposed after the last lake retreat were reworked from the basin edge and redeposited by seasonal floods. For this reason the top part of the lake sediments in the Ibrala fan sequence may have been removed by erosion. No Holocene lacustrine sediments are recorded in the fan sequence, indicating that lake levels have remained below c. 1006 m during the last 9000 years.

3.2 *Proximal fan sequence*

A less complete picture is available for the Ibrala fan apex. Cores show c. 2.5 m of sandy silt and fine sand over fluviatile gravels, plausibly correlating with the Upper and Lower fan units described above. The one substantial section available for study (4b) lay 300 m away from, and 5 m above, the modern Ibrala River. The exposed sediments consisted of colluvial red beds overlain unconformably by fluviatile gravels. Most of the colluvium was

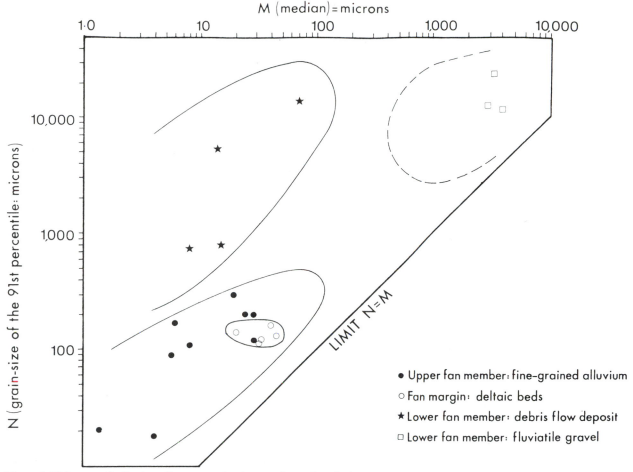

Figure 6. NM pattern for Ibrala sediment samples (see text for explanation).

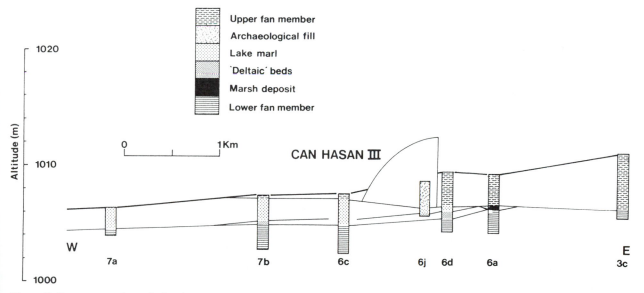

Figure 7. Core transect through distal part of Ibrala fan.

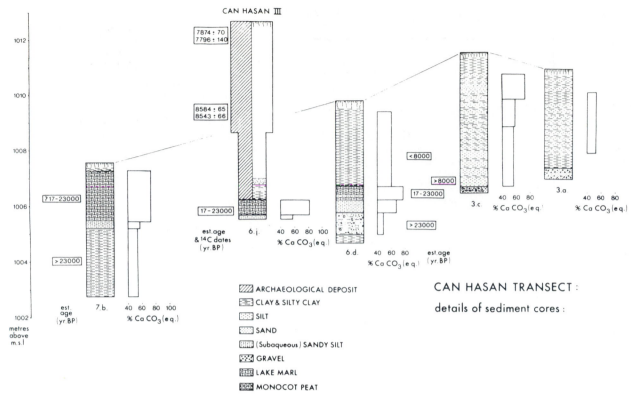

Figure 8. Detailed core stratigraphies showing correlation with sequence at Can Hasan III and estimated ages.

made of fine-grained mudflow deposits though there were occasional sheets of better sorted angular gravels.

The colluvium appears to have originated from a local side-gully rather than the Ibrala itself. Rock fragments were entirely of limestone, and the matrix fines were yellowish-red in colour, similar to the 'derived *terra rossa* soil' found elsewhere in the Konya basin by de Meester and van Schluylenborgh (1970). These authors concluded that eroded limestone hills were once covered by red Mediterranean soils (= Chromic Luvisols) formed under a different climate from that of today, and that the soil was subsequently removed downhill by mudflows. The source area for the red beds at 4b is an undulating limestone platform, which has very shallow, eroded soils today (ThB a/c de Meester 1970), and for *terra rossa* soil to form here would require considerably moister conditions than are provided by the present semi-arid climate.

The red colluvium is overlain by extensive sheets of fluviatile gravels deposited by braided channels, and at two points in the sections display a clear palaeochannel. The sub-rounded limestone gravels which this contains form large-scale trough cross-stratified bedding, and grain-size analysis shows sorting values virtually the same as those reported by Bull (1972, 71) for stream-channel deposits on Californian alluvial fans. The lithostratigraphic sequence at 4b cannot be tied directly to that at the Ibrala fan margin. However, it is clear that the sedimentary environments represented in 4b are unlike those of the Ibrala River today, which is characterised by a shallow channel and high suspended sediment load.

Nor can they be easily reconciled with the depositional environments of Holocene sediments on the Ibrala fan. This makes a Pleistocene age for the palaeochannel likely; and since it postdates the red beds, the implication must be that not merely the formation on the *terra rossa* soil but also its removal occurred before the onset of the Holocene.

4 ALLUVIAL RESPONSE TO CLIMATE CHANGE

A sharp contrast thus emerges between the nature of the Ibrala fan during the Holocene and at least part of the Late Pleistocene. Evidence therefore suggests a division into a fine-grained, moderately sorted upper member of Holocene age, and a coarse-grained and/or poorly sorted lower member of Late Pleistocene age. In core 6d, for instance, colluvium directly underlay lake beds whose inferred age is between c. 23,000 and 17,000 ^{14}C years BP. The sequence here indicates that mudflow processes operated on a steeper fan surface than at present immediately before the last major lacustral interval. Among these colluvial deposits are 'red beds', derived from *terra rossa* soils which must therefore have formed prior to c. 25,000 years BP under relatively warm, sub-humid climatic conditions (Table 3). Similar red beds are known from elsewhere in Anatolia (e.g. Atkinson 1970), and this suggests that there was widespread pedogenesis in the earlier part of the last cold stage (oxygen isotope stages 3

Table 3. Summary of late Quaternary geomorphological changes on the Ibrala fan.

Age ^{14}C years BP	Lake		Alluvial fan	Catchment	Postulated climate
pre-25,000	?			Terra Rossa soil formation	Interstadial? warm, sub-humid
				Removal of Terra Rossa soil and secondary deposition	
?25,000	? Intermediate		Lower Fan Member	Intense mechanical weathering and high sediment yield	Cold, semi-arid
22,000			(Fluviatile gravel, debris flows)		
	High lake level	Deltaic beds — — — — lake			
17,000		marl	No sediment deposition on edge of fan	Reduced sediment yield	Cold, arid
11,000	low lake level (Secondary lakes)			Low sediment yield	
				Ibrala springs active	
8500	Low lake	Can Hasan III	Upper Fan Member	Increased chemical weathering and sediment yield	Warm, semi-arid
6500	level		(Alluvial silts and clays)	High suspended sediment load	
present-day					

and 4) or during the last interglacial complex (oxygen isotope stage 5).

The deposits of the Lower Fan Member indicate a seasonal concentration of river discharge with bedload more important than suspended sediment load, and significant mechanical weathering in the catchment. Some of these processes are likely to have been periglacial, with frost-shattering active and spring snowmelt responsible for much of the sediment transport. The Ibrala catchment was not glaciated, so that the fan deposits cannot be of glacio-fluvial origin.

During the following stage, a Gilbert-type fan-delta formed at the distal edge of the Ibrala fan as lake levels rose. Continued fan sedimentation contemporary with the last phase of high lake levels indicates that sediment discharge on the fan remained high, at least initially. The climate at this time was clearly much colder than at present in the Konya basin. An estimate based on the presence of periglacial features, the then largely treeless vegetation (Bottema and Woldring 1984) and a water-balance model for the former lake (Roberts 1980) suggests a fall in mean July temperatures of 8-9°C and little change in total precipitation, although much of this would have fallen as snow. However, the fact that water levels in palaeo-lake Konya then fell before the main rise in temperatures indicates that a progressive decline in precipitation occurred between 18,000 and 13,000 years BP, after which the climate became both warmer and wetter. Significantly there seems to have been no accumulation of sediment at the fan margin between the end of this

lacustral phase (c. 17,000 years BP) and the establishment of Can Hasan III (9000 years BP), and what evidence there is points to erosion. A shift to a more arid climate and a decrease in load might be expected to have led to fan-head trenching (Schumm et al. 1987, 283), but there is no evidence for this or for the creation of a new distal fan lobe. This may have been because both sediment and hydrological fluxes were diminished during the later Pleistocene.

After the end of the Pleistocene, alluvial sedimentation Pleistocene.

After the end of the Pleistocene, alluvial sedimentation recommenced on the fan, but the silts and clays of the Upper Fan Member indicate a steady rather than a flashy river regime, with a high suspended sediment load from the catchment. The onset of alluviation was partly due to the re-activation of the springs which feed the Ibrala River. No major changes in sedimentation are indicated during the Holocene on the Ibrala fan, but this may not be true of other fans which are rain- rather than spring-fed. Given that an alluvial fan effectively acts as a closed system for particulate sediments, it is possible to calculate the net Holocene sediment flux from the Ibrala catchment. Approximately 4 cm/century of fine-grained, water-laid sediment was laid down during the last 10,000 years at the edge of the Pleistocene fan and apparently rather less than this (2-3 cm/century) in both the outer segment of the Holocene fan and at the fan apex. Overall this amounts to $2.5\text{-}3.0 \times 10^8$ m^3 of sediment deposited on the Ibrala fan during the Holocene, and a mean catchment surface lowering of 7.0-8.5 mm/century, excluding solute loss; (the Ibrala has a mean modern solute concentration of 0.26 gm/l). This is a higher rate of denudation than that calculated for the catchment of the much smaller Karaömerler fan. The Karaömerler fan has a sediment volume of c. 3.8×105 m^3, equivalent to a mean catchment surface lowering of 5.6 mm/century over the last 10,000 years, or 3.5 mm/century over the last 16,000 years (when the lake last retreated from this site). The lower values for Karaömerler are surprising given its steeper, smaller catchment and more efficient sediment delivery, and are probably due to sediment supply limitation once the topsoil had been removed (cf. Fig. 2).

5 CONCLUSION

Although the present-day alluvial sediments at the surface of major Konya fans are fine-grained and spread as a thin veneer over a large area, this was not always the case. The core transect through Can Hasan has shown that the Ibrala fan was formerly smaller in extent and that related sediments were coarse-grained or poorly sorted or both. The sediments of the Upper Fan Member indicate a floodplain environment (flood basin and levée); those of the Lower Fan Member are typical of true alluvial fans (debris flows, coarse water-laid). In other words a Pleistocene alluvial fan (area 32.4 km^2) became an alluvial plain (area 101.6 km^2) after c. 10,000 ^{14}C years BP. As the region is tectonically stable there is no reason to think that this switch related to faulting activity. In contrast to late Quaternary fan-delta deposits from other eastern Mediterranean lake basins, such as the Dead Sea (Sneh 1979) and Lake Hazar in eastern Turkey (Dunne and Hempton 1984), both of which are tectonically active, one can be confident that the changing sedimentary environment on the Konya fans was responding solely to changes in hydro-climatic conditions.

The conclusion that alluvial regimes responded to major climate changes at the Pleistocene – Holocene transition is not entirely a surprising one. On the other hand, alluvial fan sedimentation in the Konya basin has been a function of topographic relief as well as climate. Because smaller fan systems generally have steeper slopes, true alluvial fan regimes have continued during the Holocene on smaller, higher-slope angle fans elsewhere in the Konya basin (e.g. at Karaömerler). Equally, the Late Pleistocene was not characterised by a single set of morpho-climatic conditons, but witnessed significant temporal changes; in this case, fan sedimentation was active before and during the last major lacustral interval (i.e. before c. 17,000 years BP), but may not have been during the Late glacial (i.e. 17,000 – c. 12,000 years BP).

6 ACKNOWLEDGEMENTS

I am pleased to acknowledge for their assistance: Mustafa Karabıyıkoğlu, Oğuz Erol, David Alexander, Claudio Vita-Finzi and David French; MTA Institute and the British Institute of Archaeology at Ankara. Erica Millwain asssisted with drafting the diagrams.

REFERENCES

Atkinson, K. 1970. Fossil limestone soils in northwest Turkey. *Palaeogeography, Palaeoclimatology, Palaeoecology* 8: 29-35.

Bottema, S. & H. Woldring 1984. Late Quaternary vegetation and climate of southwestern Turkey II. *Palaeohistoria* 26: 123-149.

Bull, W.B. 1972. Recognition of alluvial-fan deposits in the stratigraphic record. In W.K. Rigby & J.K. Hamblin (eds), *Recognition of ancient sedimentary environments*: 63-83. Soc. Ec. Pal. and Min. Spec. Pub. 16.

Bull, W.B. 1977. The alluvial-fan environment. *Progress in Physical Geography* 1: 222-270.

Bull, W.B. 1991. *Geomorphic responses to climatic change*. Oxford University Press, Oxford.

Burleigh, R., J. Ambers & K. Matthews 1982. BM Natural Radiocarbon measurements XV. *Radiocarbon* 24: 262-290.

Deevey, E.S. 1969. Coaxing history to conduct experiments. *BioScience* 19: 40-43.

D.S.I. (Devlet Su Isleri) 1975a. *Konya-Çumra Karapınar ovası*. Hidrojeolojik etüd raporu. Ankara.

D.S.I. (Devlet Su Isleri) 1975b. *Karaman-Ayrancı ve Akşehir ovalarası*. Hidrojeolojik etüd raporu. Ankara.

Dunne, L.A. & M.R. Hempton 1984. Deltaic sedimentation in the Lake Hazar pull-apart basin, south-eastern Turkey. *Sedimentology* 31: 401-412.

Ergin, M. 1979. The Hacettepe University Radiocarbon Laboratory and chronological prospection of the archaeological sites in Turkey. *Chimica Acta Turcica* 7: 31-38.

Erol, O. 1978. The Quaternary history of the lake basins of central and southern Anatolia. In W.C. Brice (ed), *The environmental history of the Near and Middle East since the last ice age*: 111-139. London: Academic

French, D.H., G.C. Hillman, S. Payne & R.J. Payne 1972. Excavations at Can Hasan III 1969-1970. In E.S. Higgs (ed), *Papers in economic prehistory*: 181-190. Cambridge: Cambridge University Press.

de Meester, T. (ed) 1970. *Soils of the Great Konya basin, Turkey*. Wageningen: Centre for Agricultural Publishing and Documentation.

de Meester, T. 1971. *Highly calcareous lacustrine soils of the Great Konya basin, Turkey*. Wageningen: Centre for Agricultural Publishing and Documentation.

de Meester, T. & J. van Schuylenborgh 1970. Genesis and morphology of reddish-brown soils in the Great Konya basin. In T. de Meester (ed), *Soils of the Great Konya basin, Turkey*: 221-227. Wageningen: Centre for Agricultural Publishing and Documentation.

Nemec, W. & R.J. Steet (eds) 1988. *Fan deltas: sedimentology and tectonic settings*. Int. Assoc. Sed. Spec. Pub., Blackwell Scientific.

Rachocki, A. 1981. *Alluvial fans. An attempt at an empirical approach*. Chichester: Wiley.

Roberts, N. 1980. Late Quaternary geomorphology and paleoecology of the Konya basin, Turkey. Unpublished Ph.D., London University.

Roberts, N. 1982. A note on the geomorphological environment of Çatal hüyük, Turkey. *Journal of Archaeological Science* 9: 341-348.

Roberts, N. 1983. Age, palaeoenvironments and climatic

significance of Late Pleistocene Konya lake, Turkey. *Quaternary Research* 19: 154-171.

Roberts, N. 1991. Late Quaternary geomorphological change and the origins of agriculture in south central Turkey. *Geoarchaeology* 6: 1-26.

Schumm, S.A., M.P. Mosley & W.E. Weaver 1987. *Experimental fluvial geomorphology*. Chichester: Wiley.

Sneh, A. 1979. Late Pleistocene fan-deltas along the Dead Sea rift. *Journal of Sedimentary Petrology* 49: 541-552.

Tite, M.S., S.G.E. Bowman, C.J. Ambers & K.J. Matthews 1987. Preliminary statement on an error in British Museum radiocarbon dates (BM-1700 to BM-2315). *Antiquity* 61: 168.

CHAPTER 20

Facies sequences and sedimentation influences in late Quaternary alluvial deposits, southeast Cyprus

RODNEY L. STEVENS and PER O. WEDEL
Department of Geology, University of Göteborg and Chalmers University of Technology, Göteborg, Sweden

ABSTRACT: The complex response of alluvial systems is troublesome for stratigraphers wishing to isolate and evaluate individual parameters, and suggests that we need increasingly specific methods for relating sedimentary detail to the interactions of processes within such environments. We have documented facies sequences from late Quaternary alluvial settings: fan, axial stream and coastal plain. The facies include sandy gravels, diamictons, muds, calcrete and palaeosols, and subdivisions of these general types using sedimentary structures and composition. Markov-chain analyses help identify non-random facies associations that provide support for interpreting processes related to different parts of the alluvial systems. The facies characteristics, stratigraphic development and geographical distribution are further valuable for relating the sedimentological processes to known and plausible influences upon sedimentation. The main emphasis of this study is upon one alluvial sequence that is widely represented. It underlies a terrace that is correlated with the broad coastal plain in southeast Cyprus, which contains raised marine deposits of Tyrrhenian-I age (c. 200 ka) and younger. The sequence is characterised by bounding erosional surfaces, palaeosol and calcrete development, and upward trends from axial-stream to alluvial fan facies. This sequence possibly reflects one or two glacial cycles, but has probably been considerably influenced by tectonic and intrinsic factors as well.

KEYWORDS: Fluvial, Cyprus, Quaternary, facies sequence, stratigraphy.

1 INTRODUCTION

The known complexity of external influences during Quaternary sedimentation provides sobering perspective for interpreting ancient alluvial sedimentation in terms of one or even a few isolated influences. Schumm (1977) used the term 'complex response' to describe stream behaviour that did not correlate with known influences. Leopold and Langbein (1963) conclude that although the range of uncertainty will decrease, it will never be entirely removed when numerous interacting factors allow physical laws to be fulfilled by a variety of combinations of these factors. The greater time resolution and sequence preservation common within Quaternary deposits are, in combination with stratigraphical and sedimentological detail, advantageous for the stochastic modelling that may be the only realistic solution for most geological systems. Constructing local facies models, including as many architectural elements as possible with genetic significance, will systematically improve the basis for wider environmental interpretations (cf. Walker 1990).

The separation of allo- and autocyclic trends is an old problem, perhaps inherited from the Huttonian-Lyellian combination of cyclic and directional changes within geological uniformitarianism. Considering the on-going and century-old debates about fan entrenchment, river-terrace formation and landscape evolution in general (e.g. Vita-Finzi 1969; Lewin et al. 1991), these problems of randomness and causality, catastrophism and gradualism,

and periodocity and evolution are well exemplified within studies of fluvial systems. Furthermore, interpretations are often limited by a preferential emphasis upon either rational or empirical methodologies (Baker 1988).

Recently, Miall (1991) and others have again stressed concerns about extending the eustatically oriented framework for sequence-stratigraphic interpretations into fluvial environments (e.g. Posamentier and Vail 1988). Base-level changes may be isolated by thresholds or considerably modified and delayed within the adjustments of a graded system. Additionally, the climatic, tectonic and anthropogenic factors have considerable impact and must become increasingly predominant in terrestrial environments away from the shoreline. A fundamental principle within sequence stratigraphy is that eustatic changes allow for fluvial aggradation, mainly with the change from rising to falling relative sea level (Posamentier and Vail 1988). But the erosion or aggradation induced by climatic changes during the same time are not negligible, and could complement or diminish the eustatic effects upon fluvial responses, both of which probably vary within different parts of the system (Schumm 1977).

The objective of this paper is to relate logged sequences to the processes within different parts of the Quaternary alluvial systems in southeast Cyprus. Three valleys have been investigated: the Tremithios, Pouzi and

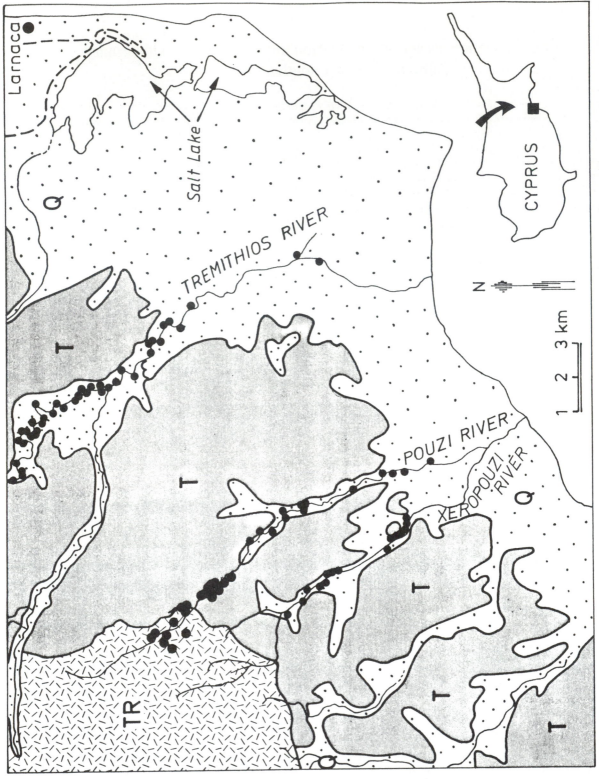

Figure 1. Geologic map (after Bagnall 1960) of the study area where Quaternary (Q) alluvial sequences and marine deposits occur along the coastal plain and the stream valleys draining the upland areas with crystalline mafic and ultra-mafic rocks of the Troodos Massif (TR) and chalk hills of the Upper Cretaceous to Tertiary

Xeropouzi Rivers (Fig. 1). Markov-chain analysis is used to identify facies associations that are greater than random. Randomness, although an essential aspect of most geological systems, often obscures those facies associations with relatively low frequency but possibly high significance for interpreting processes. Rather than providing a predictive basis where randomness and absolute frequencies are crucial, the non-random facies models resulting from Markov-chain analysis largely show genetic relationships. However, this is a complementary viewpoint when considering the known and plausible influences upon sedimentation and the general stratigraphic developments. Although a synthesis of environmental factors is necessarily limited in this present context, the complex response of natural systems may be better understood with increasingly defined facies relationships and distribution.

2 GEOLOGICAL AND GEOMORPHOLOGICAL SETTING

The emplacement tectonics of the Troodos ophiolitic sequence have received considerable attention (Wilson 1959; Robertson 1977; Varga and Moores 1985; Dilek and Eddy 1992). The island's emergence and unroofing since the early to mid-Quaternary (Robertson 1977) has exposed the mafic and ultra-mafic crystalline rocks and the circum-Troodos sedimentary sequence, both of which have provided sediments to the streams and alluvial fans that radiate from the Troodos Massif (Fig. 1).

One of the advantages of working with the sediments from contrasting bedrock sources is especially evident in this situation where the axial streams carry a mixture of mafic clasts, from the massif area to the northwest, and chalk fragments (including some chert), eroded from the early Tertiary limestone bedrock (Lefkara Group). In the areas where the Lefkara chalks are exposed, fans extend from the limestone hills toward the axial stream. Accordingly, the late Quaternary valley-fill deposits have resulted from the deposition of the axial-stream and alluvial fan systems which have had close contact within the valleys through time (Bagnall 1960; Pantazis 1967; Turner 1971; Gomez 1987; Poole and Robertson 1991). The fan material is composed only of chalk material since their drainage areas do not here include other bedrock types. The sediment composition can therefore indicate provenance and, through analogy with the modern setting, the environments of deposition. Overlying the Lefkara Group to the southeast are the softer marls and limestone of the Pakhna Formation, largely covered by the Quaternary sediments. A gently sloping, slightly hilly coastal plain replaces the upstream valley topography, in similarity with other parts of the southern coast.

River terraces have been recognized within the study area by Bagnall (1960) at approximately 5, 10 and 25 m above the present floodplain. Pantazis (1967) and Gomez (1987) investigated four terraces along river valleys to the west. Gomez also suggested a correlation between the four terraces which he identified along the Vasilikos

Valley and others elsewhere on Cyprus. Bagnall's 10 m terrace (25-35 feet) is the most extensively developed and occurs above the main sequence of valley-fill deposits. According to Bagnall this terrace may correspond to raised-beach deposits at 12 m (40 feet) above sea-level near the coast in the study area. The gravel beds and shelly sandstone of these and similar raised-beach deposits (8-11 m) along the southern coast of Cyprus contain the gastropod *Strombus bubonius* and coral *Cladocora caespitosa*, which have been uranium-dated by Poole et al. (1990) to c. 200 ka (Tyrrhenian I and inferred oxygen isotope stage 7). In exposures 3-4 km south of Larnaca (Fig. 1), reef deposits of the shelly sandstone are in the lower portion and the datings probably relate to this stratigraphic position. The correlation with these marine deposits, although tentative, is consistent with our field observation that the 10 m fluvial terrace is more-or-less continuous with the broad coastal plain southeast of the limestone hills and that it contains the raised beach deposits. The extent of the terrace and the coastal plain implies a major aggradational development, and this will be the main sequence for discussion below.

3 FACIES DESCRIPTIONS

The sedimentary facies, summarised in Table 1 and illustrated in Figure 2, were documented in the field mainly from natural exposures. The facies codes have a basic similarity with those used by Eyles et al. (1983), but are adapted for the variations and purposes of this study. In addition to the textural and structural characteristics, the lithological composition is noted in terms of two generalised components. The mafic component (M) includes clasts from the basalt (presumably from pillow lava), diabase, gabbro and ultra-mafic rock fragments of the Troodos Massif. The limestone clasts (L) are chalks from the Lefkara Group. The younger Pakhna Formation is occasionally exposed toward the coast, but these marls are too soft to be well preserved in the sand and gravel fractions.

The gravels containing both mafic and limestone components (facies sG, sGx and sG↑) are generally sandy, moderately sorted and show only poor imbrication. The limestone gravels (Gi) tend to be bimodal in that the silty, clayey matrix does not have a sandy transition to the pebble and cobble clast framework. This facies essentially always shows imbrication of the flagstone clasts, and calcareous cements are common. The matrix-supported diamicton facies (D) varies in its clast content and can be, in places, transitional to either the clast-supported gravel or gravelly mud.

The distinction between the two fine-grained facies (F and gF) would not always be evident if not for the pronounced colour contrasts imparted by the two contrasting lithological components. The overall brownish colour of the F-facies is interrupted by lighter layers with calcareous cements and grain coatings. In some instances, the layering appears rhythmic, with light-coloured portions at intervals c. 10 cm apart or less. There

Figure 2. (A) The Tremithios River showing terraces and typical exposures of alluvial sequences, an alluvial fan surface and the present day ephemeral stream channel. (B) Modern river-channel sediments similar to those of the sandy gravel (fluvial) facies, with dark, well rounded mafic clasts and white, sub-angular and somewhat larger chalk. The hammer head is 17 cm long. (C) Facies sG and sGx (in the centre), in the lower portion of the exposure in A. (D) Facies F, showing palaeosol development and calcareous banding (the vertical streaks are from surface wash-down). (E) Facies sG, D and Gi (upwards), with typically distinct contrasts in composition and sedimentary structure. Some reworking of the diamicton is evident. (F) Exposed sequence (c. 8 m) underlying the 10 m terrace and reflecting changes from fan to stream (darker) to fan deposition.

Table 1. Facies descriptions and lithologic codes.

Facies Code	Lithology	Description
sG	M + L	Sandy, clast-supported gravel. Generally with poor to moderate sorting and stratification and imbrication. Pebble and cobble sizes predominate, although boulder gravels also occur.
sGx	M + L	Sandy, clast-supported gravel with cross-bedding. Cross-bed sets are mainly planar-tabular and up to 50 cm thick. Pebble gravels and pebbly sands are most common. Moderately well sorted.
sG↑	M + L	Sandy, clast-supported gravel with fining-upward trend that often concludes with pebbly sand. Moderate to good sorting.
F	M + L	Fine-grained deposits, mainly sandy silt. Textural, colour and carbonate variations are often evident and form sub-horizontal laminations and beds. Poor sorting. Some lenses of gravelly sand. Calcareous.
gF	L (+M)	Fine-grained deposits, mainly clayey silt. Isolated chalk pebbles occur. Minor mafic components are occasionally observed within the silt and sand fraction. Strongly calcareous.
D	L	Diamicton. Generally a pebbly, cobble- and boulder-rich mud, but matrix-supported conglomerates are also common. Minor sorting and textural trends. No internal stratification or imbrication. Lateral continuity poor.
Gi	L	Clast-supported gravel with well developed imbrication. Moderately poor sorting of, predominantly, cobbles and pebbles, and with notably limited amounts of sand. Interparticular clay filling in many deposits.
K	L ± M	Calcrete. Locally termed kafkala and havara, depending upon the hardness of the calcareous cement. Occurs mainly within fine-grained lithofacies.
P	L ± M	Palaeosol (and recent soils where strongly developed). Both brown and reddish brown colours are developed. Observed only within the fine-grained lithofacies.

*M = mafic sources; L = limestone.

are some sub-vertical and bifurcating root-like holes and these are in places in-filled or associated with local carbonate precipitation. The gravelly lime muds (gF) only rarely show good clear stratification, and the light grey colour has little variation. Root holes were very rare. The isolated limestone clasts (< 10%) are usually of pebble size, but larger clasts also occur. Microscopic examination revealed a minor content of mafic sand and silt grains in some cases.

Calcrete (or caliche) occurs within both the gravelly and fine-grained facies. 'Havara' and 'kafkala' are Cypriot terms for carbonate horizons with differing hardness: approximately equivalent to the respective carbonate morphology stages 3 (extensive carbonate cement but soft) and 4 (plugged and hardened with laminar deposits on the surface) of Gile et al. (1966). The general calcrete facies (K) was often observed underlying a well developed soil layer (facies P), but has been kept separate here since some occurrences may be related to the flow and evaporation of groundwater rather than soil water. The interpretative 'soil' and 'palaeosol' terms are included with the descriptive facies as useful summaries of the iron staining and sediment structure typical of B horizons. Although the calcrete and soil layers are most commonly within two metres of terrace surfaces, some more deeply buried examples were also observed. We noted that stage 4 calcrete layers occurred close to exposed, natural cuttings but carbonate morphology was less developed in recent pits several metres away from these exposures. Better carbonate development was tentatively associated with layers on western exposures.

4 FACIES SEQUENCES

Markov-chain analysis is useful for identifying that portion of sequence trends not explicable by the random tendencies related to the frequencies of each facies (Selley 1970; Miall 1973; Cant and Walker 1976; Davis 1986). The greater than random relationships are presumed to be due to environmental influences, provided that facies definitions and field observations are consistent and non-preclusive. Application of this statistical procedure requires considerable caution since random associations can very well be those with the highest frequency and the most important in predictive modelling of the overall trends. However, geological processes are generally stochastic, containing both random and deterministic aspects. The preferential relationships between facies may be genetically important but obscured by the abundance of random components.

Table 2 gives the results of a Markov-chain analysis for the combined set of 442 facies transitions observed in 122 logged sections (Fig. 1). The Markov-chain procedure entails using the tally matrix (Table 2A) to calculate the random probabilities (Table 2C) which are then subtracted from the observed frequencies (Table 2B). The resulting differences (Table 2D), both negative and positive, have interpretative value if the nature of tested hypotheses is appropriate to the facies divisions and varieties chosen. The statistical reliability of the database is tested using a Chi squared test (lambda value) and the significance level of individual transition trends (Table 2E) should also be calculated from their binomial probabilities (Harper 1984). The details of our procedure are given in Stevens (1990). Holm et al. (1986) consider the special problems and mathematical treatment of embedded Markov chains (where only transitions between different facies are considered), as in this study.

All observed transitions from Table 2A are also shown diagrammatically in Figure 3. There are transitions between most facies, but by far the most frequent are between the abundant sandy gravel (sG) and mud (F) facies, 179 together. High absolute frequencies are also

Table 2. Results of Markov-chain analysis for all transitions.

	sG	sGx	F	sG↑	D	gF	Gi	K	P
A. Transition-count matrix									
sG	–	21	107	5	13	9	3	4	1
sGx	17	–	5	1	0	1	1	0	0
F	72	1	–	4	9	10	3	8	4
sG↑	6	1	5	–	0	0	0	0	0
D	6	0	1	0	–	3	21	1	0
gF	9	0	1	0	6	–	16	3	1
Gi	2	0	0	0	24	15	–	0	1
K	2	0	1	0	0	2	0	–	4
P	2	0	6	0	1	1	2	0	–
B. Observed transition probabilities									
sG	–	0.13	0.66	0.03	0.08	0.06	0.02	0.04	0.01
sGx	0.68	–	0.20	0.04	0.00	0.04	0.04	0.00	0.00
F	0.65	0.01	–	0.04	0.08	0.09	0.03	0.07	0.04
sG↑	0.50	0.08	0.42	↑	0.00	0.00	0.00	0.00	0.00
D	0.19	0.00	0.03	0.00	–	0.09	0.66	0.03	0.00
gF	0.25	0.00	0.03	0.00	0.17	–	0.44	0.08	0.03
Gi	0.05	0.00	0.00	0.00	0.57	0.36	–	0.00	0.02
K	0.22	0.00	0.11	0.00	0.00	0.22	0.00	–	0.44
P	0.17	0.00	0.50	0.00	0.08	0.08	0.17	0.00	–
C. Transition-probability estimates with partial independence									
sG	–	0.06	0.45	0.03	0.15	0.11	0.13	0.04	0.03
sGx	0.36	–	0.31	0.02	0.10	0.08	0.09	0.03	0.02
F	0.49	0.06	–	0.03	0.14	0.11	0.12	0.04	0.03
sG↑	0.35	0.04	0.30	–	0.10	0.08	0.09	0.03	0.02
D	0.38	0.05	0.33	0.02	–	0.08	0.09	0.03	0.02
gF	0.37	0.04	0.32	0.02	0.10	–	0.09	0.03	0.02
Gi	0.38	0.05	0.32	0.02	0.11	0.08	–	0.03	0.02
K	0.35	0.04	0.30	0.02	0.10	0.08	0.09	–	0.02
P	0.35	0.04	0.30	0.02	0.10	0.08	0.09	0.03	–
D. Difference matrix (B-C)									
sG	–	0.07	0.21	0.00	−0.07	−0.06	−0.11	−0.02	−0.02
sGx	0.32	–	−0.11	0.02	−0.10	−0.04	−0.05	−0.03	−0.02
F	0.16	−0.05	–	0.01	−0.06	−0.02	−0.09	0.03	0.01
sG↑	0.15	0.04	0.12	–	−0.10	−0.08	−0.09	−0.03	−0.02
D	−0.19	−0.05	−0.30	−0.02	–	0.01	0.56	0.00	−0.02
gF	−0.12	−0.04	−0.29	−0.02	0.06	–	0.35	0.05	0.01
Gi	−0.33	−0.05	−0.32	−0.02	0.47	0.28	–	−0.03	0.00
K	−0.13	−0.04	−0.19	−0.02	−0.10	0.15	−0.09	–	0.43
P	−0.18	−0.04	0.20	−0.02	−0.02	0.01	0.08	−0.03	–
E. Transition level of significance (binomial)									
sG	–	1.00	1.00	0.56	–	–	–	–	–
sGx	0.99	–	–	0.63	–	–	–	–	–
F	1.00	–	–	0.70	–	–	–	0.93	0.64
sG↑	0.79	0.60	0.73	–	–	–	–	–	–
D	–	–	–	–	–	0.50	1.00	0.37	–
gF	–	–	–	–	0.83	–	1.00	0.91	0.47
Gi	–	–	–	–	1.00	1.00	–	–	0.41
K	–	–	–	–	–	0.85	–	–	1.00
P	–	–	0.88	–	–	0.39	0.72	–	–

*N = 442; Lambda $\doteq$ 389.08; Degrees of freedom: 55.

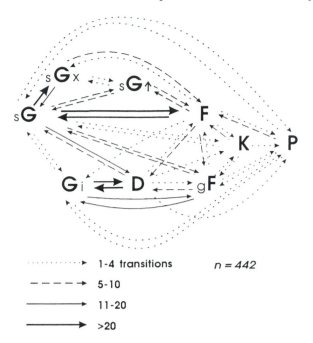

.................► 1-4 transitions   *n = 442*

- - - - → 5-10

⎯⎯⎯⎯► 11-20

⎯⎯⎯► >20

Figure 3. Flow diagram of all facies transitions (442). The fancies codes are from Table 1.

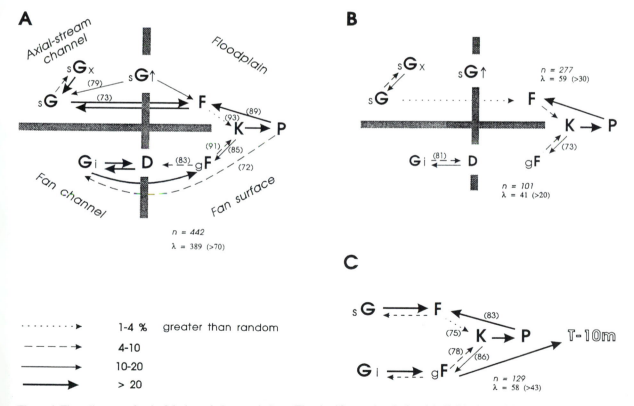

.................► 1-4 %   greater than random

- - - - → 4-10

⎯⎯⎯⎯► 10-20

⎯⎯⎯► > 20

Figure 4. Flow diagrams for the Markov-chain associations. The significance level of each individual transition is indicated if less than 90%. The lambda value is a test of the data set, and if greater than the Chi-squared value (in parentheses) the probability of obtaining these associations by chance is less than 5%. (A) Markov trends identified using the total data set. (B) Trends found by testing the facies groups interpreted to occur in fluvial and, separately, alluvial fan environments. (C) Facies associations from the sequences underlying the 10 m terrace.

observed between sG and the cross-bedded, sandy gravel (sGx) and between the imbricated gravel (Gi) and diamicton (D).

Markov-chain associations for the total transitions (Table 2D) are presented in Figure 4A, where the arrow character reflects the positive difference from randomly predicted transitions. Individual associations with <90% significance are indicated since these ideally require a larger statistical basis for improved confidence. Several of the high frequency transitions from Figure 3 remain above the random background (e.g. sG–F, sG–sGx, Gi–D). Low frequency transitions, such as K–P, show significant Markov association since these few transitions are very consistent. Although process interpretations could be suggested from these Markov associations, the analysis has involved all 9 facies, and presumably combines environmental systems that should be tested separately.

That no Markov trends occur between the upper and lower facies in Figure 4A (transitions occur but are less than predicted by chance) suggests that these facies belong to separate axial-stream and alluvial fan environments. Supported by the facies characteristics, the stream and fan environments can be further subdivided into channel and non-channel associations, as is illustrated in Figure 4B. In this diagram the Markov trends shown are the result of two separate and restricted analyses of the facies transitions interpreted, respectively, to be from axial-stream and fan environments (both including K and P facies transitions). The weak or absent Markov associations reflect the near random alternations that are apparently possible between facies within either the axial-stream or fan environments. For instance, the high frequency of the sG–F transitions (Fig. 3) is largely explained by randomness, since these facies are so abundantly represented in this fluvial environment. Testing of only the four axial-stream facies and, separately, only the three fan facies (excluding K and P) gave even fewer and weaker Markov associations than in Figure 4B. The channel origin of sG and sGx is supported by their Markov associations. The sGx facies, which has negative Markov associations (not shown) with the floodplain mud (F), is suggested to originate from the migration of channel bars with lee-face avalanching that produces large sets of planar cross-stratification. The graded gravel beds are randomly associated with all the other axial-stream facies and can develop both within the channel and on the floodplain, presumably with waning storm discharge in shallow channels. The alluvial fans are believed to have been largely characterised by gravity-flow processes, i.e. D and gF deposition and Gi by reworking of this material (cf. Blair and McPherson 1992).

Although calcrete (K) does develop with gravelly sediments, there is a Markov tendency, albeit weak, suggesting an environmental association with the fine sediments of both floodplain and alluvial fan environments. The strong trend for calcrete to be followed by buried or recent soils (P) is consistent with its commonly interpreted origin from soil leaching (Wright and Tucker 1991). Soil development improves away from the chan-

nel source of disturbing overbank sedimentation (Bown and Kraus 1987) and this is evident from the lack of positive Markov trends with the channel facies of the axial streams and alluvial fans. The less than random, but notable, occurrences of K in gravels may be related to the conductivity of these layers that might allow soil and groundwater to collect and concentrate carbonate, especially if transported laterally to free surfaces for easier evaporation. When not followed by P, there is a trend from K to gF that can be explained by the carbonate composition on the alluvial fan which would hinder soil development but still allow the downward translocation and carbonate cementation. The poor soil development on the fans is also an underlying reason for mainly floodplain fines to be observed above the buried soils, leading to a strong Markov trend in Figure 4B.

Figure 4C shows the Markov associations from sequences underlying the 10 m terrace, which is well developed in the study area and of special interest for the discussion below. The terrace surface is treated here as a facies even though there are some definitional inconsistencies. The axial-stream gravel facies and the muddy fan facies have been generalised and included in sG and gF due to their restricted number of occurrences. Strong autocyclic trends from channel gravels to finer sediments are consistent with expectant tendencies for aggrading fluvial systems in either axial-stream or alluvial fan settings. The fining-upward relationship in fan facies may also be related to the increased importance of colluvial, aeolian and surface flow processes with abandonment following lobe shifting or entrenching. Our observations of the overall stratigraphic developments, below, also support these Markov-chain results for the 10 m terrace sequences. The associations between the F, gF, K and P facies are the same as noted above. Floodplain and fan-lobe abandonment allows for the diminished sedimentation that would be favourable for calcrete, and for soil development on the floodplain where fluvial supply of non-carbonates occurs. A strong Markov trend from palaeosols to floodplain muds, but not to fan mud, is also logical given the noted compositional differences in soil material.

The 10 m terrace has a greater than random association with the gravelly muds and diamicton (included in gF here) from fan deposition. There are, of course, other facies that occur at the top of the terrace sequences, mainly the axial-stream sG and F facies, but these have a less than random occurrence with respect to their high frequencies. This suggests that there were environmental factors favouring fan facies in the upper portion of these sequences, despite the overall dominance of axial-stream facies in the regional environment. It is also possible to have continued sedimentation on the terrace surfaces after river down-cutting since the connection to valley-side sources for alluvial fan materials could allow fan deposition to persist and extend over the former floodplains.

5 STRATIGRAPHIC DEVELOPMENT AND INFLUENCES UPON SEDIMENTATION

The discussion below will primarily concern the sequences underlying the 10 m terrace that occurs throughout the three investigated valleys and is more-or-less continuous with the raised, broad coastal plain. Figure 5 gives a schematic summary of the most important observations and interpretations.

An angular unconformity separates, where observable, the 10 m terrace sequences from the Lefkara chalks. In the lowermost portions of these sequences the axial-stream gravels predominate and seem to reflect lateral migration of channels with good competence for transport of the coarse-grained sediment. Alluvial fan deposits were notably observed below the interval of axial-stream dominance at a few localities in upstream areas. The floodplain mud becomes better represented upward and, in turn, is stratigraphically succeeded in many sections by increasing amounts of alluvial fan facies. As suggested from the Markov-chain analysis, there is a certain Markov tendency for the 10 m terrace to be directly underlain by alluvial fan deposits, where these occur near limestone hills and adjacent areas towards the coast (cf. Fig. 2A and 2F). Floodplain sediment with soil and calcrete development (P and K) more commonly conclude this terrace sequence in the coastal plain area.

Although tectonic uplift of Cyprus provided the basic conditions for erosion, tectonic activity was seemingly greatest in the early to mid-Quaternary and decreased during the Late Pleistocene and Holocene (Robertson et al. 1991; Poole and Robertson 1991), during which time eustatic and climatic influences may have gained considerable importance. The only age control presently available is the tentative correlation of these deposits with the Tyrrhenian-I dating of raised marine sequences near the coast (Poole and Robertson 1991). This, combined with the number of environmental factors still poorly defined, makes the presentation of alternative hypotheses speculative, but nevertheless helpful for identifying remaining problems and avenues for investigation.

If the angular unconformity of the alluvial sequence corresponds with that separating the marine sequence from the underlying Athalassa Formation (Pliocene; Bagnall 1960; Poole and Robertson 1991), then the correlation of these two sequences is strengthened. In addition to the Tyrrhenian-I date for the marine deposits, water temperature of 18 degrees or higher that would be conducive to a thriving coral fauna (Butzer and Hansen 1968) suggest an interglacial age for at least part of this sequence. Although uncertain, increased interglacial precipitation in the Mediterranean areas (cf. Gates 1976) might explain a trend toward fluvial dominance in the lower alluvial stratigraphy. This, together with the eustatically high sea level would also allow for the succession of reef to shelly sandstone to uppermost beach gravels in the Larnaca area. A change to drier conditions perhaps caused a relative shift toward increased alluvial fan deposition, decreased floodplain sedimentation and soil and calcrete formation. The diminished supply of sandy fluvial sediments may also have aided the concentration of beach gravels near the top of the marine sequence. A glacially induced lowering of the sea level would then favour erosion and isolation of both the 10 m terrace and the raised beach deposit. With this scenario, the sequence between the two erosional episodes would represent one glacial cycle. In the upper valley cycles, the eustatic changes are considered to have been important for erosion, whereas climatic effects are most influential with respect to aggradation. Both factors would cause regionally similar trends, as is consistent with the terrace correlations both on Cyprus and, possibly, elsewhere.

If the 10 m terrace sequence represents Tyrrhenian-I deposition, a corresponding sequence of aggradation for

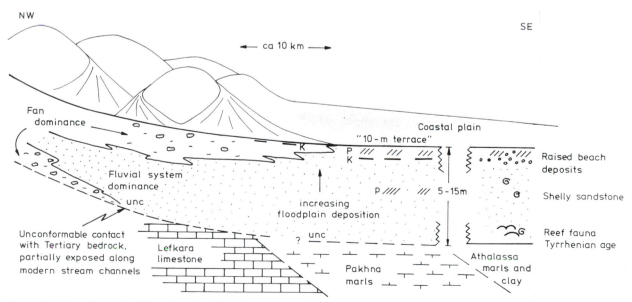

Figure 5. Schematic section illustrating the alluvial sequences underlying the 10 m terrace. P = palaeosol/soil, K = calcrete.

the Tyrrhenian-II period (c. 100 ka, inferred isotope stage 5e) appears to be absent since the lower terrace in the study area is related to much younger, perhaps Neolithic age (c. 5 ka BP) deposition, as indicated by pottery and fire-pit datings (Gillford 1978; Gomez 1987; Wedel and Stevens, unpubl. documentation). An alternative hypothesis, unresolved because of the lack of stratigraphic control, is that the alluvial (and marine) sequences could represent two glacial cycles. The observation of palaeosols within the sequence (Fig. 5) at several localities may reflect a period of diminished overbank sedimentation. This could have occurred with limited down-cutting and/or climatic shifts during an earlier glacial phase than that interpreted to have caused down-cutting and isolation of the 10 m terrace.

Low eustatic sea levels are interpreted to have had a major influence upon fluvial erosion, creating the lower unconformity and the terrace isolation. However, this effect may have been delayed or strongly diminished in the upstream areas where alluvial fan deposits commonly underlie the fluvial (axial-stream) portions of the 10 m terrace sequence (Figs 2A and 5). If down-cutting was restricted upstream, these may be glacial-period fan deposits from a previous cycle, the equivalent of the fan facies concluding the 10 m sequence and possibly reflecting transition to drier glacial climates. Aggradation, on the other hand, has been interpreted with both rising and falling relative sea levels (cf. Posamentier and Vail 1988; Miall 1991). However, climatic conditions, presumably an increase in runoff and sediment yield during certain time periods (early interglacial?), were perhaps most important considering the regional and seemingly concurrent changes (cf. Butzer and Hansen 1968; Schumm 1977).

Finally, the autocyclic mechanisms that can be expected within axial-stream and fan environments may have been crucial to the deposition and preservation of these alluvial sequences. The transitions from axial-stream to alluvial fan predominance in the deposits along the upper stream reaches might merely reflect the change expected as the channel migration shifts the concentration of erosional processes and fluvial deposition. The fine-grained, high carbonate content of the fan deposits may have made them relatively resistant to erosion. Confined stream flow with increased energy would be able to down-cut and possibly produce sequence cycles difficult to distinguish from those influenced by allocyclic factors. A similar effect is plausible in the coastal plain areas where increasingly abundant floodplain sediments could also stabilise channels and increase erosion. Calcrete development would help resist erosion in both axial-stream and fan environments, especially on the eastern banks where best development occurs.

6 CONCLUSIONS

The study of facies sequences suggests that highly frequent transitions between the most common facies are, despite their importance for predictive modelling, not necessarily process related. The facies themselves reflect individual processes, but their transitions may largely be induced by the availability of grain sizes and the random repetition of certain depositional processes. In fact, facies-transition frequencies between the axial-stream and alluvial fan facies groups are nearly random. Some stronger Markov associations suggest genetically related processes within channels and on the floodplain and fan surfaces.

The facies sequences underlying the widely developed 10 m terrace indicate a certain influence favouring alluvial fan deposition prior to terrace formation. This trend was also observed in the general stratigraphy of the terrace exposures, where interpreted axial-stream gravels are often followed by increasing floodplain and then fan deposits. The terrace sequence, including the underlying disconformity, the upper calcrete and soil layers, and the isolated terrace surface itself, are perhaps related to the probable changes in eustatic and climatic influences that were related to glacial cycles, specifically the Tyrrhenian-I to Tyrrhenian-II, and even possibly during the latest glaciation. Autocyclic influences within the alluvial environments should also be expected.

7 ACKNOWLEDGEMENTS

We are indebted to the students, spread out over several years of field courses, who have participated in the sedimentologic documentation. The Swedish Natural Science Research Council has funded our work. The manuscript has been improved by suggestions from Mark Johnson and the anonymous reviewers.

REFERENCES

Bagnall, P.S. 1960. *The Geology and Mineral Resources of the Pano Lefkara-Larnaca Area.* Geological Survey Department, Cyprus, Memoir 5.

Baker, V.R. 1988. Geological fluvial geomorphology. *Geol. Soc. Amer. Bull.* 100: 1157-1167.

Blair, T.C. & J.G. McPherson 1992. The Trollheim alluvial fan and facies model revisited. *Geol. Soc. Amer. Bull.* 104: 762-769.

Bown, T.M. & M.J. Kraus 1987. Integration of channel and floodplain suites, I. Developmental sequence and lateral relations of alluvial palaeosols. *J. Sed. Petrology* 57: 587-601.

Butzer, K.W. & C.L. Hansen 1968. *Desert and River in Nubia. Geomorphology and Prehistoric Environments at the Aswan Reservoir.* Madison: Univ. Wisc. Press.

Cant, D.J. & R.G. Walker 1976. Development of a braided-fluvial facies model for the Devonian Battery Point Sandstone, Quebec. *Canadian J. Earth Sci.* 13: 102-119.

Davis, J.C. 1986. *Statistics and Data Analysis in Geology.* New York: Wiley.

Dilek, Y. & C.A. Eddy 1992. The Troodos (Cyprus) and Kizildag (S. Turkey) ophiolites as structural models for slow-spreading ridge segments. *J. Geology* 100: 305-322.

Eyles, N., C.H. Eyles & A.D. Miall 1983. Lithofacies types and vertical profile models: an alternative approach to the description and environmental interpretation of glacial diamict and diamictite sequences. *Sedimentology* 30: 393-410.

Gates, W.L. 1976. Modeling the ice-age climate. *Science* 191: 1131-1144.

Gifford, J.A. 1978. Palaeogeography of archaeological sites of the Larnaca lowlands, south eastern Cyprus. Univ. Minnesota, PhD. Thesis.

Gile, L.H., F.F. Peterson & R.B. Grossman 1966. Morphological and genetic sequences of carbonate accumulation in desert soils. *Soil Science* 101: 347-360.

Gomez, B. 1987. The alluvial terraces and fills of the Lower Vasilikos Valley, in the vicinity of Kalavasos, Cyprus. *Trans. Inst. Br. Geographers* 12: 345-359.

Harper, C.W. 1984. Improved methods of facies sequence analysis. In R. Walker (ed.), *Facies Models*, 11-13. Geoscience Canada Reprint series 1.

Holm, S., I. Isaksson & R.L. Stevens 1986. A test of independence for stratigraphic sequences with respect to embedded Markov chains. *Math. Geology* 18: 551-561.

Leopold, L.B. & W.B. Langbein 1963. Association and indeterminancy in geomorphology. In C.C. Albritton (ed.), *The Fabric of Geology*:184-192. London: Addison-Wesley.

Lewin, J., M.G. Macklin & J.C. Woodward 1991. Late Quaternary fluvial sedimentation in the Voidomatis basin, Epirus, northwest Greece. *Quaternary Research* 35: 103-115.

Miall, A.D. 1973. Markov chain analysis applied to an ancient alluvial plain succession. *Sedimentology* 20: 347-364.

Miall, A.D. 1991. Stratigraphic sequences and their chronostratigraphic correlation. *J. Sed. Petrology* 61: 497-505.

Pantazis, Th.M. 1967. *The Geology and Mineral Resources of the Pharmakas-Kalavasos Area*. Geol. Survey Dept., Cyprus. Memoir 8.

Poole, A.J. & A.H.F. Robertson 1991. Quaternary uplift and sea-level change at an active plate boundary, Cyprus. *J. Geol. Soc. London* 148: 909-921.

Poole, A.J., G.B. Shimmield & A.H.F. Robertson 1990. Late Quaternary uplift of the Troodos ophiolite, Cyprus: Uranium series dating of Pleistocene coral. *Geology* 18: 894-897.

Posamentier, H.W. & P.R. Vail 1988. Eustatic controls on clastic deposition II – sequence and systems tracts models. In Wilgus, C.K., B.S. Hastings, C.G.St.C. Kendall, H.W. Posamentier, C.A. Ross & J.C. Van Wagoner (eds.), *An intergrated Approach*. Soc. Econ. Mineral. Paleontol. Sp. Publ. 42: 125-154.

Robertson, A.H.F. 1977. Tertiary uplift of the Troodos Massif, Cyprus. *Geol. Soc. Amer. Bull*. 88: 1763-1772.

Robertson, A.H.F., E. Eaton, E.J. Follows & J.E. McCallum 1991. *The role of local tectonics versus global sea-level change in the Neogene evolution of the Cyprus active margin*. Spec. Publ. Intern. Assoc. Sedimentologists 12: 331-369.

Schumm, S.A. 1977. *The Fluvial System*. New York: Wiley.

Selley, R.C. 1970. Studies of sequence in sediments using a simple mathematical device. *Q. J.Geol. Soc. London* 125: 557-581.

Stevens, R.L. 1990. Markov-chain analysis as a pedagogic tool. *J. Geol. Education* 38: 288-293.

Turner, W.M. 1971. Quaternary sea levels of western Cyprus. *Quaternaria* 15: 197-202.

Varga, R.J. & E.M. Moores 1985. Spreading structure of the Troodos ophiolite, Cyprus. *Geology* 13: 846-850.

Vita-Finzi, C. 1969. *The Mediterranean Valleys: Geological Changes in Historical Times*. Cambridge: Cambridge Univ. Press.

Walker, R.G. 1990. Facies modeling and sequence stratigraphy. *J. Sed. Petrology* 60: 777-786.

Wilson, R.A.M. 1959. *The Geology of the Xeros-Troodos Area*. Geological Survey of Cyprus Memoir 1.

Wright, V.P. & M.E. Tucker 1991. *Calcretes*. Oxford: Blackwell.

CHAPTER 21

Quaternary wadi and floodplain sequences of Tripolitania, northwest Libya: A synthesis of their stratigraphic relationships, and their implications in landscape evolution

J.M. ANKETELL
Department of Geology, University of Manchester, UK

S.M. GHELLALI
Faculty of Earth Science, Al Fateh University, Tripoli, Libya

D.D. GILBERTSON
Institute of Earth Studies, UCW Aberystwyth, Aberystwyth, Dyfed, UK

C.O. HUNT
Department of Geographical and Environmental Sciences, University of Huddersfield, UK

ABSTRACT: This paper demonstrates the multiphase history of Tripolitanian valley systems and argues for their initiation in or before the mid Tertiary. Correlation of erosion surfaces, palaeosols and sediment bodies in Quaternary deposits of Tripolitania shows that they can be traced throughout the region. Major correlatable events allow subdivision of lithologically monotonous sequences into discrete units and allow interpretation of the deposits in terms of an interplay of erosion, deposition and landscape evolution. In the coastal region a subdivision of the fluvio-aeolian deposits of the Jifarah Formation into six aggradational units Q1-Q6 forms the basis of the sequence stratigraphic framework. The units can be correlated with wadi gravel fill and alluvial fan gravel sequences of the Jabal Nafusah, and with gravel and sand fills in major wadis of the Beni Walid region to the south of the Jabal which display similar relationships, albeit in coarser terrigenous lithofacies. A seventh aggradational unit between Units Q4 and Q5 is found in upland valley areas. A major phase of downcutting and terrace formation occurred in the early Holocene beginning with the erosional surface which defines the base of unit Q5. Modern erosion of Q5 wadi fill produces a characteristic upper and lower terrace morphology. The age of the older units is as yet in doubt but there is evidence to assign unit Q1 to the Early Pleistocene, unit Q2 to the late Middle Pleistocene, and units Q3 and Q4 to the Late Pleistocene. Radiometric dates on valley basalts indicate, however, that valley formation commenced in the Early Eocene following uplift of the region in the Late Palaeocene. The valleys continued to form conduits for lava flows in the Oligocene, Miocene, Pliocene and Pleistocene. However, no terrestrial sediments older than Late Pliocene-Early Pleistocene have as yet been identified. It is likely that older valley deposits were flushed out during each rejuvenation event.

KEYWORDS: Libya, Tripolitania, wadi, floodplain, stratigraphy, erosion surfaces, palaeosols, lava flows, terrace morphology.

1 INTRODUCTION

Detailed studies of Quaternary deposits in northwest Libya have been concentrated in two regions, the Jifarah Plain, including the foothills of the Jabal Nafusah (Vita Finzi 1971; Ghellali 1977; Anketell 1989; Anketell and Ghellali 1983; 1988; 1991a; 1992a, 1992b, 1992c), and the major valleys of the Beni Walid region on the gently dipping south-facing slope of the Jabal (UNESCO Libyan Valleys Project – Barker and Jones 1981, 1982; Barker et al. 1983; Gilbertson et al. 1987; Gilbertson and Hunt 1988; Hunt et al. 1986) (Fig. 1). General accounts of the Quaternary sequences in Tripolitania and Sirt are recorded in regional geological surveys by the Industrial Research Centre, notably the Nalut and Mizdah regions to the west of the Beni Walid area (Antonovic 1977; Mijalkovic 1977a, 1977b; Novovic 1977a, 1977b) and the Al Qaddahiyah area of the Sirt coastal region to the east (Zivanovic 1977).

Stratigraphic sequences recorded in these studies are summarised in Table 1. Correlation is based on field comparison of lithological associations in the major aggradational units, correlation of the erosional and soil forming events which bound them, together with limited archaeological and palynological data.

Early field surveys (Anketell and Ghellali 1983; Barker and Jones 1981, 1982; Barker et al. 1983) noted the extent to which it was possible to map in many wadis a 'classic sequence' of coarse gravel infill, erosion and incision, and shallow infill with finer clastic materials – recalling the Older Fill/Incision/Younger Fill sequences recorded by Vita Finzi (1969). Subsequent examinations have tended to confirm the frequency with which the topographic pattern associated with this sequence recurs. These later investigations have, however, also resulted in more detailed analysis of the depositional environments of the sediments involved (Anketell and Ghellali 1992b) together with the relationships that exist between the deposits, regional tectonic evolution, volcanicity, large scale and long term landscape development, and Quaternary climatic fluctuations as well as the activities of people (Anketell 1989; Gilbertson et al. 1984; 1987;

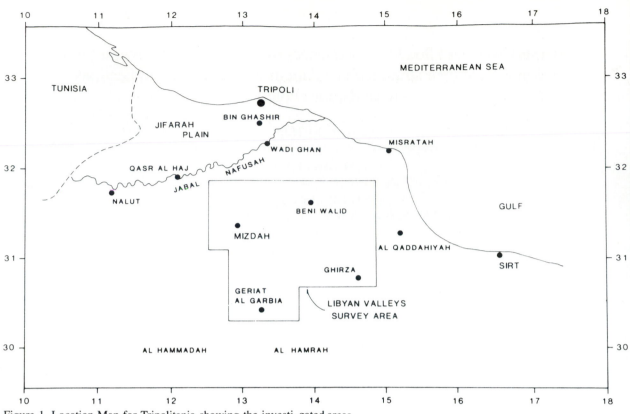

Figure 1. Location Map for Tripolitania showing the investi-gated areas.

Table 1. Proposed correlation of Late Tertiary-Quaternary deposits in northern Libya.

	A. Jifarah Plain	B. Qasr Al Haj	C. Wadi Ghan	D. Nalut/Mizdah	E. Beni Walid.	F. E. Sirt	G. Cyrenaica
Q6	Modern wadi gravels aeolian dune sands on interfluves	Modern wadi fill/	Modern wadi fill/	Modern wadi fill/ active dunes	Modern wadi fill/ active dunes (S17b/S18) Colluvium (S19)	Modern wadi fill active dunes	Modern wadi fill
E6					---------------- Erosion ----------------		
Q5	Lower Terrace **Lebda Alluvium** Historic gravels/sands Reddened fixed dunes on interfluves	Gravel terrace	Gravel and sand	Fluvio-aeolian wadi fill. Older aeolian sheet sands	Soil (S14). Romano Libyan wadi fill (S17a) Older climbing dunes (S16)	Fluvio-aeolian sands	**Bel Ghadir Alluvium**
					Colluvial silts of Gasr Banat Lacustrine deposits of Grerat D'nar Salem (S13)		Upper Palaeolithic scree
E5			---------- Erosion ----------				
	Thin impersistent calcrete <<<<<<<<<<<<<<<<<<<<<<<<<<<<<<<<<<<<<<<<<<				Soil (S11)		
Q4	Upper Terrace columnar 'loessic' silts	Gravel and columnar silts Terrace	Gravel and columnar silts	'Old Wadi Terrace' Weakly cemented gravels and sands	'Upper Cobbly Fill' weakly cemented gravels (S8). Colluvium (S9)	Fluvio-aeolian sands/silts	**Kuf Alluvium** ('Younger gravel')
E4			---------Erosion---------				
	Very thin calcrete <<<						<<<<<<<cementation<<<<<<
Q3	Grey lacustrine silts/ white aeolian sands	--------------- non deposition and/or erosion ---------------					**Ain Mara Tufa** and and 'Older Gravels
E3			---------- Erosion ----------				
	Upper Calcrete <<<<<<<< <Laminar calcrete <<<<<<<<<<Thin laminar<<<< <<<<<<<Calcite-filled joint <<<< <<Thick calcrete (S7b)<<<<<<<<<<<Calcrete<<<<<<<<<<<<<<<<Calcrete <<<<<<<<						
	laminar		**Basalt flow.**	in basalt flow top **Basalt flow**			
Q2	Red silt with calcrete nodules	Conglomerate well cemented conglomerates	Sandy conglom. Locally comprised of calcrete pebbles	'Old Wadi Terrace' well cemented conglomerates	'Lower Cobbly Fill' (S7) Well cemented conglomerates	Red silts with *Helix*/ Old Wadi Terraces'	Beach gravels
E2			---------- Erosion ----------				
	Lower Calcrete <<<<<<<<Thick calcreted cong.<<<<<<<<Calcite-filled joints<<<<<<<<<<Calcrete<<<<<<<<<<<<< ??????????????????? <<<<< Thick calcrete<<<<<<<<<<<<Thick calcrete<<<<<<						
			in basalt flow top **Basalt flow**				
Q1	Heavily calcreted red silts and breccia	Strongly cemented conglomerate/ cong. mudstone	Strongly cemented conglomerate/ cong, mudstone	Calcreted breccias and conglomerates	Lower Cobbly Fill?	'Old Wadi Terrace'' Strongly cemented conglomerates	
E1			---------- Erosion ----------				
	Al Assah Fm.		**Basalt flow**		**Basalt flow (S3)** -------- Erosion (S2) ---------- Karst/cave collapse (S1)	Al Hishah Fm.	
E0			---------- Erosion ----------				
	Mioc./Cret.	Cret./Jur	Cret./Trias.	Cret.	Cret.	Mioc.	Tert./Cret

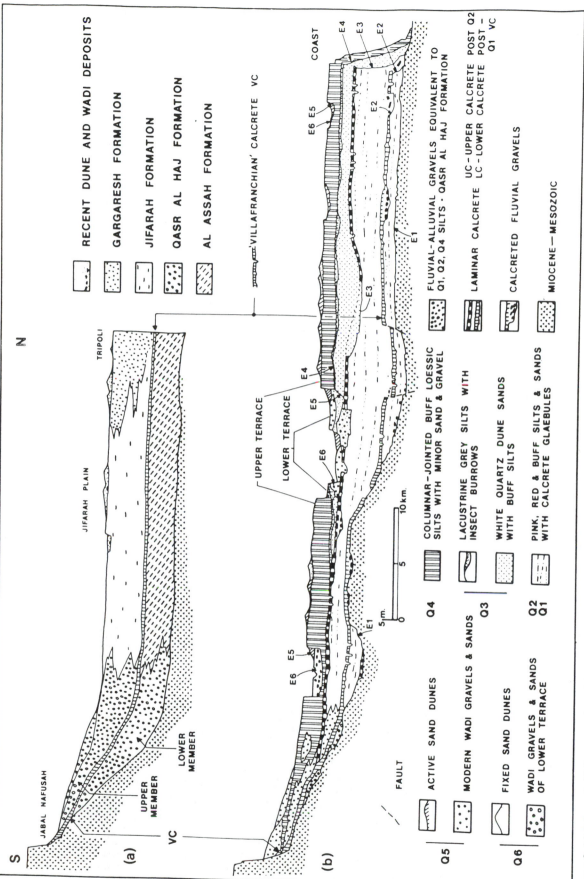

Figure 2. (a) Schematic N-S cross section of Quaternary deposits in the Jifarah Plain showing relationships between formations. (b) Schematic N-S cross section of the Eastern Jifarah Plain showing subdivisions of the Jifarah Formation (Q1-Q6), regional erosion surfaces (E1–E6) and major palaeosols.

Gilbertson and Hunt 1988; Hunt et al. 1985; 1986; 1987). Work has proceeded to a point where it is possible to set out and elaborate a multi-aspect geomorphological history of the Libyan Valleys in the Beni Walid region and to assemble a lithostratigraphic sequence which can be correlated with the better known sequences of the Jabal Nafusah and Jifarah Plain (Anketell 1989; Anketell and Ghellali 1983, 1991a, 1992b).

2 QUATERNARY DEPOSITS OF THE JIFARAH PLAIN

Quaternary sediments have been more intensively studied in the Jifarah Plain-Jabal Nafusah region of Tripolitania than elsewhere in Libya (Vinassa de Regny 1901, 1902; Crema et al. 1913; Stella 1914; Parona 1915; Desio 1940, 1970, 1971; Desio et al. 1963; Lipparini 1940; Jones 1969; El Hinnawy and Cheshitev 1975; Ghellali 1977; Anketell and Ghellali 1983, 1987, 1988, 1990) and these display a stratigraphic sequence which forms the basis for correlation with other areas of Tripolitania and elsewhere in Libya (Anketell 1989, 1991)

They are characterised by three major lithofacies associations: calcarenite aeolian dunes with marine gravels, terrigenous fluvio-aeolian silts and sands and alluvial-fan gravels. Evaporite sabkha deposits form a significant sub-facies of fluvio-aeolian and aeolianite sequences in western Jifarah coastal regions (Klen 1974; Ruhlich 1974). El Hinnawy and Cheshitev (1975) sub-divided the terrigenous fluvio-aeolian silts and sands and alluvial fan gravels. Evaporite sabkha deposits form a significant but cropping out locally in the western Jifarah Plain only, this formation comprises gypsum-rich sands and silts with occasional gravel lenticles near the top. It is locally capped by, or contains in its uppermost part, a thick calcrete which El Hinnawy and Cheshitev (1975) correlate with a 'Villafranchian' calcrete in southern Tunisia (Burollet 1960; Coque and Jauzein 1967; Banerjee 1980), thus placing the formation broadly in the Plio-Pleistocene.

2. Qasr Al Haj Formation: largely confined to the foothills and wadis of the Jabal Nafusah and comprising fanglomerates, gravel wadi fill and screes in two members. The lower member, capped by a thick calcrete is comprised of strongly-cemented conglomerates and conglomeratic mudstones whereas the upper is largely unconsolidated. El Hinnawy and Cheshitev (1975) correlate the calcrete with the 'Villafranchian' calcrete, the lower member with the Al Assah Formation and the upper member with the Jeffara Formation (Anketell and Ghellali 1992b) (Fig. 2a).

3. Jeffara Formation – renamed Jifarah Formation by Anketell and Ghellali (1992c): underlying the plain between the coast and Jabal Nafusah and comprised of weakly to well-cemented terrigenous silts and fine sands deposited and reworked by fluvial, colluvial and aeolian agencies

4. The Jifarah Formation passes northwards into the Gargaresh Formation (El Hinnawy and Cheshitev 1975):

aeolianite calcarenite dunes with interbedded lenticles of marine gravels and coquinas, occupying a coastal zone up to 4 km in width. The calcarenites are comprised of oolites and abraded marine shells, and form part of the sequence of regressional aeolianite deposits which fringe much of the Mediterranean coastline (Yaalon 1967; Butzer 1975; Hey 1978).

Stratigraphic relationships within and between the terrigenous formations have been very poorly documented due to their monotonous lithologies and lack of diagnostic fauna. Within the silts and gravels, however, erosional surfaces and palaeosol horizons define stratigraphic discontinuities which attest to a complex episodic history of degradation, landscape stability and aggradation (Anketell and Ghellali 1983, 1987, 1991a). Recognition of regional discontinuities enables subdivision of an otherwise lithologically monotonous sequence into discrete depositional and non-depositional phases and establishes a framework for correlation across major facies boundaries (Anketell 1989; Anketell and Ghellali 1991a).

2.1 *Jifarah Formation*

The sands and silts of the Jifarah Formation are well exposed in the eastern Jifarah Plain in coastal cliffs and in the banks of deeply incised wadis extending from the jabal to the sea. Vinassa de Regny (1901, 1902), Crema et al. (1913), Parona (1915), El Hinnawy and Cheshitev (1975), Mann (1975a, 1975b) and Smetana (1975), observed lithological variations, but accorded them little stratigraphic significance giving instead an impression of a monotonous sequence of red silts and fine sands with calcrete levels and abundant *Helix* sp., devoid of mappable units lower than formation in rank.

Recognition of erosional surfaces and palaeosols with regional expression allows subdivision of the formation into six members, designated Q1-Q6 (Fig. 3) (Anketell and Ghellali 1983, 1992b, 1992c). These aggradational units, although comparable in lithology and consisting of fine-grained fluvial, fluvio-aeolian, aeolian and colluvial deposits, are clearly defined and separated one from the other by the erosional surfaces. Termination of units Q1 and Q2 by major palaeosols provides an additional basis for definition and correlation (Fig. 3). Units Q1-4 are Pleistocene and Units Q5-6 are Holocene.

2.1.1 *Palaeosols*

Subdivision of lithologically monotonous, non-fossiliferous sequences using palaeosols is well documented (Allen 1977, 1986; Atkinson 1986; Kraus and Bown 1986).

Calcrete, first described from the area by Crema et al. (1913), and displaying a wide range of morphologies, is a significant feature of the Jifarah Formation. Stage IV profiles (Gile et al. 1966) are developed at two horizons. The older calcrete, capping Unit Q1, and reaching up to 1.2 m in thickness, is the 'Villafranchian' calcrete recorded by El Hinnawy and Cheshitev (1975). The

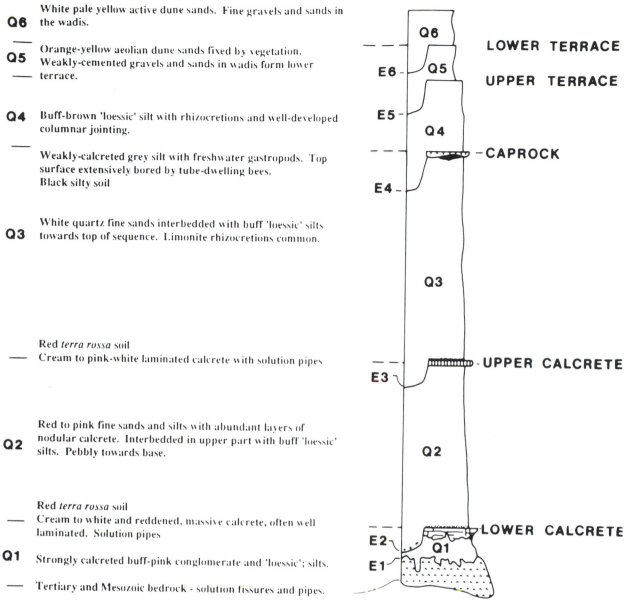

Q6 White pale yellow active dune sands. Fine gravels and sands in the wadis.

Q5 Orange-yellow aeolian dune sands fixed by vegetation. Weakly-cemented gravels and sands in wadis form lower terrace.

Q4 Buff-brown 'loessic' silt with rhizocretions and well-developed columnar jointing.

Weakly-calcreted grey silt with freshwater gastropods. Top surface extensively bored by tube-dwelling bees. Black silty soil

Q3 White quartz fine sands interbedded with buff 'loessic' silts towards top of sequence. Limonite rhizocretions common.

Red *terra rossa* soil
Cream to pink-white laminated calcrete with solution pipes

Q2 Red to pink fine sands and silts with abundant layers of nodular calcrete. Interbedded in upper part with buff 'loessic' silts. Pebbly towards base.

Red *terra rossa* soil
Cream to white and reddened, massive calcrete, often well laminated. Solution pipes

Q1 Strongly calcreted buff-pink conglomerate and 'loessic'; silts.

Tertiary and Mesozoic bedrock - solution fissures and pipes.

Figure 3. Composite section of the Jifarah Formation displaying typical lithofacies of units Q1-Q6 and positions of regional erosion surfaces and palaeosols.

younger, capping Unit Q2, seldom exceeds 40 cm. Both are highly indurated, crudely laminar, and are formed of close-packed, irregular glaebules capping honeycombed and more openly nodular levels (Netterburg 1967; Anketell and Ghellali 1984). The older locally displays pisolitic structure and micro-lamination at its top and along joint surfaces (Kornicker 1958; James 1972; Anketell and Ghellali 1983). Tepee structure (Price 1925; Reeves 1976) is widely developed.

Both calcretes, developed on a variety of substrates ranging from limestone bedrock to silts and gravels, can be traced throughout the region. Other than an increase in thickness on gravel deposits they display little lateral variation. Where intervening sediment pinches out, they amalgamate to form a layer up to 1.5 m in thickness

which locally blankets the bedrock (Anketell and Ghellali 1983). Laminar calcrete locally caps units Q3 and Q4, but is impersistent and rarely thicker than 3 cm (Fig. 3).

Nodular calcretes (stage II, Gile et al. 1966), predominate in units Q1, and Q2 and occur locally in Q4. They form horizons which can be followed for tens of metres and can, with care, be used to correlate between closely adjacent outcrops. Glaebules range from irregular to roughly equant. Some, oriented bedding-normal, are clearly rhizocretions.

Terra rossa is much less common than calcrete and only patchily preserved, much having been reworked into immediately overlying deposits. It is found in association with solution pipes and enlarged joints, and is particularly well developed in pipes in the calcrete formed prior to deposition of Unit Q3 where it is accompanied by a marked reddening of the sediments.

2.1.2 *Solution features*

Karst phenomena generated during surface or near-surface exposure occur in association with the two main calcrete horizons as well as on limestone bedrock. Two types are common:-

1. Solution pipes: These are generally cylindrical but occasionally ovate in cross-section and range from 0.4 to 0.7 m in diameter to 0.9 to 2.5 m in depth. Within the succession they may either be infilled with the immediately succeeding aggradational deposits or by *terra rossa* silts, commonly with well-developed root systems preserved as casts. The latter association argues for pipe formation by leaching of the calcrete beneath and around tree root systems (Jennings 1969; Perkins 1977). In wadis and on marine platforms floored by calcrete, pipes, either empty or filled with modern marine or fluvial sands and gravels (Anketell and Ghellali 1983), may have formed by solution potholing. However, their morphological similarity to, and close association with, rooted soil-filled types argues for the majority having a similar origin (Anketell and Ghellali 1992b). The pipes appear to be independent of joint systems.

2. Enlarged joints: the thick lower calcrete capping Q1 is locally invaded by sub-parallel fissures a few centimetres in depth which may have developed by solution of tectonic joints. Enlargement of tectonic joints in the Miocene and Mesozoic limestone has locally produced a complex network of fissures reaching depths in excess of 1 metre and now filled with sediment.

2.1.3 *Erosion*

Erosional surfaces in the Jifarah Formation are mainly fluvial in origin. Their geometry changes rapidly from near the escarpment where they are deep cut, irregular linear zones marking fluvial valleys along which all or part of older units have been removed, to sensibly planar surfaces in interfluve areas and downvalley. On these surfaces erosion is relatively minimal, commonly only involving the upper part of soil profiles so that successive units follow each other with apparent conformity. In the valleys complete removal of the palaeosols by the erosional phase defining the base of the succeeding unit commonly results in successive units displaying closely similar lithofacies across contacts. The erosion surfaces are usually sufficiently obvious to distinguish between members, particularly where modified by features formed by solution and/or bio-erosion processes.

Marine erosional surfaces (fossil cliffs, cliff notches and wave-cut platforms with associated beach deposits) are well developed in the eastern Jifarah. They are broadly coeval with the fluvial surfaces and allow relative dating of units within and between the Jifarah and Gargaresh Formations (Anketell and Ghellali 1987).

At several localities in the coastal region, the upper surface of the lowermost calcrete displays a *trypanites* ichnofauna dominated by *Gastrochaenolites*. The borings are filled with sediment and indicate that the horizon

has a long and complex history, ranging from soil formation to marine erosion and bio-excavation. Although marine borings are the most widely documented in sediment sequences (Warme and McHuron 1978), those of insects are common in a variety of terrestrial environments (Ratcliffe and Fagerstrom 1980). The surface of the lightly calcreted lacustrine silts capping unit Q3, followed in interfluve areas with apparent conformity by unit Q4, is intensely excavated by borings of tube-dwelling bees (Anketell and Ghellali 1984). This allows recognition of a substantial depositional hiatus between the two members.

2.2 *Lateral variation in the Jifarah Formation*

Variation in lateral relationships between members of the Jifarah Formation result from the interplay of episodes of aggradation, soil formation and erosion, together with thickness variation inherent to the mode of sedimentation. Typical relationships, based on sections in the eastern Jifarah (Anketell and Ghellali 1983) are summarised in Figure 2b.

Erosional surfaces and palaeosols, defining the six major aggradational phases, can be followed throughout the region. The pre-Q1 erosion surface (E1) forms the contact between Mesozoic-Miocene rocks and Quaternary silts. The surface is irregular and unit Q1 varies rapidly in thickness filling hollows and pinching out over highs. The unit consists of silts with well-developed nodular calcrete levels resting on, and passing southwards into, more pebbly deposits. Anketell and Ghellali (1992b) interpret the deposits as fluvio-aeolian in origin, with fluvial gravel wash followed by floodplain fine sands and silts reworked by wind and sheetwash. The deposits are capped by the thick lower calcrete of supposed 'Villafranchian' age which was faulted prior to deposition of Unit Q2 (Anketell and Ghellali 1983).

The pre-Q2 surface (E2) is marked by extensive erosion of the 'Villafranchian' calcrete so that Unit Q2 occasionally rests directly on the Miocene substrate or on Unit Q1 to which it is lithologically very similar. Comprising red sandy silts and buff 'loessic' silts, both with well-developed levels of calcrete glaebules, it too passes southwards into gravels and again is interpreted as fluvio-aeolian in origin. An erosional episode at the top of Q2 is displayed locally by a valley topography preserved by the 'upper' calcrete (Fig. 3b). Locally, particularly towards the east and in inter-wadi areas near the Jabal Nafusah, Unit Q2 pinches out and the upper calcrete rests directly on the lower.

The pre-Q3 erosional surface (E3) is particularly obvious as local, broadly linear depressions cut through the Q2 calcrete and where cliffs, cut in Q2 by marine erosion, are blanketed by the distinctive white quartz sand deposits of Q3. Elsewhere, the sands rest on the Q2 calcrete hardpan, thinning rapidly in interfluve areas near the coast and tapering out southwards to a depositional feather edge. The deposits represent a small aeolian dune field (Anketell and Ghellali 1992b). Subsequent drowning of the dunes with the development of lacustrine silts following an increase in rainfall, was probably facilitated by ponding behind a cemented calcarenite dune barrier (Ghellali 1977; Anketell and Ghellali 1992a).

Local stripping of the lacustrine deposits, planation of earlier dune topography, and blanketing of marine cliff sections cut in Garagaresh aeolianite equivalent to unit Q3 (Anketell and Ghellali 1983), attest to the erosional surface (E4) which defines the base of Unit Q4. This unit, comprised of 'loessic' silts is interpreted as due to reworking of loess deposits by sheet flood and colluvial processes. The silts extend southwards into the valleys of the Jabal Nafusah, locally passing into and blanketing thick wadi gravel deposits. The top of the unit is generally planar, and dips very gently northwards from the jabal to form the surface of the Jifarah Plain.

Strong downcutting along well-delineated wadi channels by the erosional episode (E5) that defines the base of unit Q5 formed an 'upper' terrace (Vita Finzi 1969, Ghellali 1977) while erosion of the Q5 wadi fill by 'modern' erosional episodes (E6) has formed a well defined 'lower' terrace (Vita Finzi *ibid*) and recent wadi fill making up unit Q6 (Fig. 2a). The fluvial sands and gravels that make up units Q5 and Q6 overlie and interleave with aeolian facies, which are well developed on interwadi areas as well as in the wadis. Q5 dunes are deep ochre to red with a weakly developed red palaeosol. They are commonly fixed by vegetation whereas Q6 dunes are yellow-white and active (Anketell and Ghellali 1983, 1990; Anketell 1989). The strong colour contrast and invasion by root systems allows clear discrimination between the two sequences.

3 QUATERNARY DEPOSITS OF THE JABAL NAFUSAH

Towards the scarp base and in the deep valleys of the Jabal Nafusah (Fig. 1) the Jifarah Formation passes into alluvial fan, slope wash and wadi fill gravels and conglomerates which make up the Qasr Al Haj Formation. A composite section, built up from exposures in the alluvial fan apron to the east of Qasr Al Haj village (Anketell and Ghellali 1988), shows a depositional sequence closely comparable to that of the Jifarah Formation. The lowermost unit (Q1; Anketell and Ghellali 1988), capped by the thick 'Villafranchian' calcrete (El Hinnawy and Cheshitev 1975), consists of strongly-cemented, framework-supported conglomerate overlying conglomeratic mudstone. It rests on an erosional surface cut in Mesozoic sandstones and is separated from overlying, less well-cemented conglomerates, by another erosional surface (E2). These conglomerates (Q2; Anketell and Ghellali 1988) are capped by a thinner calcrete (30 cm), locally with small solution pits, which is closely comparable to the calcrete that caps unit Q2.

Sediments equivalent to Unit Q3 of the Jifarah Formation are absent from the Qasr Al Haj Formation at Qasr Al Haj and have not been recognised anywhere in the jabal region. They thin out to a depositional feather edge in the eastern Jifarah (Anketell and Ghellali 1983) while in the

area south and west of Bin Ghashir (Fig. 1) their absence is due to a combination of intraformational erosion and wedging out (Anketell and Ghellali 1991a).

As a result, the upper calcrete is succeeded by a sequence of loosely cemented gravels and silts equivalent to unit Q4 which forms the surface of the plain to the north, sweeping south and upwards towards the foothills where, incised by wadis, it forms the upper terrace. The characteristic columnar-jointed silts, capped by a patchily preserved and thin calcrete initially overlie and then interfinger with the weakly cemented gravels. These are capped by a thin (14 cm) laminar calcrete and overlie the two older conglomerate units near Qasr Al Haj (Table 1).

Sediments of the lower terrace, making up Unit Q5, and the modern wadi gravels of Unit Q6 are much coarser and thicker than 'downstream' Jifarah equivalents with which they interdigitate. Both deposits rest on strongly erosional bases and are easily identifiable throughout the Qasr Al Haj region although the lower terrace is commonly fragmentary.

At Wadi Ghan (Fig. 1) the succession consists of a valley fill sequence of gravels and conglomerates containing two levels of valley lava flows (Table 1C) (Anketell and Ghellali, 1988). The lavas occur at the top of Unit Q1 and Q2 and are capped respectively by the 'Villafranchian' and Upper calcretes (Anketell and Ghellali 1988; Anketell 1989). The sequence rests on Cretaceous limestones and sandstones at Wadi Ghan dam but to the south Q1 conglomerates are banked against a valley lava flow which rests on the limestones.

The columnar-jointed terrace silts and gravels which succeed the older units are equivalent to Q4, the silty upper part recording a significant southwards incursion of the Jifarah Formation into the jabal. The lower surface is strongly erosional, and the deposits fill a valley deeply cut in the Q1 and Q2 gravels, and calcretes. The silts of the terrace have yielded Aterian artefacts from near the base and a Neolithic assemblage from the upper surface (McBurney and Hey 1955; Hey 1963).

Aggradational sequences equivalent to Units Q5 and Q6 in the Wadi Ghan area are easily identified and correlated along the lengths of the wadis, particularly in the eastern Jifarah region. Anketell and Ghellali (1988) show that the type section of the upper member of the Qasr Al Haj Member, as established in Wadi Ghan by El Hinnawy and Cheshitev (1975), corresponds only to Unit Q5 of the Jifarah Formation.

4 THE NALUT AND MIZDAH REGIONS

Stratigraphic sequences recorded by Antonovic (1977) and Novovic (1977a, 1977b) for the Nalut Mizdah region to the south of the jabal (Fig. 1) are summarised in Table 1 (Anketell 1989). The deposits are thickest in closed depressions and in wadis where a number of terraces, 'Old Wadi Terraces' (Novovic 1977a), rest locally on strongly cemented 'breccia Sheets' and are succeeded by several aggradational units with strongly erosional bases. In the Mizdah area the well cemented conglomerates of the Old Wadi Terraces and the underlying breccia sheets are overlain by basaltic lava flows. Anketell and Ghellali (1990) confirmed correlation of the succession with the lower part of the sequence at Wadi Ghan proposed by Anketell (1989) (Table 1). In this area, the breccia unit, comprising sheets of strongly calcreted conglomerates and breccias, is capped by a laminar calcrete crust and occurs mainly as erosional remnants left on hill tops following dissection of the area by a major phase of erosion which cut the valleys in which the younger, less strongly indurated conglomerate sequences making up the Old Wadi Terraces rest. Elsewhere the younger conglomerates directly overlie the breccia and are capped by the lava flow (Antonovic 1977). Joints in the top of the lava flow are lined with laminar calcrete, remnants of a calcrete crust. Anketell and Ghellali (1990) correlate the pre-basalt calcrete with the 'Villafranchian' calcrete; the underlying breccias and conglomerates with the debris flow deposits of Q1 of the Qasr Al Haj Formation; the younger conglomerate sequence with Unit Q2 and the lava with calcrete-filled fissures with the upper lava-upper calcrete association at Wadi Ghan. Thick calcrete which locally blankets the country rock in inter-wadi areas is probably an amalgam of 'Villafranchian', post Q2 and possibly younger calcrete, similar to that described from Wadi Gerrim in the eastern Jifarah by Anketell and Ghellali (1983).

The younger deposits characteristically comprise three units (Anketell 1989). The lowermost, resting on the deeply incised erosional base is lithologically similar to Unit Q4 in the Jabal foothills area and yields Middle to Upper Palaeolithic artefacts. It is terraced by the erosional surfaces which define the bases of the two younger aggradational sequences. The older of these sequences ranges from gravel fill in the wadi axis to partially cemented, reddened climbing dunes banked against the terrace walls. The erosional surface defining the youngest fill locally forms a lower terrace in the immediately preceding unit. Wadi fill and colluvial deposits again pass laterally into active climbing dunes preserved along the wadi margins. The similarity in erosional, lithological and geomorphological expression of the sequence to the topmost units of the Jifarah Formation supports their correlation with units Q4, Q5 and Q6.

5 THE BENI WALID REGION – UNESCO LIBYAN VALLEYS SURVEY

The Beni Walid region to the east of Mizdah (Fig. 1) includes the largest wadi system in Tripolitania centred on the Wadi Suf Al Jin (Soffegin). This flows WSW to ENE with a number of southeasterly flowing tributaries, notably Wadis Tininai, Merdum and Ghobeen. The wadis contain a complex of fluvial, colluvial and fluvio-aeolian deposits. The UNESCO Libyan Valleys Survey (Barker and Jones 1981, 1982; Barker et al. 1983; Hunt et al. 1986; Gilbertson et al. 1987; Gilbertson and Hunt 1988) has shown that terrestrial erosional and depositional episodes in the valleys during the Quaternary incorporate a number of distinct elements whose relationships, illustrated by reference to a composite section (Gilbertson and Hunt 1988, their Fig. 21) are here updated (Fig. 4) to include the results of sedimentological and palynological research in deposits of the Grerat D'nar Salem hammada basin (Fig. 1). For consistency, the original distinction between event and deposit, as well as the numbering sequence employed in Gilbertson and Hunt (1988) are maintained Table 1.

The valley systems of the Beni Walid region can be shown to be of considerable antiquity. An early topography (S2) was partly flooded by basalt lava flows (S3) during the Pliocene (Piccoli 1971). At some time before this, solutional collapse of evaporite horizons and karstic solution of limestones (S1) took place (Hunt et al. 1985). Following the valley lava flows, the valleys were re-excavated by one or more significant downcutting events (S6), probably equivalent to episode E1 in the Jifarah (Table 1). This interpretation of these early phases supercedes that by Gilbertson and Hunt (1988).

The depositional sequence in the valleys characteristically commences with a complex of gravels which have been termed the 'Cobbly Fill' (Gilbertson et al. 1984) in order to avoid confusion between, as well as unintended correlations with, 'Older fill' of valleys at the coast (Vita Finzi 1969). In Wadi Merdum/Beni Walid, as well as in other wadis of the area, deposition of coarse clastic deposits by powerful, typically braided rivers (S7) formed large dunes which now rise through the alluvium and colluvium of the modern valley floor. Locally, they

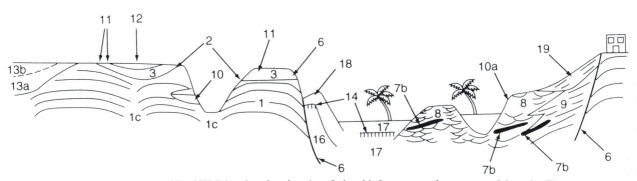

Figure 4. Schematic cross section of Beni Walid region showing the relationship between major events and deposits. The numbers refer to the scheme by Gilbertson & Hunt (1988), as described in the text.

pass into low angle alluvial fan gravels and scree which enter the wadi floors laterally from adjacent plateaux (e.g. in the Wadi Suf Al Jin near Ras el Qattar – Dorsett et al. 1984, their Figs 1 and 4). Some of these gravels are associated with Acheulian hand axes. The coarse clastic deposits may be separated in a general way from similar but younger facies (S8) by their partial induration in calcretes of various thicknesses and possibly different ages (S7b).

The younger gravel facies (S8) lack significant induration but indicate a return to powerful braided rivers draining seawards through the wadis to the Mediterranean coast. Exposures in Wadi Mansur (Fig. 1) indicate that there is complex and significant lateral interdigitation between fluvial deposits S8 and alluvial fan with colluvial deposits (S9) in 'terrace gravels'(Hunt et al. 1986) which had earlier been described as 'Older Fill' (Barker and Jones 1981, 1982, 1983; Jones and Barker 1980, 1983). Alluvial fans and associated colluvial deposits are dominant features along the steep wadi edges throughout the Beni Walid region.

Anketell (1989) and Anketell and Ghellali (1991) correlate the indurated older Cobbly Fill with Q2 of the Jifarah sequence and the overlying younger Cobbly Fill with unit Q4. It is possible that part of the older sequence may be equivalent to Q1 but this remains to be established (Table 1).

In the Gasr Banat region (Fig. 1), Gilbertson et al. (1987) record a minor colluviation phase following the 'Cobbly Fill' but pre-dating the Holocene. This phase has not as yet been identified in the Jifarah region where associated deposits may have been included in younger or older parts of the sequence. Anketell (1989) does, however, correlate this event with an Upper Palaeolithic scree deposit which lies between the Kuf and Bel Ghadir Alluvium in Cyrenaica (Hey 1962).

The Cobbly Fill is succeeded by a complex sequence of deposits with weathering profiles which have been assigned to the Holocene (Gilbertson et al. 1984, Gilbertson and Hunt 1988). Understanding the significance of the accumulation of cave deposits (S10), the weathering profiles on the limestones and basalts (S11), and of basin fills on the basalts (S12) is complicated by the possibility of their being the composite result of Holocene and earlier events.

Understanding the nature of the wadi sequence has, however, recently been greatly improved by sedimentological and palynological studies of a large isolated basin, Grerat D'nar Salem, northwest of Beni Walid (Gilbertson et al. 1994). This area is shown to have been a large permanent lake (2-3 km wide) in the early to mid-Holocene (S13a). This then became ephemeral, and more recently a dry open basin accumulating wind-blown sand (S13b), and only flooded by occasional storm runoff. The desiccation cannot be precisely dated at this site but has been shown to have happened well before the basin became the site of floodwater farming of olives and cereals, currently known from the Romano-Libyan period on. Independent evidence for this wet period is found in well developed palaeosols and inferred greater biolo-

gical activity attributed to the early and mid-Holocene for the region (Gale et al. 1986; Gilbertson et al. 1984), features which are widely recorded from many areas of northern Africa at this time (Bradley 1985; Barker 1986; Gilbertson et al. 1994). The evidence indicates that the wadis are likely to have been occupied by some sort of permanent stream or river flow in more densely vegetated floodplain than has occurred since the mid-Holocene. At present, land forms or deposits of this episode (designated here S11b) have not been recognised in wadis of the Libyan Valleys study area.

The types of deposits that accumulated in the wadis during the mid to late Holocene are now reasonably well known, but their detailed sequence is not. Aeolian sands, collecting as climbing dunes banked against wadi walls (S16), developed a thin palaeosol (S14) in the Wadi Merdum (Gilbertson and Hunt 1988) at some point in the pre Romano-Libyan Holocene. This pre Romano-Libyan sequence is almost certainly the equivalent of the older aeolian sand/fluvial wadi fill sequences of the Nalut area and probably of the reddened older dunes/wadi fill of the Jifarah region (Q5) (Table 1). The sequence points to an episode of aeolian erosion and deposition, presumably with a drier climate and probable aeolian reworking of sands laid down by the inferred wadi-floor rivers of the early to mid Holocene followed by episodes of fluvial reworking of the dune sands by intermittent wadi flows. The Romano-Libyan floodwater farming wall systems which play such a dominant role in the hydrology, geomorphology and biogeography of the wadi floors are underlain by a mix of deposits which essentially reflect the present depositional environment with its floodwater loams which become re-worked in spring and summer by aeolian processes (S17). These contain weak palaeosols (S14). This sequence may also be correlated with the lower terrace deposits making up unit Q5 in the Jifarah region.

Aeolian dune sands are presently developing from sand exposed on an ancient wadi floor by ploughing where mid-Holocene dunes collected in the Wadi Merdum (S18). This appears to have happened several times in the Romano-Libyan and subsequent period and is indicated in the complexes of interbedded aeolian, floodloam and slope deposits in wadi edge exposures. These modern dunes and associated fluvial reworked deposits are equivalent to Unit Q6 of the Jifarah Formation and to the younger dunes and wadi fill of the Nalut and Mizdah regions.

The universal occurrence of substantial sediment traps in Roman water supply and floodwater farming systems, as well as detailed investigations of Romano-Libyan sediment accumulating within them, suggest that soil erosion was relatively rapid during this period of agrohydrological farming (Hunt et al. 1987). The most interesting exposures occur beneath the post Medieval archaeological site of Gasr Abzam in the Wadi Merdum (S19). Here a combination of radiocarbon and archaeological dating has shown that erosion of the plateau and the wadi edge has occurred at a surprisingly rapid rate in the past with a sedimentary column of breccias and aeolian sands 5.9 m thick having accumulated within the period c. 439 to c. 390 years ago demonstrating a rapid and significant supply of sediment to the wadi floor. Examination of the modern plateau has frequently revealed a 'lichen line' 2-10 cm above the modern spoil sediment surface both on the rocky hammadas and basin edges, pointing to a high, if unquantified, rate of surface stripping at the present day – presumably as a result of modern grazing practices (Hunt et al. 1986; Pyatt et al. 1990).

6 EASTERN TRIPOLITANIA COASTAL REGION

To the east of Misratah along the western and southern coast of the Gulf of Sirt (Table 1) the Plio-Pleistocene Al Hishah and Qarat Weddah Formations (Mijalkovic 1977a), equivalent to the Al Assah Formation (Mijalkovic 1977a; Innocenti and Pertusati 1977), are overlain unconformably by 'Old Wadi Terrace' conglomerates capped by a thick calcrete assigned to the 'Villafranchian' by Hecht et al. (1963), Desio (1935) and Mijalkovic (1977a) (Table 1). The calcrete is overlain by a thin development of red silts with Helix. The silts, strongly calcreted and capped by a stage IV calcrete, are correlated with unit Q2 and the upper calcrete of the Jifarah coast (Crema et al. 1913; Parona 1915; and Anketell 1989) while the 'Old Wadi Terrace gravels' are correlated with unit Q1 of the Qasr Al Haj Formation.

The red silts of Unit Q2 are succeeded by a sequence of fluvio-aeolian deposits similar to that of the Jifarah region, comprising three aggradational units bounded by erosion surfaces. The oldest unit, columnar-jointed loessic silts, rests on a strongly erosional surface and displays a well-defined terrace resulting from erosion at the base of the succeeding deposit. This deposit forming a palaeowadi fill and named the Lebda Alluvium by Vita Finzi (1971), is incised by modern wadi systems to form a lower terrace. The three units are clearly equivalent to Q4, Q5 and Q6 (Table 1).

7 REGIONAL CORRELATION AND CHRONOSTRATIGRAPHY

Absence of reliable dating of the Tripolitanian deposits precludes accurate correlation with Quaternary depositional sequences in neighbouring regions, however, useful comparison can be made on the basis of lithology and stratigraphic sequence. Tentative correlation of the Tripolitanian sequence with Quaternary deposits of Morocco, western Italy and Majorca and with marine and Alpine glacial stages are shown in Table 2.

Assignment of the major calcrete capping Q1 to the 'Villafranchian' has not been independently established in Libya and although it can be followed more or less continuously from southern Tunisia into the Jifarah and beyond, some question must remain about its age. Its assignment to the 'Villafranchian' places the lower member of the Qasr Al Haj Formation (Q1) and its equivalent in the Jifarah Formation in the Lower Pleisto-

cene or older. In the Al Qaddahiyah/western Sirt area, Q1 conglomerates underlying the calcrete rest on Plio-Pleistocene deposits of the Al Hishah Formation so it seems likely that unit Q1 is largely Lower Pleistocene in age. The valleys that cut the Jabal Nafusah are thus at least Lower Pleistocene in age and probably older (Ghellali 1977, Ghellali and Anketell 1991). Ghellali (1977) proposed that an ancestral Jabal Nafusah and valleys were initiated in the early Miocene by continental erosion and retreat of a complex fault scarp related to the Aziziyah Fault system follwing uplift of the region earlier in the Tertiary. The valley-fill lavas within the Q1 gravels at Wadi Ghan are thus Pleistocene in age, as noted by Christie (1955) and Hey (1962). At the same time dating of other valley lava flows in the Gharyan volcanic province (Piccoli 1971) show that the Lower Pleistocene valleys were probably developed by rejuvenation of older valley systems. In Wadi Ghan the basalt flow underlying the gravel sequence is Oligocene in age as is a flow in Wadi Guasem. Flows in Wadi Beni Walid range from 5.7-3.5 m.y. in the Pliocene while the oldest valley flow, occurring in Wadi Hashim to the northeast of Mizdah is dated at 52 m.y. in the Early Eocene. Emergence and uplift of the Jabal Nafusah thus appears to date from the Eocene. Following its emergence, south-eastwards draining rivers carved valleys and cave systems in the Cretaceous and Lower Tertiary limestones on the dip slope of the jabal following solution and collapse of evaporite horizons (Hunt et al. 1985). These valleys became the sites of valley lava flows throughout the Eocene, Oligocene and probably the early Miocene. The region was partially flooded during the Middle Miocene following a eustatic sea level rise. The Jifarah region was drowned up to the level of the remnant Aziziyah fault scarp, while to the east and southeast, brackish and shallow water marine deposits overstepped the Eocene onto the Cretaceous. Mann (1975a) and Ghellali (1977) show that the surface had considerable local relief and Ghellali argues for extensive stripping and reworking of pre-Middle Miocene Tertiary continental facies, mainly gravels and sands, by the advancing Middle Miocene sea.

In Late Miocene times, the Messinian sea level fall led to re-emergence of the region that had been flooded by the Middle Miocene seas, probably with rejuvenation of the valleys, some of which were again invaded by lava flows. Subsequent rise in sea level in the Pliocene led to drowning of the valley mouths and the northern part of the Jifarah Plain with reworking of the fluvial input in sabkha and estuarine environments (Mijalkovic 1977b). This period was terminated by a major rejuvenation with deeply incised valleys and braided gravel sedimentation recorded in the Q1 sequences. Anketell and Ghellali (1992b) relate this event to onset of glacioeustatic sea level fall with associated climatic change to semi-arid conditions.

Anketell (1989, 1991) argues for correlation of the lacustrine deposits capping unit Q3 with the Ain Mara Tufas of Cyrenaica and Units Q4 and Q5 with the Kuf Alluvium and Bel Ghadir Alluvium of the same region. In Cyrenaica, correlation of wadi fills with cave deposits

Table 2. Proposed ages of Tripolitanian Quaternary deposits and their regional correlation.

Northwest Libya	Morocco		Western Italy		Majorca		Marine	Europe/ Alpine	Age
Units	Stage Continental	Marine	Stage Continental	Marine	Hemicycle Continental	Marine	Stage		
Q6									
Q5									HOLOCENE
		Mellahian		Versilian	A	Z			
Q4	Soltanien		Pontinian		B		4		
		Ouljian ll		Strombus lll		Y3	5a	WURM	
Q3	Soltanien		Maspinian				5b		LATE
Q2?						Y2	5c 5d	EEM	PLEISTOCENE
		Ouljian l		Strombus ll		Y1	5e		
Q2?	Presoltanien		Maspinian		C				
Hiatus	Tensiftien	Harounian	Rianan	Strombus l	D	X1-2		RISS	
	?	Anfatian ll	Pariolian	Torrimpetran		W4			MIDDLE
		Anfatian l		Tarquinian		W1			PLEISTOCENE
Hiatus	Amirian		Galerian		E			MINDEL	
		Maarifian		Portuensan		V			
Q1	Saletian		Upper Villafranchian		F				
		Messaudian	Emilian					GUNZ	
	Moulouyan		Upper Villafranchian						EARLY
		Fouaratian		Santernian					PLEISTOCENE

dated by Higgs (1961) and with temperature minima in Mediterranean ocean cores (Emiliani 1955), led Hey (1963) to assign an age of c. 50,000 BP to the Ain Mara Tufas and c. 18,000 BP to overlying Upper Palaeolithic screes sandwiched between the Kuf and Bel Ghadir deposits. The screes occupy a similar stratigraphic position to the colluvial silts which succeed the upper Cobbly Fill in the Tripolitania valleys (Anketell 1989). This correlation suggests that unit Q4 and probably Q3 also, are Late Pleistocene in age.

If the calcrete capping Q1 is 'Villafranchian' and thus equated with the later Lower Pleistocene (Anketell and Ghellali 1991) then Q2 lies in a time slot ranging from the middle to Late Pleistocene. Lipparini (1940) records that 'red silts with Helix' i.e. Q2, in Tripolitania overlie a coquina with a warm water Senegalese fauna and that this, although Strombus bubonius, has not been found (Hey 1963; Anketell and Ghellali 1987), places the deposits in the Tyrhennian. If the age of the oldest strombus beach in the Mediterranean is accepted as 200,000-170,000 BP (Strombus I of Ambrosetti et al. 1972) (Table 2), then Unit Q2 dates from at most the late Middle Pleistocene interglacial. As a consequence it would seem that an appreciable part of the sediment record equivalent to the Middle Pleistocene is absent or as yet unrecognised in the Tripolitanian region (Table 2).

Evidence for placing post Q4 erosion in the early Holocene and Units Q5 and Q6 in mid-Holocene onwards is fairly well established and it is expected that this will be refined in ongoing studies.

8 CONCLUSIONS

This essentially historical account demonstrates that there has been considerable progress in identifying the Quaternary sequences in northwest Tripolitania, particularly in the Jifarah and Beni Walid regions. In the Jifarah region, fluvio-aeolian deposits which make up the Jifarah Formation are subdivided into six aggradational units Q1-Q6, by erosion surfaces. The two oldest units are capped by significant calcrete and terra rossa soil developments. Units Q3 and 4 display patchy calcrete only.

Strongest erosion tends to be concentrated along linear fossil wadi trends. Inter-wadi erosion is limited and generally resisted by the well-cemented Ca-rich horizons of the calcrete profile. As a result inter-wadi areas tend to be planar or, where the top of the previous unit is dominated by aeolian dune sediment, gently undulate.

Correlation of the younger deposits throughout Tripolitania and the Sirt region appears to be fairly straightforward. Erosional surfaces and associated calcretes can be traced into contemporaneous gravel-dominated formations in the Jabal Nafusah scarp and valleys, onto the south facing dip slope of the jabal with its minor wadis, and into the major wadi system of the Beni Walid region leading to the coast in eastern Sirt.

The major erosional surfaces delineate and bound a similar number of depositional events in each of these regions. The older, cemented Cobbly Fill is equivalent to unit Q2 and possibly unit Q1 of the Jifarah region. The younger Cobbly Fill/fluvio-aeolian deposits of the valleys, western jabal top and Sirt (Table 1) are equivalent to the upper terrace loess and gravel deposits of Unit Q4 in the Jifarah region (Table 1). Unit Q3 is absent in the Jabal region, which may have remained as stony desert during accumulation of the aeolian sand field in the eastern Jifarah Plain. Unit Q5, together with the older aeolian sheet sands and older climbing dunes, corresponds to the mid-Holocene and historic Romano-Libyan wadi fills of the valleys (Gilbertson and Hunt 1988; Anketell 1989) and the lower terrace of other areas. Modern wadi fills and active sand dunes correspond to Unit Q6. Only the slope colluvium sandwiched between and separated from units Q4 and Q5 by erosional surfaces is as yet unrecognised in the Jifarah region and elsewhere in Tripolitania.

It is clear that the 'Old Wadi Terrace' deposits (Antonovic 1977; Novovic 1977a, 1977b; Mijalkovic 1977a, 1977b) consist of more than one aggradational episode ranging from pre- 'Villafranchian' in the lower reaches of the Tripolitanian valleys to gravels equivalent to Q2 and probably Q4 elsewhere (Table 1).

Lava flows occupying valleys which radiate out from the Gharyan volcanic province, furnish a range of dates ranging from the Eocene up to the Pliocene. These indicate that the Tripolitanian valleys are of considerable antiquity and were initiated on a land mass, the Jifarah Arch, uplifted in Late Palaeocene times. Partial submersion of the arch in the Middle Miocene was followed by Late Miocene sea level fall, probably accompanied by rejuvenation and incision of the valleys. Subsequent sea level rise in the Pliocene drowned the valley mouths before major downcutting with coarse braided gravel deposits commenced in the Early Pleistocene. The severity of fluvial erosion processes related to this event was probably instrumental in the removal of valley deposits formed in earlier events. There appears to be a major gap in sedimentation following deposition of Unit Q1. Faulting of the 'Villafranchian' calcrete and evidence for uplift in neighbouring areas of Tunisia suggests that the absence of post-'Villafranchian' pre-Tyrhennian deposits

may be due to erosion following renewed tectonic uplift of the Jifarah Arch.

9 AKNOWLEDGEMENTS

We wish to express our thanks to Br Abdullah Shaiboub, then President of the Department of Antiquities in Tripoli, Nr Muammar M Boggar of Beni Walid, and our other Libyan colleagues in the field. Our thanks are due to the University of Al Fateh and the National Drilling Company, Tripoli for logistical help during fieldwork in the Jifarah. J. M. Anketell gratefully acknowledges travel grants from the Council for Libyan Studies and The Royal Society of London. We also wish to acknowledge the help of many others in the Universities of Manchester, Sheffield and Tripoli.

REFERENCES

Allen, J.R.L. 1977. Wales and the Welsh Borders, In M.R. House (ed), *A correlation of the Devonian Rocks of the British Isles*:40-54. Spec. Rep. geol. Soc. London. 8.

Allen, J.R.L. 1986. Pedogenic calcretes in the Old Red Sandstone facies (late Silurian-early Carboniferous) of the Anglo-Welsh area, southern Britain. In V.P.Wright (ed), *Palaeosols – Their recognition and Interpretation*:58-82. Blackwell, London.

Ambrosetti, P., A. Azzaroli, F.P. Bonadonna & M. Follieri 1972. Scheme of Pleistocene Chronology for the Tyrrhenian side of Central Italy. *Boll. Soc. Geol. Ital.* 91:169.

Anketell, J.M. 1989. Quaternary deposits of northern Libya -lithostratigraphy and correlation. *Libyan Studies* 20:1-27.

Anketell, J.M. & S.M. Ghellali 1983. Stratigraphic studies on Quaternary floodplain deposits of the eastern Jeffara Plain. S.P.L.A.J. *Libyan Studies* 14:16-36.

Anketell, J.M. & S.M. Ghellali 1984. Nests of a tube-dwelling bee in Quaternary sediments of the Jeffara Plain. *Libyan Studies* 15:137-141.

Anketell, J.M. & S.M. Ghellali 1987. Stratigraphic aspects of the Gargaresh Formation. Tripolitania S.P.L.A.J. *Libyan Studies* 17:123-131.

Anketell, J.M. & S.M. Ghellali 1988. Stratigraphic aspects of the Qasr Al Haj Formation, Tripolitania S.P.L.A.J. *Libyan Studies* 18:115-127.

Anketell, J.M. & S.M. Ghellali 1990. Stratigraphic relationships of basalt lava flows to the Pleistocene sedimentary sequence in the Mizdah region, Tripolitania, S.P.L.A.J. *Libyan Studies* 21.

Anketell, J.M. & S.M. Ghellali 1991a. Quaternary fluvio-aeolian sand/silt and alluvial gravel deposits of northern Libya – event stratigraphy and correlation. *Journal of African Earth Sciences* 13:457-469

Anketell, J.M. & S.M. Ghellali 1992a. The Karawah and Qarabulli Members of the Jifarah Formation – Late Pleistocene Aeolian and Lacustrine Deposits – Northwest Libya. *Libyan Studies* 23.

Anketell, J.M. & S.M. Ghellali 1992b. Quaternary sediments of the Jifarah Plain – stratigraphy and sedimentary history. In M J. Salem (ed), *Third Symposium of the Geology of Libya. Tripoli 1987*:1987-2013

Anketell, J.M. & S.M. Ghellali 1992c. The Jifarah Formation –

Aeolian and Fluvial deposits of Quaternary age – Jifarah Plain, S.P.L.A.J. A redefinition in terms of a composite stratotype. In M. J. Salem (ed) *Third Symposium on the Geology of Libya. Tripoli 1987*:1967-1985

Anketell, J.M. & S.M. Ghellali 1992d. A palaeogeologic map of the pre-Tertiary surface in the region of the Jifarah Plain and its implications to the structural history of northern Libya. In M.J. Salem (ed), *Third Symposium on the Geology of Libya. Tripoli 1987*:2381-2406

Antonovic, A. 1977. Geological map of Libya, Mizdah. (NI33-1). Explanatory Booklet. Industrial Research Centre. S.P.L.A.J.

Atkinson, C.D. 1986. Tectonic control on alluvial sedimentation as revealed by an ancient catena in the Capella Formation (Eocene) of northern Spain. In V. P.Wright, *Palaeosols – Their recognition and Interpretation*:139-174. Blackwell.

Banerjee, S. 1980. Stratigraphic Lexicon of Libya. *Dept. of Geol. Res. and Mining. Bull. I.R.C. Libya.* 13. 300p.

Barker, G.W.W. 1986. Prehistoric rock art in Tripolitania. *Libyan studies* 17:69-86.

Barker, G.W.W. & G.D.B. Jones 1981. The UNESCO Libyan Valleys Survey 1980. *Libyan Studies* 12:9-48.

Barker, G.W.W. & G.D.B. Jones. 1982. The UNESCO Libyan Valleys Survey 1979-1981; Palaeoeconony and environmental archaeology in the pre-desert. *Libyan Studies* 13:1-34.

Barker, G.W.W., D.D. Gilbertson, C.M. Griffin, P.P. Hayes & D.A. Jones 1983. The UNESCO Libyan Valleys Survey V: Sedimentological properties of Holocene Wadi Floor and Plateau Deposits in Tripolitania, North-West Libya. *Libyan Studies* 14:69-85.

Burollet, P.F. 1960. Lexique Stratigraphique International. 4. Afrique, No. 4a, Libye. Names and Nomenclature Committee, Petrol. Explor. Soc. Libya. 20th International Geol. Congr., Mexico (1956) – Stratigraphic Commission. Cent. Nat. Resch. Sci. Paris. 62p.

Butzer, K.W. 1975. Pleistocene littoral sedimentary cycles of the Mediterranean Basin: A Mallorquin view. In K.W. Butzer & G.L. Izaac, *After the Australopithecines*:25-71.

Christie, A.M. 1955. Geology of the Gharian Area, Tripolitania, Libya. United Nations Report No. TAA/LIB/2. [Republished 1966 by Geol. Section. Kingdom of Libya, Ministry of Industry, Bulletin No. 5.]

Coque, R & A. Jauzein 1967. The geomorphology and Quaternary geology of Tunisia. In L. Martin, *Tunisia. Ninth Ann. Field Conf. P.E.S.L.*:227-258.

Crema, C., S. Franchi & C.F. Parona 1913. Descrizione fisica e geologica della regione. *Commissione per lo Studio Agrocolico della Tripolitania. La Tripolitania settentrionale* 1:3-42. Roma.

Desio, A. 1935. Studi geologici sulla Cirenaica, sul Deserto Libico, sulla Tripolitania e sul Fezzan Orientale. *Missione Scientifica dell R. Accad. d'Italia a Cufra (1931)* 1:480p.

Desio, A. 1940. Sulla Posizione Geologica e sull' Origine delle Falde Artisiane della Gefara Tripolina e del Misuratino. *La Ricerca Scientifica*, Anno 11 Roma.

Desio, A. 1970. Outlines of the Geomorphological Evolution of Libya from the early Tertiary. *Atti. della Accad. Naz. dei Lincei* 10:19-65. Roma.

Desio, A. 1971. Outlines and Problems in the Geomorphological Evolution of Libya, from the Tertiary to the Present Day. In C. Gray (ed), *Symposium on the Geology of Libya*:11-36. Faculty of Science, University of Libya, Tarabulus. S.P.L.A.J.

Desio, A., C.R. Ronchetti, R. Pozzi, F. Clerici, G. Invernizzi, C.

Pisoni & P.L. Vigano 1963. Stratigraphic Studies in the Tripolitanian Jebel, Libya. *Riv. Ital. di Paleont. Strat.* 9:1-126.

Dorsett, J.E., D.D. Gilbertson, C.O. Hunt & G.W.W. Barker 1984. The UNESCO Libyan Valleys Survey VIII: Image analysis of Landst data for archaeological and environmental surveys. *Libyan Studies* 15:71-80.

El Hinnawy, M. & G. Cheshitev 1975. Geological Map of Libya. Tarabulus (NI33-13). Explanatory Booklet, Industrial Research Centre, S.P.L.A.J.

Emiliani, C. 1955. Pleistocene temperature variations in the Mediterranean. *Quaternaria* 2:87-98. Rome.

Etourbi, I.Y. 1989. Study of the main facies of the Farwah Group, late Palaeocene to Middle Eocene, of the northwest Libya Offshore. M.Sc. Thesis University of Manchester, England.

Gale, S.J., C.O. Hunt & D.D. Gilbertson 1986. The infill sequence and water carrying capacity of an ancient irrigation channel: Wadi Gobbeen, Tripolitania. *Libyan Studies* 17:1-5.

Ghellali, S.M. 1977. On the Geology of the eastern Gefara Plain, North West Libya. Ph.D. Thesis. University of Manchester, England. 296p.

Ghellali, S.M. & J.M. Anketell 1991. The Suq al Jum'ah Palaeowadi; an example of a Plio- Quaternary Palaeovalley from the Jabal Nafusah, Northern Libya. *Libyan Studies* 22:1-6.

Gilbertson, D.D., P.P. Hayes, G.W.W. Barker & C.O. Hunt 1984. The UNESCO Libyan Valleys Survey VII: An Interim Classification and Functional Analysis of Ancient Wall technology and Land Use. *Libyan Studies* 15:45-70.

Gilbertson, D.D., C.O. Hunt, D.J. Briggs, G.M. Coles & N.M. Thew 1987. The UNESCO Libyan Valleys Survey XVIII: The Quaternary geomorphology and calcretes of the area around Gasr Banat in the pre-desert of Tripolitania. *Libyan Studies* 18:15-27.

Gilbertson, D.D. & C.O. Hunt 1988. The UNESCO Libyan Valleys Survey XIX: A Reconnaisance Survey of the Cenozoic Geomorphology of the Wadi Merdum, Beni Ulid, in the Libyan Pre-desert. *Libyan Studies* 19:95-121.

Gilbertson, D.D., C.O. Hunt, N.R.J. Feiller & G.W.W. Barker 1994. The Environmental Consequences and Context of Ancient Floodwater Farming in the Tripolitanian Pre-Desert. In: A.C. Millington & K. Pye (eds), *Environmental Change in Drylands: Biogeographical and Geomorphological Perspectives*. Chichester, John Wiley, 229-251.

Gile, L.H., F.F. Peterson & R.B. Grossman 1966. Morphological and genetic sequences of carbonate accumulation in desert soils. *Soil Science* 101:347-360.

Hecht, F., M. Furst & E. Klitzsch 1963. Zur Geologie von Libyen. *Geol. Rundsch.* 53:413-470.

Hey, R.W. 1962. The Quaternary and Palaeolithic of northern Libya. *Quaternaria* VI:435-449.

Hey, R.W. 1963. Pleistocene scree in Cyrenaica (Libya). *Eiszeitalter und Gegenwart*. Ohringen. 14:77-84.

Hey, R.W. 1978. Horizontal shorelines of the mediterranean. *Quaternary Research* 10:197-203.

Higgs, E.S. 1961. Some Pleistocene faunas of the Mediterranean coastal areas. *Proc. Prehist. Soc.* 27:144-154.

Hunt, C.O., S.J. Gale & D.D. Gilbertson 1985. The UNESCO Libyan Valleys Survey IX; Anhydrite and limestone karst in the Tripolitanian Pre-desert. *Libyan Studies* 16:1-13.

Hunt, C.O., D.D. Gilbertson, R.D.S. Jenkinson, M. Van der Veen, G. Yates & P.C. Buckland 1987. The UNESCO Libyan Valleys Survey XVI: The palaeoecology and agriculture of an abandonment phase at Gsur Mnio, Wadi Mimoun, in the Tripolitanian pre desert. *Libyan Studies* 18:1-13.

Hunt, C.O., D.J. Mattingly & D.D. Gilbertson 1986. ULVS XIII: interdisciplinary approaches to ancient farming in the Wadi Mansur, Tripolitania. *Libyan Studies* 17:7-47.

Hunt, C.O., D.J. Mattingly, D.D. Gilbertson, G.W.W. Barker, J.N. Dore, J.R. Burns, A.M. Fleming & M. Van der Veen 1986. The UNESCO Libyan Valleys Survey XIII: Interdisciplinary Approaches to Ancient Farming in the Wadi Mansour, Tripolitania. *Libyan Studies* 17:23-63.

Innocenti, F & P. Pertusati 1984. Geological map of Libya, Al Aqaylah. (NH34-5). Explanatory Booklet. Industrial Research centre. S.P.L.A.J.

James, N. 1972. Holocene and Pleistocene calcareous crust (caliche) profiles: Criteria for subaerial exposure. *Jour. Sed. Petr.* 42:817-836.

Jennings, J.N. 1968. Syngenetic Karst in Australia. In P.W. Williams & J.N. Jennings (eds), *Contributions to the study of Karst. Res. School of Pacific Studies. Dept. Geogr, Publ. G/5/1968*, Austral. Nat. Univ. Canberra.

Jones, J.R. 1969. Ground Water in Libya – A Summary. Hydrogeology of the Southern Mediterranean Littoral and the north central Sahara. Open File report. June 1969. *Water Resources Division U.S. Geological Survey* (U.S.A.I.D.)

Jones G.D.B. & W.W. Barker 1980. Libyan Valleys Survey. *Libyan Studies* 11:11-36.

Jones G.D.B. & W.W. Barker 1983. The UNESCO Libyan Valleys Survey IV: the 1981 season. *Libyan Studies* 14:39-68.

Klen, L. 1984. Geological map of Libya, Benghazi. (NI34-14). Explanatory Booklet. Industrial Research Centre. S.P.L.A.J.

Kornicker, L.S. 1958. *Bahamian limestone crusts.* Gulf Coast Assoc. Geol. Soc. Trans. 8, 167-170.

Kraus, M.J. & T.M. Bown 1986. Palaeosols and time resolution in alluvial stratigraphy. In V.P. Wright (ed.), *Palaeosols – Their recognition and Interpretation*:180-201. Blackwell.

Lipparini, T. 1940. Tettonica e geomorfologia della Tripolitania. *Boll. Soc. Geol. Ital.* 59:222-301.

Mann, K. 1975a. Geological map of Libya, Al Khums. (NI33-14). Explanatory Booklet. Industrial Research Centre. S.P.L.A.J.

Mann, K. 1975b. Geological map of Libya, Misratah. (NI33-15). Explanatory Booklet. Industrial Research Centre. S.P.L.A.J.

McBurney, C.B.M. & R.W. Hey 1955. *Prehistory and Pleistocene Geology in Cyrenaican Libya.* Cambridge University Press. 308p.

Mijalkovic, N. 1977a. Geological map of Libya, Al Qaddahiyah (NH33-3). Explanatory Booklet. Industrial Research Centre. S.P.L.A.J.

Mijalkovic, N. 1977b. Geological map of Libya, Qasr Sirt. (NH33-4). Explanatory Booklet. Industrial Research Centre. S.P.L.A.J.

Netterburg, F. 1967. Some roadmaking properties of South African calcretes. *Proc. 4th Regional Conference Soil Mech. and Eng.* Cape Town.

Novovic, T. 1977a. Geological map of Libya, Nalut. (NI32-4). Explanatory Booklet. Industrial Research Centre. S.P.L.A.J.

Novovic, T. 1977b, Geological map of Libya, Djeneien. (NH32-3). Explanatory Booklet. Industrial Research Centre. S.P.L.A.J.

Parona C.F. 1915. Impressioni di Tripolitania – Note geomorfologiche sulla Gefara. *'Natura'* VI:217-248.

Perkins, R.D. 1977. Depositional Framework of Pleistocene Rocks in South Florida. *Geol. Soc. Am. Mem.* 147:131-198.

Piccoli, C. 1971. Outlines of volcanism in northern Tripolitaia. In C. Gray (ed), *Symposium on the Geology of Libya.* Faculty of Science, University of Libya, Tarabulus :323-321.

Price, W.A. 1925. Caliche and pseudoanticlines. *Am. Assoc. Petr. geol. Bull.* 9:1009-1017.

Pyatt, F.B., D.D. Gilbertson & C.O. Hunt. 1990 ULVS XXII: Crustose Lichen Affecting the Geological Interpretation of Digital Landsat Imagery on the Tripolitanian Pre-desert. *Libyan Studies* 21:43-47.

Ratcliffe, C.F. & J.A. Fagerstrom 1980. Invertebrate Lebensspuren of Holocene Floodplains: Their morphology, origin and palaeoecological significance. *Journ. Palaleont.* 54:614-630.

Reeves, C.C. 1976. *Caliche.* Estacado Books. P.O. Box 4515. Lubbock, Texas.

Ruhlich, P. 1974. Geological map of Libya, Al Bayda. (NI34-15). Explanatory Booklet. Industrial Research Centre. S.P.L.A.J.

Smetana, R. 1975. Geological map of Libya, Ra's Jdeir. (NI32-16). Explanatory Booklet. Industrial Research Centre. S.P.L.A.J.

Stella, A. 1914. Geologia. In: La Missione Franchetti in Tripolitania, II Gebel. Milano. 81-151.

Vinassa de Regny, P. 1901. Note Geologiche sulla Tripolitania. *Rendic. Sess. R. Acc. Istituto di Bologna*:177-186.

Vinassa de Regny, P. 1902. Note Geologiche sulla Tripolitania. *Rendic. Sess. R. Acc. Istituto di Bologna*:1-12.

Vita-Finzi, C. 1969. *The Mediterranean Valleys.* Cambridge: Cambridge University Press, 131p.

Vita-Finzi, C. 1971. Alluvial History of Northern Libya since the last Interglacial. In C. Gray (ed), *Symposium on the Geology of Libya.* Faculty of Science, University of Libya, Tarabulus. 408-429

Warme, J.E. & E.J. McHuron 1978. Marine Borers; Trace Fossils and their significance, In P.B. Basan, *Trace Fossil Concepts, S.E.P.M. Short Course* 5:67-118.

Yaalon, D.H. 1967. Factors affecting the lithification of eolianite and interpretation of its environmental significance. *Journ. Sed. Petr.* 37:1189-1199.

Zivanovic, M. 1977. Geological map of Libya, Bani Walid. (NH33-2). Explanatory Booklet. Industrial Research Centre. S.P.L.A.J.

CHAPTER 22

Soils of Quaternary river sediments in the Algarve

P.A. JAMES & D.K. CHESTER
Department of Geography, University of Liverpool, Liverpool, UK

ABSTRACT: The soils of younger and older valley fills and those of their chief source area, the upland hillslopes, are analysed. The relationships between these groups of soils are examined using a sequence of multivariate statistical techniques. Some evidence of possible sediment-source linkage between valley fill and catchment is presented, but there is no clear chronosequence within the older fill soils. It appears that the younger fill largely comprises the former topsoil of the chief catchments, the erosion of which was caused by human disturbance during the last three millennia.

KEYWORDS: Mediterranean soils, river sediments, Holocene erosion, multiple discriminant analysis, factor analysis, cluster analysis.

1 INTRODUCTION

Information derived from the analysis of alluvial soils has several uses in the study of Quaternary river sediments. In particular, soil profiles may be used to help differentiate land surfaces and to evaluate their relative age (e.g. McFadden and Hendricks 1985). Soil properties may also hold the key to the identification of catchment sediment sources (e.g. Macleod and Vita-Finzi 1982) allowing discrimination either in terms of catchment location or of topsoil as opposed to subsoil source. Knowledge of the dominant alluvial sediment sources in turn throws light upon the nature of catchment erosion.

It is to the questions of sediment age and of catchment erosion that this analysis turns. In the Algarve, there is a clear geomorphological distinction between a valley floor alluvium and an older alluvial fill which occurs upon spur-tops and as terrace remnants. It has been established that the valley floor sediment is largely of late Holocene age, and that the closing stages of deposition of the older sediments were in the very early Holocene (Devereux 1983; Chester and James 1991). Thus in the Algarve there exists the clear dichotomy of valley fills as proposed by the Vita-Finzi model for the Mediterranean lands at large (Vita-Finzi 1969). For convenience, we call the two groups of sediment in the Algarve the 'younger fill' and the 'older fill'. The distinction between the two is very clear, though we do not imply by the use of these terms that the alluvial stratigraphy of the region is simple.

In soil chronosequences of Mediterranean climatic environments one of the most distinctive pedogenetic trends involves iron oxi-hydroxide transformations and an increase in redness of hue (rubification) (Torrent et al. 1983). An apparent distinction between a generally brown younger fill and a redder older fill of the Algarve

valleys is consistent with this model. However, in the catchments of the region at the present day, brown soils are very restricted. The possible reason for the colour difference between younger fill and present catchment soils may be diagenesis in the fill, or the fact that the chief sediment source was a former brown topsoil which erosion has comprehensively removed. The extensive truncation of hillslope soils would have implications for the nature of the catchment disturbance which forced the erosion.

2 THE STUDY AREA

The regional relief and geology are shown in Figure 1. The geological boundaries delimit three major landscapes. In the north, the syenite intrusion of the Monchique Massif (Rock 1978) rises to 902 m. This lies within a Hercynian upland of intensely dissected flysch facies, slightly metamorphosed and comprising fairly thinbedded shales, iron-rich phyllites and greywackes (Oliveira 1983; Ribeiro and Brandão-Silva 1983). We use the term 'metasediments' for these rocks. There are occasional minor intrusions of syenite within this zone. The Serra is the chief source of river sediments throughout the Algarve. The vegetation of much of these hills is a relatively uniform mattoral dominated by *Cistus ladanifer* L. Widespread removal of *Quercus* forest occurred during the settlement of Silves (Fig. 1) by Moorish people from the ninth to twelfth centuries AD (Chester and James 1991; Stanislawski 1963: 183). The marked decline of Silves after the 12th century appears to have been related in part to the silting of the Arade River, navigable to the sea and a route for timber export at the time of the Moors (Chester and James 1991). From the early 20th century to the 1950's a shifting cultivation of

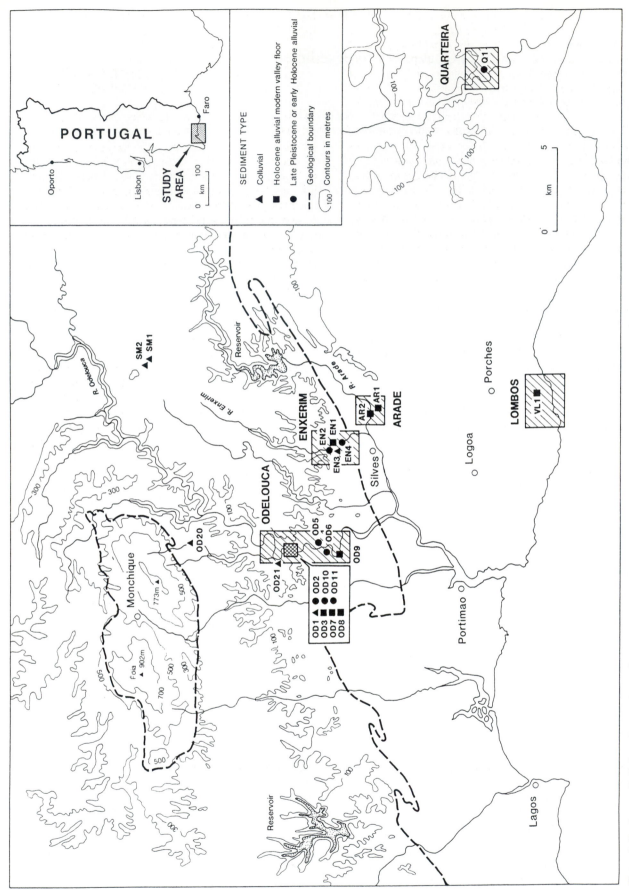

Figure 1. Study area: Relief, geology and sampling areas.

grain was practised in the thinly populated hills (Stanis-lawski 1963: 189-203; *Carta Agricola e Florestal de Portugal*, 595). The most recent and continuing phase of vegetation stripping is catastrophic: slope soils are ploughed or terraced by bulldozer for the planting of Eucalyptus. Soil, ecology and landscape are transformed. In the larger valleys, the valley floor and lower surfaces of older fill support crops, particularly irrigated Citrus.

The chief rocks of the lowland to the south of the Serra are Jurassic and Cretaceous limestones. An important formation are the unconsolidated sands which occur across parts of interfluves in the lowland and littoral zones from west of Lagos to east of Faro. In the east of the region, Manupella et al. (1987) name them the Faro-Quarteira sands and date them as Pleistocene. We interpret the deposits as a piedmont fluviatile sediment derived in part from the Serra. In the littoral zone are Miocene rocks and Pleistocene marine and terrestrial sediments (Rocha et al. 1983; Manupella et al. 1987). The lowlands are an area of traditional Mediterranean agriculture undergoing intensification, particularly for irrigated Citrus.

The climate of the Algarve is portrayed as thermomediterranean with shorter dry season in the lowlands; and mesomediterranean with longer dry season in the Serra, and with shorter dry season in the Monchique Massif (UNESCO/FAO 1963). Feio (1949) quoted mean annual rainfall figures of 421 mm for Praia da Rocha (2.5 km to the south of Portimão, Fig. 1) and 1205 mm for Monchique (395 m). At Faro, only 6% of the mean annual total of 363 mm fell in the period June to September; mean monthly temperature for January was 11.6°C and 24.1°C for August. In the *Atlas do Ambiente* (1974), a mean annual temperature of > 17.5°C is mapped for the eastern portion of the area of the Serra shown in Figure 1. This is higher than the mean annual temperature shown for the lowland to the south (15.0-17.5°C).

3 FIELD SAMPLING, LABORATORY METHODS AND MULTIVARIATE TECHNIQUES

The variables analysed on the soil samples (Tables 1A and 1B) relate to chief soil-forming processes in the Mediterranean environment, particularly transformations of iron oxi-hydroxides, including rubification, and the pedogenetic redistribution of clay and carbonate. In addition to the physical and chemical analyses, mineral magnetic properties of the soils were determined because of their potential for differentiating between topsoil and subsoil sources of sediments (Maher 1986; Fine et al. 1989; Fine et al. 1992). The distribution of sample sites is shown in Figures 1 and 2; sedimentary and pedological characteristics of the three groups of soils/sediments are shown in Figures 3 to 6. The older fill was sampled in the reaches of the valleys of the Odelouca and Enxerim which have received sediment only from the Serra, and in the Quarteira Valley, to which lowland rocks and sediments (particularly the Faro-Quarteira sands) have been important contributors. Soils of the younger fill were sampled in the Odelouca and Enxerim valleys; in the valley of the R. Arade, which receives flow chiefly from the Serra but also from the limestone region to the south; and in the valley of Lombos, where the present ephemeral stream has a small, entirely lowland catchment and in which the interfluve sand deposits are important.

A multivariate statistical perspective on relationships between the soils is gained with factor, cluster and discriminant analysis. Because of the need to consider variations within as well as between profiles, the cases exa-

Table 1A. Soil physical and chemical analysis (< 2 mm fraction).

Property	Method
pH	Electrometrically in a 2:1 water:soil suspension
% loss on ignition (LOI)	850°C for 30 minutes; where carbonates are present: 450°C for 4 hours
Particle-size	On < 2 mm fraction by sieve and pipette method after H_2O_2 digestion and dispersion in 'Calgon'. Sand = .063-2.0 mm; Silt = .002-.063 mm; Clay = < .002 mm
Pedogenetic oxyhydroxides	Fe determined by atomic absorption spectrophotometer
Total pedogenetic (Fe_d)	Dithionite-citrate-bicarbonate (Mehra and Jackson 1960)
Paracrystalline + organically complexed (Fe_o)	Acid ammonium oxalate (Schwertmann 1964)
Organically complexed (Fe_p)	Sodium pyrophosphate (McKeague 1967)
Redness rating	Based on Munsell colours of moist soil and using the Hurst index as modified by Torrent et al. 1980
% Calcium carbonate	Gasometric determination

Table 1B. Soil mineral magnetic properties analysed after methods of Thompson and Oldfield (1986).

Low frequency susceptibility ($10^{-8}m^3 kg^{-1}$)
Frequency-dependent susceptibility (%)
Anhysteretic remanent magnetisation ($10^{-6}A m^2 kg^{-1}$)
Saturated isothermal remanent magnetisation (SIRM) ($10^{-6}A m^2 kg^{-1}$)

'Hard' and 'Soft': reverse field demagnetisation measures of isothermal remanent magnetisation (IRM) ($10^{-6}A m^2 kg^{-1}$):
'Soft' is the difference in the IRM at SIRM and that at −20mT
'Hard' is the difference in the IRM at SIRM and that at −300mT

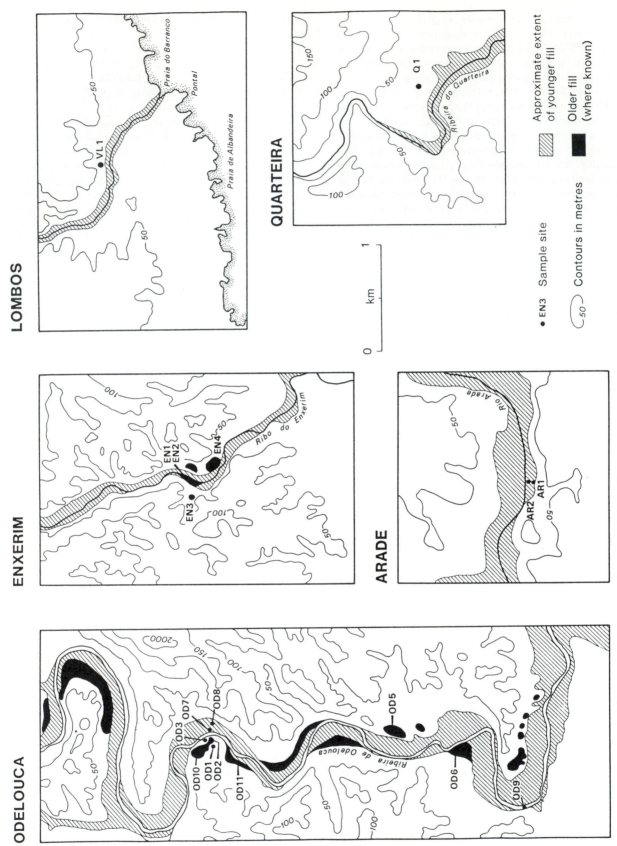

Figure 2. Location of sampling sites. (For OD20, OD21, SM1 and SM2, see Fig. 1).

mined are the 166 individual soil samples which were taken from 22 profiles. The variables on each case are those listed in Tables 1A and 1B, though $CaCO_3$ was excluded because of its limited occurrence in the profiles analysed. The distribution of many of the variables was made more nearly normal by a logarithmic transformation. The physical and chemical variables (Table 1A) – referred to as 'soil' variables – were processed separately from mineral magnetic variables because an initial exploration showed that the two sets of data appeared to reflect different spatial controls.

The factor analytical method used was that of principal axis factoring (Harman 1967) with varimax rotation. For cluster analysis, squared Euclidean distance was the similarity measure, and Ward's the method of grouping. For clustering samples on the basis of soil properties, the cases were defined in terms of their scores on the first three factors derived in the factor analysis. As only six variables were selected for the mineral magnetic data, cluster analysis was applied to the original data rather than to factor scores. For the classification procedure in multiple discriminant analysis, an equal probability of group membership is assumed (i.e. 0.333).

4 THE RIVER SEDIMENTS

Knowledge of Quaternary geology in the Algarve was based formerly on the work of Zbysewski (1958). A lithostratigraphic study of river sediments of the Algarve was completed by Devereux (1983). In addition to the younger and older fills, fluviatile sediments of the Algarve include the Faro-Quarteira sands referred to above (and called the Porches Sands by Devereux). We have also identified a Pleistocene or older sediment lying at 80 m a.s.l. on the western seaboard which has the characteristics of a *raña* soil (Espejo 1987), but we have not as yet defined the stratigraphic relationship between this and the Faro-Quarteira sands. These two sediments are excluded from the present analysis.

The distribution of the younger and, where known, older valley fills in the areas sampled is shown in Figure 2. The older fill survives in major valleys of both Serra and lowland. In the former they occur between 10 m and 40 m elevation and in many cases cover rock-cut valley spurs (Chester and James 1991). The sediment is everywhere reddened and may be correlated geomorphologically between the valleys studied and, tentatively, with similar sediments described from other valleys and named the Purgatorio alluvium by Devereux (1983). He dated the closing stages of deposition at 7450 ±90 BP. They are similar, sedimentologically and geomorphologically, to older fills described from many sites throughout the Mediterranean region (Vita-Finzi 1969; 1976) and generally ascribed to the Pleistocene. They represent deposition in conditions of high energy, probably episodic, fluvial discharge. Sedimentological characteristics at the sites chosen for detailed analysis are shown in Figure 3.

The younger fill is browner and commonly contains abundant charcoal. Sedimentological details are given in Figure 4. On the grounds of geomorphological situation and age, the younger fill we describe is clearly correlated with the Odiaxere formation of Devereux (1983). He dated late stages at 520 ±60 BP and 780 ±50 BP. He also distinguished coarse and fine facies, which can be recognized in the valleys we study. In some locations the coarse facies passes upward into the fine. Charcoal from 2.8 m below the present ground surface of AR2 yielded a date of 2690 ±70 BP (OxA-2267); the base of the sediment lay deeper than 3.6 m. A Roman floor lies 1.25 m below the present ground surface in a position between AR1 and AR2. Sherds in AR1 were identified by Ms M. Vita-Muller as Roman at 1.1 m and Moorish at 0.9 m below the surface (though part of the sediment at AR1 may have mudflowed off the valley-side). These data, therefore, suggest a sediment accumulation of 2.8 m since the Bronze Age; most of the one metre or so of post-Roman sediment has been deposited since Moorish times.

5 THE SOILS

5.1 *Catchment soils*

In the lowlands, red Mediterranean soils predominate, particularly in association with limestone (*Carta dos Solos de Portugal* 49D, 52B). The Serra rocks have been the chief source of river sediments throughout the Algarve. Soils in metasediment colluvium were sampled between 20 m and 290 m above sea level in the catchments of the Odelouca and Enxerim river systems (Fig. 2). One profile, OD20, was sampled on syenite. The Serra hillslope soils are mapped throughout as lithosols by the *Carta dos Solos de Portugal* (49C) and as being prone to erosion (*Carta de Capacidado de Uso do Solo de Portugal*, 49C). However, they are by no means everywhere lithosols, a thickness of 2 m occurring on many slopes of moderate gradient (Fig. 5). The erosion risk reflects the nature of the terrain rather than any inherent erodibility, as the textures are generally very fine (% clay reaches 81.6 in OD21). The Serra soils are acid with pH almost everywhere < 5.0, and are iron-rich; values of Fe_d generally being above 4%, and reaching 7.8% in OD20. However, high Fe_o amounts give high Fe_o/Fe_d ratios ('activity ratios': Blume and Schwertmann 1969), especially in OD21 (0.13 to 0.37). Torrent et al. (1980: 192) found that the Fe_o/Fe_d ratio for red soils in Spain, irrespective of parent material and environment, rarely exceeds 0.15 and is usually lower than 0.07.

Evidence of slight clay illuviation with cappings of clay on clasts and clay bridges was observed in OD1 (40 to 100 cm depth), the syenite soil, OD20, (30 to 65 cm), and SM2 (13 to 18 cm). Particle-size curves give a stronger suggestion of clay illuviation, with clay increasing markedly with depth whilst the ratio of silt to sand tends to remain fairly constant. The strongly seasonal rainfall would favour clay illuviation, but its effects would suggest a degree of soil age, and therefore of

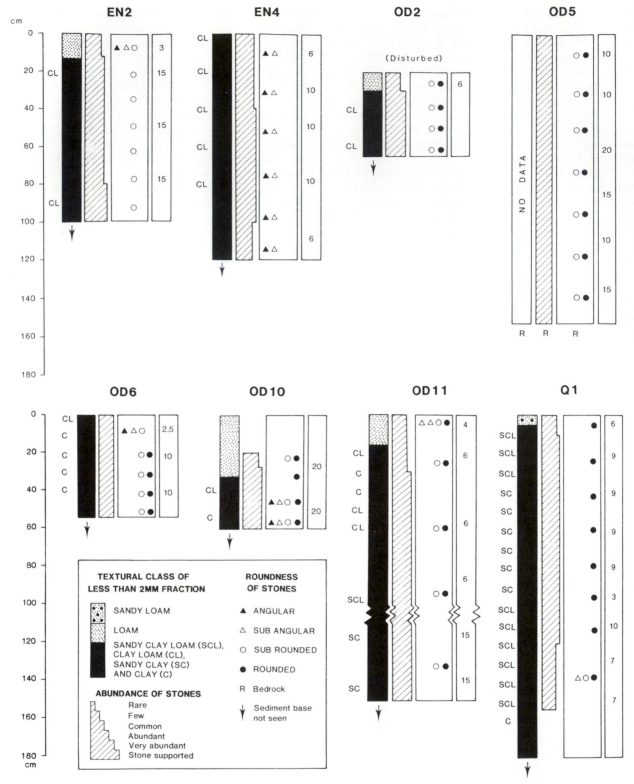

Figure 3. Older fill profiles: sedimentology.

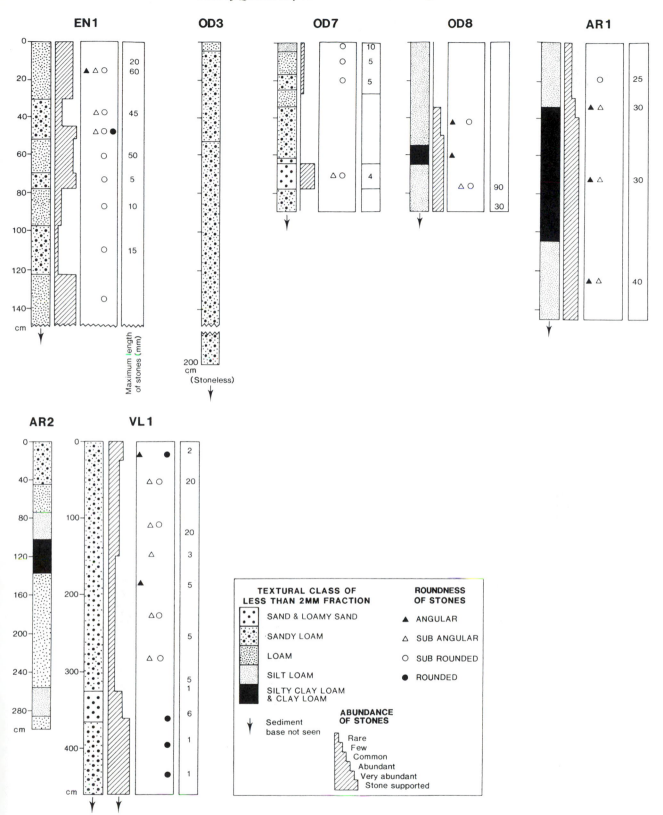

Figure 4. Younger fill profiles: sedimentology.

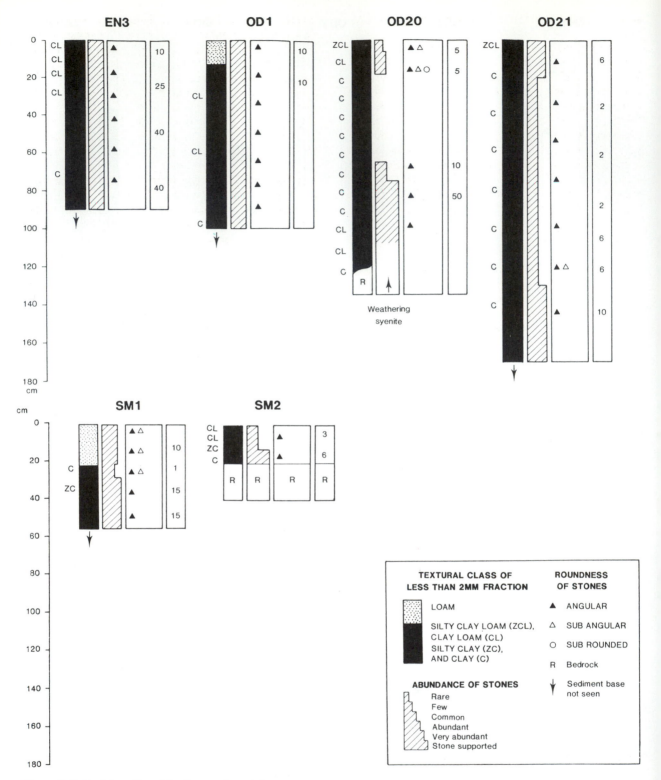

Figure 5. Colluvial profiles of the Serra: sedimentology.

stability of at least the subsoil upon certain slopes of the Serra.

The high iron content of the metasediment soils is reflected in their reddish colours. These vary between 7.5YR7/2 (dark brown: redness rating 6.6) and 10R4/6 (red: redness rating 15). Mottles of 10R3/6 (dark red: redness rating 20) occur at 1.2 m depth in the syenite soil, OD20. The surfaces of quite freshly exposed metasediment rock (as well as clasts in the soils) are commonly very red. A hue of 10YR (redness rating 0) is uncommon in the Serra soils, occurring in organic surface horizons and shallow (< 10 cm) A1 horizons (e.g. 10YR4/2 – dark greyish brown) beneath dense scrub woodland, but not beneath the Cistus mattoral.

5.2 *Soils of the Older Fill*

The older fill soils are significantly redder than those of the younger fill, but there is no statistically significant difference between the redness ratings of older fill and catchment soils (Table 3). Clasts are smaller in maximum size and more rounded than in the younger fill sites examined, though subangular and angular clasts occur in some of the older fill profiles, particularly in EN4 (Fig. 3)

Table 2. Percent weight of calcium carbonate. Sample depths shown in Figure 9.

A. Soils of the Arade and Lombos Younger Fill.

AR1		AR2		VL1	
1	0.84	1	1.08	1	13.03
2	0.00	2	1.20	2	6.34
3	0.28	3	3.01	3	9.18
4	0.88	4	3.54	4	5.62
5	4.30	5	2.36	5	15.19
6	5.73	6	3.97	6	0.93
7	5.56	7	3.38	7	5.58
8	6.49	8	0.41	8	3.34
9	9.23	9	0.30	9	1.90
10	6.32	10	0.10	10	1.84
11	6.19	11	0.62	11	5.61
12	8.15			12	7.27
13	0.00			13	7.43
				14	12.11
				15	0.92
				17	10.67
				18	2.74
				19	8.49
				20	4.87

B. Soils of the Enxerim and Quarteira Older Fill.

EN4		Q1	
1	0.60	1	0.00
2	0.79	2	0.19
3	0.49	3	0.21
4	0.39	4	0.39
5	0.20	5	0.20
6	0.58	6	0.00
		7	0.00
		8	0.00

where they were recorded throughout the profile exposed and may be attributed to movement off the valley side-slope. Clast lithology is dominated by the metasediments, with quartz and quartzite being more abundant than in the younger fill. Sandstone occurs in OD6 and Q1, and syenite cobbles are common in OD5. In general, the matrix particle-size is much finer than in the younger fill, clays and clay loams being predominant. The lowland zone profile, Q1, contained a high proportion of sand in the lower horizons, as high as 77% in the sandy clay. The sand is derived from the Faro-Quarteira formation.

Soils of the older fill are typically Mediterranean: they are reddened and contain redistributed clay and, in some cases, calcium carbonate. There is an increase in clay content (which is mostly < 1 mm) with depth in all profiles (Fig. 6). In a number, clay deposition features in Bt horizons are pronounced and comprise inter-grain bridges, void fillings and coatings of aggregates, sand grains and clasts. The upper surfaces of platy clasts commonly have clay cappings of 2 mm maximum thickness. Judging from the depth curves for the clay fraction (Fig. 6), the upper boundary of the Bt horizon lies at 20 to 40 cm below the present ground surface. This may reflect either erosional stability of the older fill surfaces since clay illuviation took place, or recent continuation of clay illuviation.

Low amounts of free calcium carbonate in the older fill soils (limited to profiles Q1 and EN4) reflect the acid soils of the chief sediment source, the Serra. EN4 was the only soil sampled in the Serra where pH was near or above neutral. The source of calcium may have been atmospheric, though apatite is present in the sand fraction of this soil.

5.3 *Soils of the Younger Fill*

The younger fill is brown and is coarser than the older fill, both in clast and matrix size. Brown colours (10YR3/3 to 7.5YR4/2) predominate, with reddish browns (5YR3/3 to 4/4) occurring more frequently in the Arade and Lombos than in the Serra valleys. Yellowish-red (5YR4/6) was recorded at one level in the middle of the VL1 profile. The mean Fe_d and Fe_o/Fe_d ratio for the younger fill samples is 2.45 and 0.031 respectively (Table 3), values for both variables being significantly lower than for older fill and Serra colluvial soils in the Mann-Whitney U test (Siegel 1956).

Charcoal is abundant, particularly in OD7 (from zero to > 70 cm depth); 70 cm and 80 cm in OD8; throughout EN1 down to 114 cm; to 111 cm in AR1; between 280 cm and 300 cm in AR2; and common to 80 cm depth in VL1. This amount of charcoal, particularly in sediments deposited under fairly high energy conditions, reflects very extensive burning over a period of centuries (at least since the Bronze Age in the case of the Arade catchment).

The coarsest sediments occur in the more confined valleys of the Enxerim and Lombos with clast-supported units in EN1 and VL1 (Fig. 4). One profile, OD3, contained no clasts. Though largely subrounded, subangular and angular clasts are locally important, especially where

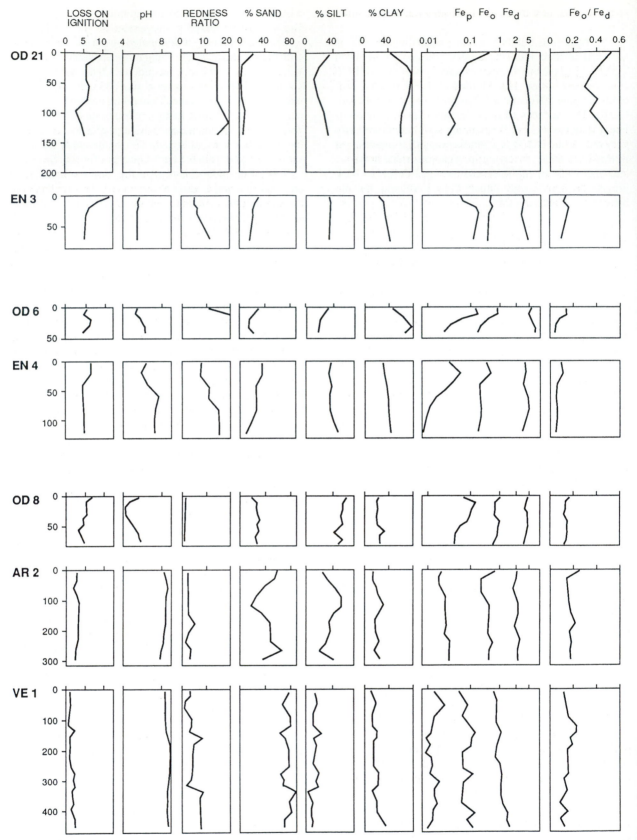

Figure 6. Physical and chemical data for selected catchment and valley fill soils. OD21, EN3: catchment; OD6, EN4: older fill; OD8, AR2, VE1: younger fill.

Table 3. Means of soil variables in the three sediment groups.

	Sand	Silt	Clay	pH	% LOI	Fe_p	Fe_o	Fe_d	Fe_o/Fe_d	RR
Younger Fill	46.2	34.7	19.0	3.65-9.1	3.12	.031	.296	2.45	.031	2.56
Older Fill	39.7	23.7	35.2	4.1-7.6	4.04	.057	.247	3.69	.072	10.78
Catchment colluvial	17.6	36.5	45.8	4.6-6.0	6.4	.145	.949	4.56	.208	9.35

there is likely to have been inputs from the valley-sides (EN1, OD8, AR1). The lithological composition included shales, quartzites and quartz of the Serra Carboniferous rocks, with limestone in AR1, AR2 and VL1. The particle-size distribution of the < 2 mm matrix varies from fine loams in the Odelouca and Arade valleys (maximum values of < 2 mm are 32% in AR1 and 30% in OD8) to sands and sandy loams elsewhere. Appreciable variation in particle-size of sediments occurs across valleys, as in the transects of OD3 – OD7 – OD8 and AR1 – AR2.

Evidence of pedogenesis in the younger fill sediments is important to the question of the origin of the brown colours of the soils. Cessation of deposition at most younger fill sites sampled has permitted pedogenetic features to be superimposed upon the sedimentary. A relatively rapid process is bioturbation, particularly by earthworms. Casts and channels dominated the structure of a 30 cm-deep horizon in OD3 and the uppermost 5 cm of OD8. Worm casts and channels were observed to depths of up to 110 cm (AR1). The effect of worm activity is much less in OD7 than in OD8, the latter lying on the older surface of a terrace which stands 1 m higher than the floodplain at OD7. A weak subangular blocky soil structure has developed in all the younger fill profiles with the exception of OD7, which was massive below a 5 cm surface layer of recently deposited laminae of silty fine sand. In AR1 a prismatic structure occurred in the silty clay loam between 40 and 110 cm depth. Given the structural development in A horizons of these soils, greater near-surface values of loss on ignition (Fig. 6) are probably pedogenetic rather than sedimentary. However, incorporation of organic matter since deposition is not the reason for the colours which occur throughout the depths examined.

Gleyed horizons were encountered in AR1 below 110 cm and in AR2 below 260 cm. Other profiles appear not to have been dug sufficiently deep to reach horizons affected by the water table. There was evidence of slight clay illuviation in AR1, with clay deposits in the voids and lining the prism faces of the fine loam between 70 and 90 cm. Judging from more pronounced illuviated clay deposits in the basal unit of VL1, and from the change in other characteristics at its upper boundary (340 cm), it appears that this unit may be an older sediment underlying the younger fill in the Lombos valley.

The alkalinity of the Arade and Lombos sediments contrasts with the acidity of valley floor soils in the Serra and reflects the influence of the limestones of the southern geological zone. Amounts of calcium carbonate are highest in the younger fill of the Arade and Lombos (Table 2), and only in VL1 is secondary carbonate visible

in the form of thin veins between 25 cm and 230 cm depth. At no depth is the sediment cemented. The bulk of carbonate in these profiles is not pedogenetic, occurring in the detrital clasts and smaller fractions. As there was no visible evidence in the field of redeposited pedogenetic carbonate in the Arade profiles, the carbonate content probably reflects the changing relative contributions of calcareous and metasediment sources. In AR1, values are low in the uppermost 40 cm and highest (> 8%) in the Moorish and Roman horizons. One possible reason for this distribution is greater land disturbance in the lowland limestone catchments of the Arade River catchment system at these times than in more recent centuries. However, the disturbance may have been only local, as the occurrence of angular and subangular clasts in much of AR1 suggests that the source of the limestone was the near valley-side. In AR2, the profile nearer to the present river course, the limestone areas of the catchment appear to have been of little importance in the earlier stages of deposition (c. 2960 BP and later). Their contribution then increased, and later decreased toward the present surface.

One reason for deriving the mineral magnetic data was the tendency for the development in topsoils of iron oxides which carry easily measured signatures that persist through erosion, transport and deposition. These magnetic characteristics therefore have potential both for defining A horizon development in soils and for identifying a topsoil provenance in alluvial sediments (Thompson and Oldfield 1986). The differentiation comprises a magnetic 'enhancement' caused by development of fine ferrimagnetic minerals during fermentation processes (Le Borgne 1955; Mullins 1977), though burning has an even more marked effect (Longworth and Tite 1977; Maher 1986), a fact of particular significance in a Mediterranean climatic environment. This topsoil distinction, reflected in values of low frequency susceptibility, does not occur in the young soils of the Holocene fill (Fig. 7), but is a feature of the older fill and colluvial soils. It is particularly marked in the latter. The magnetic susceptibility values throughout the depths of the younger fill profiles, even though they vary appreciably between VL1 and EN1, are not inconsistent with a colluvial topsoil source (Fig. 7).

6 MULTIVARIATE ANALYSIS

In order to analyse relationships between groups of sediments, two stages of multivariate analysis were undertaken. Multiple discriminant analysis was used, first, to evaluate the difference between soils grouped on the

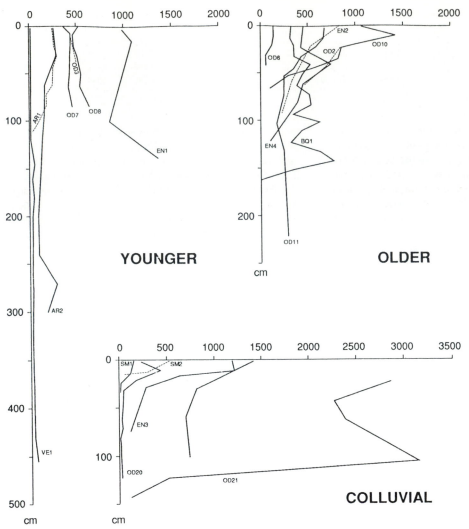

Figure 7. Low frequency susceptibility profiles for younger and older fill and colluvial soils. Units are 10^{-8} m^3 kg^{-1}.

basis of sediment (the younger and older valley fills and the catchment colluvium), and on the basis of catchment; and, second, to find which are the key characteristics that determine the differences. Cluster analysis, which is able to define smaller sample subsets than can be treated in multiple discriminant analysis, is more appropriate for considering variations within the major groups which might reflect a soil chronosequence.

6.1 Multiple discriminant analysis

Results of the discriminant analysis of the sediment groups on the basis of 'soil' variables are shown in Table 4. The two discriminant functions derived explain 81.19% and 18.81% of the variance respectively. With the first, the higher correlations with the discriminant function (> 0.5) are, in descending order, redness, clay, sand, pH (negative) and % loss on ignition. The first two variables are associated with Mediterranean red soils, including those expected to form on older fill. The fourth and fifth are associated with acid, organic-rich soil condi-

tions. The discriminant analysis, therefore, is underlining a basic construct in the data which combines the climatic regional ('zonal') soil characteristics with the local very important acid soil-forming influence of the Serra rocks. The second function is defined as one of 'activity ratio' by the dominance of Fe$_o$ (negative) and Fe$_d$.

The classification procedure shows that 91.52% of the sediment group samples are 'correctly' classified, with the most discrete group being the younger fill (97.2% correctly classified). Only two younger fill samples are reallocated, these being OD8/3 and OD8/4 which have a greater affinity for the colluvial soils. OD8 is one of the valley floor sites where angular colluvial material appears to have mudflowed off the near valley-side. One sample, OD1/1, is imported from the catchment into the younger fill group, and two samples are reallocated from the older fill: OD10/2 and Q1/1. That these three are all near-surface samples gives a hint of a topsoil source for the younger fill. There is a much less clear-cut difference between older fill and catchment samples, with 9 being reallocated.

The plot of cases by their scores on the two functions (Fig. 8) shows how the three sediment groups separate. On the first function, that involving the red and acid plus organic soil characteristics, the brown younger fill samples group clearly on the negative side and the older fill and colluvial on the positive. On the second, 'activity ratio', function, the younger fill overlaps in the positive values with the colluvial samples, and in the negative with the older fill. The separation of the majority of colluvial and older fill soils on this function is marked.

The plot is an efficient summary of the elements of variation in soil redness: the older fill and colluvial soils are distinctly redder than the younger fill but are themselves different on the basis of crystallinity of iron oxides (the second function). The clear implication is that the redness of the older fill is a factor of age: whatever the original state of the sediment, its iron oxides are now aged. The colluvial soils are very red, but contain iron oxides which are less aged.

For the discriminant analysis of soil variables by catchment, the samples are grouped into valleys of the Serra, of lowland (Lombos and Quarteira) and of the Arade, which catches sediment from both Serra and lowland. The analysis yields a first function which explains 96.43% of the variance. From the variable coefficients it is clear that soils are distinguished on the basis of particle-size and reaction. In contrast with the distinction between soils of sediment groups, iron oxide fractions and especially redness have little if any importance as discriminating variables. Thus, as may be expected, soils are distinguished between catchments on the basis of properties which reflect sediment source, but are distinguished between the groups of sediment by age-related properties.

In the discriminant analysis of the mineral magnetic data the functions derived are difficult to interpret, though the properties measured appear to reflect influences associated more with catchment (and therefore reflecting a lithological control) than with sediment group. In the analysis of the sediment groups, more than half of the colluvial samples are reallocated, the greater part of these to the younger fill group. There is here a hint of a sediment-source linkage.

6.2 *Factor and cluster analysis*

The first factor extracted from analysis of the 'soil' data explains 40.1% of the variance. It is associated with organic-rich and acid topsoil horizons and reflects the importance in the sample of pedogenetic conditions associated with the Serra: variables with high loadings are percent loss-on-ignition, depth (negative), pH (negative) and Fe_d and Fe_p. The second factor (explaining 15.4% of the variance) is a 'Mediterranean red soil' factor: redness, clay and Fe_d being the only variables with high loadings.

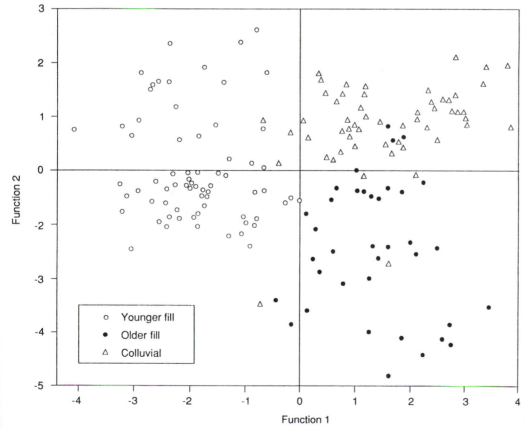

Figure 8. Multiple discriminant analysis of three sediment groups (using 'soil' data): plot of samples by scores on functons 1 and 2 (see Table 4).

Table 4. Multiple disciminant analysis: Soil variables by sediment group.

A. Discriminant functions.

	Function 1 (81.19% variance)	Function 2 (18.81% variance)
Redness index	* .7466	.2508
% Clay	* .7197	– .0801
% Sand	* .6526	.3440
pH	* –.6249	.1523
% LOI	* .5266	.0671
% Fe_p	.3545	.3097
% Fe_o	–.3363	* –1.2082
% Fe_d	–.2348	* .5511
% Silt	.1573	.2172

B. Classification.

Sediment group	No. of cases	Predicted group membership		
		1	2	3
Younger fill	71	69	0	2
		97.2%	0%	2.8%
Older fill	55	2	51	2
		3.6%	92.7%	3.6%
Colluvial	39	1	7	31
		2.6%	17.9%	79.5%

91.52% of cases 'correctly' classified.

Table 5. Characteristics of six clusters, based on means. YF = younger fill; OF = older fill; COL = colluvial.

1	Brownest; least clay; sandy; mod. Fe_d; rel. low activity ratio; strongly acid to neutral; mod. organic	YF: 74%; OF: 13%; COL: 13%
2	Brown; siltiest; rel. low Fe_d; 2nd highest activity ratio; strongly acid to alkaline; mod. organic	YF: 80%; OF: 6%; COL: 14%
3	Mod. red; clayey; silty; high Fe_d; low activity ratio; strongly acid to neutral; highest organic content	YF: 0%; OF: 56%; COL: 44%
4	Red; least silty; sandy; clayey; mod. high Fe_d; lowest activity ratio; strongly acid to neutral; low to mod. organic content	YF: 0%; OF: 100%; COL: 0%
5	Reddest; least silty; sandy; highest clay; highest for all three iron oxide fractions; highest activity ratio; mod. organic	YF: 0%; OF: 0%; COL: 100%
6	Slightly red; most sandy; low silt and clay; lowest for all three iron fractions; 3rd highest activity ratio; alkaline; least organic	YF: 100% – profile VL1

The third factor explains only 8.7% of the variance, but is of theoretical interest in that it is essentially an 'actvity ratio' factor, the highest loadings being for Fe_o/Fe_d and sand (negative). The analysis emphasizes that the soils of the Algarve are typical of the Mediterranean, but that the influence of the Serra rocks in providing acid parent materials is of great importance.

The cluster analysis of the 'soil data' using the scores of each sample on the first three factors produces an efficient grouping of the 166 samples into six clusters. Figure 9 shows the sites in their geomorphological context and maps each sample according to its cluster membership. The characteristics of the six clusters are summarized in Table 5. It is not feasible to present a dendrogram for this sample size.

The distinctiveness of the younger fill and the comparative similarity between the older fill and colluvial catchment soils highlighted by discriminant analysis are clear. The majority of younger fill profiles are relatively uniform with depth. Within this group there emerges a subdivision on the basis of catchment (sediment source). That the Arade and Lombos younger fills are slightly redder than those of the valleys which are wholly within the Serra may have implications for the history of pedological change in the lowland catchments, particularly A horizon development in relation to the timing and nature of erosion during the Holocene. The occurrence of VL1 uniquely in cluster 6 reflects the relatively iron-deficient sands and the limestones of this catchment.

Apart from the concentration of the older fill soils into clusters 3 and 4, there is little pattern evident in the clustering within this group. The exception is the separation of samples of several profiles into two or more clusters, with the samples from upper horizons falling into clusters 1, 2 or 3. This division reflects pedogenesis, the first three clusters having greater affinity with topsoil conditions (and also with the younger fill). More fully developed soils with greater differentiation within the profile could be expected to fall across several clusters. This aspect of cluster membership of samples within single profiles could be used to consider the existence of a chronosequence, a tool of some potential value given the need to differentiate stratigraphically within the older fill of the Algarve. However, in this case there is nothing which suggests an age-dependent trend, though the number of profiles sampled is small, and erosion is likely to have truncated some of the older fill soils.

The important distinction of 'activity ratio' separates some of the older fill and colluvial soils between clusters 4 and 5. Cluster 4, comprising exclusively (and mostly subsoil) older fill, has the lowest ratio; cluster 5, containing a number of colluvial samples only, has the highest ratio. Thus although the colluvial samples of cluster 5 have the reddest colour of all clusters, their secondary iron oxides are evidently less aged than those of the older fill. This distinction reflects the two controls of parent rock and age upon the redness of Algarve soils which are highlighted in the discriminant analysis.

The clustering of the Serra catchment soils shows there to be a degree of similarity between near-surface samples and the sediments of the younger fill throughout the depths sampled (Clusters 1 and 2). This similarity suggests a sediment – source linkage, though the horizons of the Serra soils which do cluster with the younger fill are very shallow, varying between 2 cm and 10 cm in thickness.

With the aim of investigating the mineral magnetic reflection of pedogenesis which could define a chronosequence within the older fill and a possible sediment-

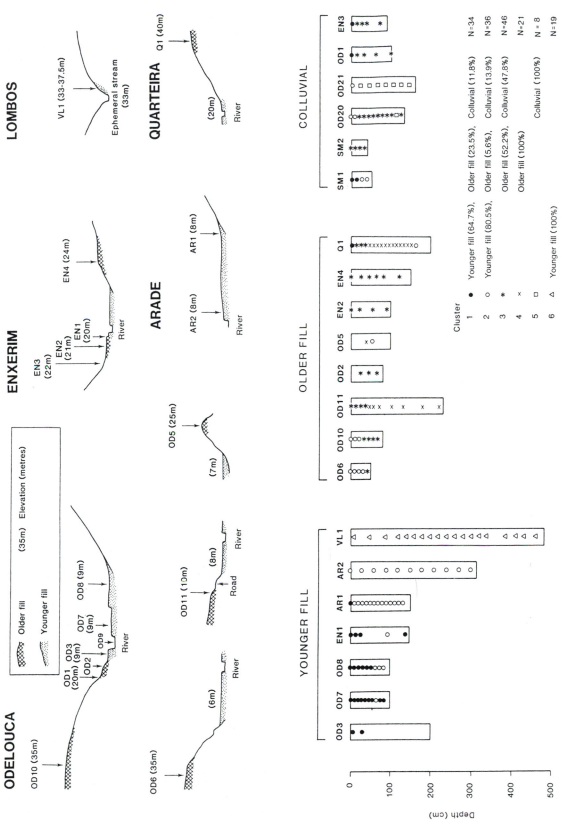

Figure 9. Cluster analysis of samples (using 'soil' data): distribution and composition of six clusters.

source linkage, a cluster analysis was also performed on the basis of the mineral magnetic variables listed in Table 1B. Sixty two of the 166 samples form a cluster which has means for parameters which fall within the middle of their range, except for the 'hard' parameter which is high. This pattern of data reflects a mixture of magnetic mineral type and grain size. The properties of the younger fill relate to a greater catchment control than is evident in the older fill and colluvial soils. Profiles in the first group form 'tighter' clusters with less spread across cluster boundaries than occurs in the second and third groups, a feature of the cluster analysis based on the 'soil' variables. The greater degree of pedogenesis in both older fill and colluvial soils is here defined by the higher amounts of ferrimagnetic minerals in the A horizon. This 'softer' magnetic character may also reflect a greater influence of fire upon the older fill surfaces and hillslopes.

There is no clear pattern of magnetic similarity between younger fill sediments and the topsoils of their catchments. The exception are the Enxerim profiles, though the number of sites is too small for any firm inferences to be drawn. If anything, the clustering of the Odelouca samples suggests a dissimilarity between the younger fill and catchment topsoils. This could have arisen from selective deposition of, or diagenesis within the former, or it may reflect the fact that the present catchment topsoils may not be representative of those which contributed to the fill. It may, on the other hand, reflect a complexity in controls on mineral magnetic properties which is difficult to 'unscramble'.

A cluster analysis was undertaken of the mineralogical composition of four sand fractions determined for a number of horizons in older and younger fill and colluvial soils of the Odelouca, Enxerim and Arade valleys by Ashton (1989). The results show a dominance of catchment (lithological) control over age-related weathering patterns.

7 DISCUSSION AND CONCLUSION

The distinctiveness of the relatively brown younger fill which is evident in the landscape is reinforced at all stages of the analysis. Multiple discriminant analysis of the soil data separates this sediment group from the others by a function combining variables which relate to both rubification and the development of acid soils. These sets of processes are represented respectively in the first two factors extracted in the factor analysis and clearly define the two most essential trends which characterize the soil geography of the Algarve river sediments and Serra hillslopes.

The distinction between colluvial and older fill soils, though less marked than that between these and the soils of the younger fill, is brought out in both cluster and multiple discriminant analysis. Both groups of soils are red, but are clearly distinguished by the ratio of Fe_o/Fe_d which defines the second discriminant function in the analysis of sediment groups using the soil variables, and describes the third factor in the factor analysis. The lower

ratios of the older fill are inferred to reflect the greater age of this sediment group.

Within the separate identity of the younger fill there is a secondary division by catchment, defined on the basis of soil physical, chemical and mineral magnetic data, and reflecting the differing sediment sources of the Serra, the limestones in the southern portion of the Arade catchment, and the limestones and fluvial sands of the lowland valley systems. This secondary, but important catchment division also applies to the older fill and colluvial soils and is defined in multiple discriminant analysis by one function based on soil particle size distribution and pH (rather than redness and secondary iron oxide fractions which are more important in distinguishing the sediment groups). Whereas the mineral magnetic properties of the younger fill vary with catchment, no such variation appears in the older fill and colluvial groups.

The soils of older and younger fill owe many of their characteristics to their relative age, and within the latter group there are differences in soil development which reflect the length of time since the sediment surface has been stable. However, the number of profiles sampled in the older fill is insufficient to detect any soil chronosequence within the patterns that the analysis has depicted.

Despite a suggestion of a sediment-source linkage between younger fill and the colluvial soils of its catchments, it is the contrast between this Holocene fill and the hillslope soils evident in the field and borne out by the multivariate analyses that is striking. The colour, abundance of charcoal and magnetic susceptibility of the fill all point to a topsoil source which is poorly represented in the present soils of the Serra. Thus one hypothesis is that a former brown topsoil has been almost entirely stripped from the hills during the last three millennia and now lies, in part, in the valley bottoms. The lack of subsoil characteristics in the matrix of the younger fill of at least the Odelouca and Enxerim valleys may indicate that during the late Holocene, erosion in the Serra has been more superficial than deep-seated, more consistent with processes of overland flow or rilling than with gullying. The extensive stripping of near-surface soil has been more important than selective erosion by increased downcutting of stream channels, the former a more likely outcome of widespread human disturbance of the vegetation than of relatively minor climatic change. We have concluded earlier that an anthropogenic forcing of erosion would be in keeping with the documented evidence of land disturbance at least since the Moorish settlement of the region (Chester and James 1991).

Two further hypotheses for the brown colours of the younger fill are diagenetic change and selective removal of a redder fine component of the sediment during fluvial transport. The 'yellowing' of red Mediterranean soils in imperfectly drained situations has been attributed to the preferential reductive dissolution of very fine haematite over that of goethite (Boero and Schwertmann 1987). However, hydromorphic features in the brown and reddish-brown younger fill soils were restricted to depth in the profile. Generally, magnetic susceptibility does not decrease with depth which would be expected in gleyed horizons (Fig. 7). In their upper portions they are, at the present day, well aerated throughout the year. That there has been a loss of fine redder material from the younger fill is certain: it contains significantly less clay and significantly more sand than the colluvial soils; there has been no relative depletion of silt (Table 3). There is a significant correlation between clay content and redness throughout the whole sample set collected from all three sediment groups ($r = 0.6549$; $p < 0.001$), but there is no indication from material in the field or laboratory that coarser fractions from the metasediment soils have the brown hue of the younger fill.

In their analysis of recent alluvial deposits in Epirus, Greece, Macleod and Vita-Finzi (1982) attribute the brown colours of the alluvial sediments partly to silt and clay depletion and partly to hydrous iron oxides derived from brown Mediterranean soils which occur with *terra rossa* soils in the catchment. We favour a similar explanation for the colour of the Algarve younger fill, but in this case the brown soil of the catchments must be a feature of the past. In many areas of the Mediterranean lands, the difference in colour between Holocene and Pleistocene river sediments has been attributed to age-related rubification. In the Algarve, that process has contributed to the redness of Pleistocene river sediments, but a more essential reason for the colour contrast may lie in the history of soil development and erosion in the catchments during the Holocene.

8 ACKNOWLEDGEMENTS

We wish to thank the University of Liverpool for financial support, Alan Henderson for undertaking the laboratory analyses, Sandra Mather for drawing the diagrams, Ian Qualtrough and Suzanne Yee for their photographic work and Bob Jude and Derek France for assistance with the mineral magnetic analyses. We are also indebted to our colleagues, John Dickenson, Phil Lister and Bob Pullan for their support and to Dr Jack Ashworth, Ms Marjalena Vita-Muller and Mr Peter Harris for their kind help in the Algarve.

REFERENCES

Ashton, J.M. 1989. Sand mineralogy in a selection of soils from the Algarve region of southern Portugal. Unpublished B.Sc. thesis, University of Liverpool.

Atlas do Ambiente, Portugal 1:1,000,000 1974. Presidência do Conselho de Ministros Secretaria de Estado do Ambiente, Commissão Nacional do Ambient, Lisboa.

Blume, H.D. & U. Schwertmann 1969. Genetic evaluation of profile distribution of aluminium and manganese oxides. *Soil Sci. Soc. Am. Proc.* 33: 438-444.

Boero, V. & U. Schwertmann 1987. Occurrence and transformations of iron and manganese in a colluvial terra rossa toposequence of northern Italy. *Catena* 14: 519-531.

Carta Agricola e Florestal de Portugal 1:50,000 1957. Serviço de Reconhecimento e de Ordenamento Agrario, Secretaria de Estado da Agricultura, No 595.

Carta de Capacidado de Uso do Solo de Portugal 1:50,000 1959. Serviço de Reconhecimento e de Ordenamento Agrario, Secretaria de Estado da Agricultura, No 49D.

Carta dos Solos de Portugal 1:50,000 1959. Serviço de Reconhecimento e de Ordenamento Agrario, Secretaria de Estado da Agricultura, Nos 49C, 49D, 52B.

Chester, D.K. & P.A. James 1991. Holocene alluviation in the Algarve, southern Portugal: The case for an anthropogenic cause. *J. Arch. Sci.* 18: 73-87.

Devereux, C.M. 1983. Recent erosion and sedimentation in southern Portugal. Unpublished Ph.D. thesis, University of London.

Espejo, R. 1987. The soils and ages of the 'raña' surfaces related to the Villuercas and Altamira mountain ranges (Western Spain). *Catena* 14: 399-418.

Feio, M. 1949. Le bas Alentejo et l'Algarve (Livret-guide de l'excursion E). *Congrès Int. Géographie, Lisboa.*

Fine, P., M.J. Singer, R. Laven, K. Verosub & R.J. Southard 1989. Role of pedogenesis in the distribution of magnetic susceptibility in two California chronosequences. *Geoderma* 44: 287-306.

Fine, P., M.J. Singer & K.L. Verosub 1992. Use of magnetic susceptibility in chronosequence studies. *Soil Sci. Soc. Am. J.* 56: 1185-1192.

Harman, H.H. 1967. *Modern factor analysis.* Chicago: University of Chicago Press.

Le Borgne, E. 1955. Susceptibilité magnétique anormale du sol superficiel. *Ann. Géophys.* 11: 399-419.

Longworth, G. & M.S. Tite 1977. Mossbauer and magnetic suscepbility studies of iron oxides in soils from archaeological sites. *Archaeometry.* 19: 3-14.

Macleod, D.A. & C. Vita-Finzi 1982. Environment and provenance in the development of recent alluvial deposits in Epirus, N.W. Greece. *Earth Surf. Proc. Landf.* 7: 29-43.

Maher, B.A. 1986. Characterisation of soil by mineral magnetic measurement. *Phys. Earth Planetary Interiors.* 42: 76-92.

Manupella, G., M. Ramalho, M. Telles Antunes, & J. Pais 1987. Notiça explicativa da Folha 53A: Faro. Serviços Geologicos de Portugal, Lisboa. 52p.

McFadden, L.D. & D.M. Hendricks 1985. Changes in the content and composition pedogenic iron oxyhydroxides in a chronosequence of soils in southern California. *Quaternary Research* 23: 189-204.

McKeague, J.A. 1967. An evaluation of 0.1M pyrophosphate and pyrophosphate-dithionite in comparison with oxalate as extractants of the accumulation products in podzols and some other soils. *Can. J. Soil Sci.* 47: 95-99.

Mehra, O.P. & M.L. Jackson 1960. Iron oxide removal from soils and clays by a dithionite-citrate-bicarbonate system buffered with sodium bicarbonate. *Clay Clay M.* 7: 317-327.

Mullins, C.E. 1977. Magnetic susceptibility of the soil and its significance for soil science: a review. *J. Soil Sci.* 28: 223-246.

Oliveira, J.T. 1983. The marine Carboniferous of south Portugal: A stratigraphic and sedimentological approach. *Mem. Serv. Geol. Port.* 29: 3-37.

Ribeiro, O. & J. Brandão-Silva 1983. Structure of the south Portuguese zone. *Mem. Serv. Geol. Port.* 29: 83-89.

Rocha, R.B., M.M. Ramalho, M.T. Antunes & A.V.P. Coelho 1983. Noticia explicativa da Folha 52-A: Portimão. Carta Geologica de Portugal, Lisboa: Serviços Geologicos de Portugal. 57p.

Rock, N.M.S. 1978. Petrology and petrogenesis of the Monchique alkaline complex, Portugal. *J. Petrol.* 19: 171-214.

Schwertmann, U. 1964 Differnzierung der Eisenoxide des Bodens durch Extraktion mit Ammoniumoxalat-Losung. *Z Pflanzenernaehr. Bodenkd.* 105: 194-202

Siegel, S. 1956. *Non-parametric Statistics for the Behavioral Sciences*. New York: McGraw-Hill.

Stanislawski, D. 1963. *Portugal's other kingdom – the Algarve.* Austin: University of Texas.

Thompson, R. & F. Oldfield 1986. *Environmental Magnetism.* London: Allen and Unwin.

Torrent, J., U. Schwertmann, H. Fechter & F. Alferez 1983. Quantitative relationships between soil color and hematite content. *Soil Sci.* 136: 354-358.

Torrent,J., U. Schwertmann & D.G. Schultz 1980. Iron oxide mineralogy of some soils of two river terrace sequences in Spain. *Geoderma* 23: 191-208.

UNESCO/FAO 1963 Bioclimatic Map of the Mediterranean Zone 1:5,000,000. Arid Zone Research XXI. Paris: UNESCO.

Vita-Finzi, C. 1969. *The Mediterranean Valleys.* Cambridge: Cambridge University Press.

Vita-Finzi, C. 1976. Diachronism in Old World alluvial sequences. *Nature* 263: 218-219.

Zbyszewski, G. 1958. Le Quaternaire du Portugal. *Bull. Soc. Geol. Portugal* 13: 3-227.

CHAPTER 23

Quaternary soil and river terrace sequences in the Aguas/Feos river systems: Sorbas basin, southeast Spain

A.M. HARVEY and S.Y. MILLER
Department of Geography, University of Liverpool, Liverpool, UK

S.G. WELLS
Department of Earth Sciences, University of California, Riverside, California, USA

ABSTRACT: The river system of the Sorbas basin, southeast Spain, was initiated by uplift and emergence during the Pliocene. The early drainage, the proto-Feos, was superimposed, flowing southwards across the southern margin of the basin and later became antecedent as further deformation caused uplift of the Sierras Cabrera and Alhamilla. During the Pleistocene, uplift of the basin caused episodic incision, but also the development of the aggressive strike-orientated Aguas system draining to the east. This captured the Feos, causing further incision of the drainage in the basin centre, complicated locally by previously documented diapiric activity.

The sequence is clearly recorded in the Quaternary river terraces, whose characteristics differ between the three parts of the system; upstream of the capture, downstream along the captor river and downstream along the beheaded system.

Relative age, and correlations both within and between reaches, have been established on the basis of differential degrees of soil development. A chronosequence is presented using evidence from soil profile development, soil colour, carbonate characteristics, iron oxide chemistry and mineral magnetics.

KEYWORDS: River terraces, soil chronosequences, tectonic geomorphology, southeast Spain.

1 INTRODUCTION

Over the past decade advances in soil chronosequence studies have enhanced our understanding of Quaternary landscape evolution (see Knuepfer and McFadden 1990 for review). Much of the recent work relates to the drylands of the American Southwest or of Israel, where soil chronosequences are an established part of the evidence used to reconstruct Quaternary landscape evolution (see Weide 1985; Birkeland 1984, 1985; for reviews). In such regions the progressive development and increasing rubification of argillic B horizons (Hurst 1977), and the subjacent accumulation of pedogenic carbonate (Gile et al. 1966; Machette 1985) are primary morphological features of soil development over time. These soil profile characteristics typically form the basis for correlation of geomorphic surfaces (Harden et al. 1985), and may be used as indicators of relative surface age (Harden 1982; Harden and Taylor 1983) in the establishment of a chronology of landform development, or to indicate past climatic regimes (Gerson 1982).

Despite this recent research in other dry regions, there has been relatively little work applying soil chronosequence analysis to geomorphic sequences in the Mediterranean region. Exceptions are work by Torrent et al. (1980), Pope and van Andel (1984), Ajmone-Marson et al. (1988), and James and Chester (this volume). Otherwise the most common approach in this region has been restricted to using the degree of calcrete development for correlation of geomorphic surfaces, for example on alluvial fans (Harvey 1990) or on pediments (Dumas 1969).

A major field within geomorphology for the application of pedological evidence is the correlation, and at least the relative dating, of fluvial depositional surfaces. Soils on river terraces have a high potential as relative age indicators. Terrace surfaces have a similarity of origin over relatively large areas. They are flat surfaces, and especially when the parent material is gravel, soil development generally starts at the time of deposition without much inheritance from previous soils. Once former floodplains have been transformed into terraces by incision, they tend to be isolated from further fluvial deposition, resulting in simple, age-related soil profile development (Bull 1990, 1991; Harrison et al. 1990). Furthermore, river terraces also may preserve the clearest evidence for the development of the fluvial system through Pleistocene and Holocene time. Thus the correlation and relative dating of river terrace surfaces may provide the key to understanding regional geomorphic evolution.

The purpose of this paper is to elaborate the geomorphic evolution of the Sorbas basin, southeast Spain, during its dissection in the Quaternary, by analysing the terrace sequence of the Aguas/Feos river systems, using the evidence provided by soil development on the terrace surfaces. A simplified soil chronosequence was first described by Harvey and Wells (1987) primarily for soils on terraces near Los Molinos. The work presented here was undertaken to extend the regional cover, and to assess whether more detailed analyses could provide further information towards a chronosequence. It builds on the earlier work that established the importance of river capture in the overall sequence (Harvey and Wells 1987). The initiation and early evolution of the network has been described elsewhere (Mather 1991, 1993; Mather and Harvey, this volume). This paper deals with landform development during the period of incision that followed.

2 THE AGUAS/FEOS RIVER SYSTEM

The Sorbas basin is a small Neogene sedimentary basin within the eastern part of the Betic Cordillera. It originated in mid-Miocene time, with the sedimentary basin fill largely comprising Upper Miocene (Tortonian to Messinian) marine sediments, including evaporitic gypsum beds (Weijermars et al. 1985). Basin filling culminated in the Pliocene, with terrestrial sedimentation in a coastal plain environment (Zorreras Member, after Ruegg 1964; Mather 1991), passing up into coarse, braidplain and alluvial fan deposits (Gochar Formation, described in detail by Mather 1991).

Uplift and deformation initiated the main drainage, the Aguas/Feos system, from the Sierra de los Filabres (Fig. 1), north-south across the basin to an outlet through a structural low between the Sierras Cabrera and Alhamilla and into the Almeria/Carboneras basin to the south (Harvey and Wells 1987; Mather 1991). This transverse fluvial course across the southern margin of the basin was superimposed onto the underlying basement rocks, but

became antecedent with the later uplift of the Sierras de Cabrera and Alhamilla (Harvey and Wells 1987; Mather 1991). Differential epeirogenic uplift of the Sorbas basin at the end of the Gochar Stage (Early Pleistocene) led to the incision of the drainage. The incision, occurring during the Quaternary, was episodic, probably reflecting both tectonic and climatic controls (Harvey 1987). Downcutting was punctuated by periodic aggradation, resulting in a fluvial landscape dominated by a sequence of river terraces. Local modifications to the drainage pattern occurred by stream capture, whereby subsequent streams captured original consequents which drained towards the basin centre, and by continued tectonic and diapiric activity (Mather 1991; Mather and Harvey, this volume).

The most dramatic modification of the network was a major capture of the south-flowing proto Aguas/Feos (Fig. 1) by the east-flowing lower Aguas, an aggressive subsequent stream which was developing by headwards erosion along the outcrop of the weak Abad marls (terminology after Reugg, 1964). The capture took place near Los Molinos (Fig. 1), and resulted in the main drainage of the basin exiting to the east into the Vera basin, rather than to the south (Harvey and Wells 1987). The local results of the capture were the deep incision of the upper Aguas, and the beheading of the former transverse drainage (the main stem of the proto Aguas/Feos) in the southern part of the basin. Previous work (Harvey and Wells 1987;

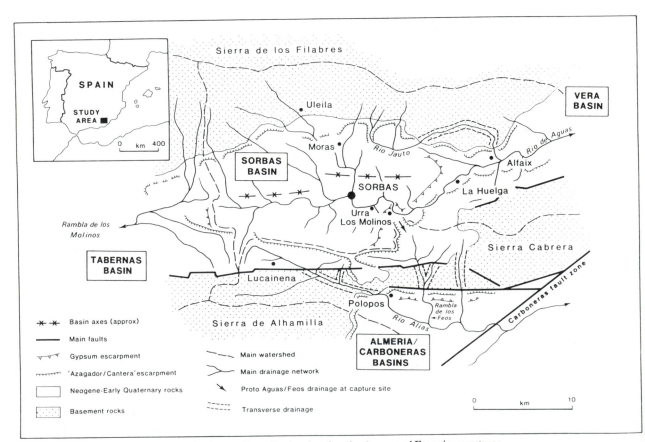

Figure 1. Location map of the Sorbas basin, southeast Spain, showing the Aguas and Feos river systems.

Harvey 1987) had identified the terrace sequence in relation to the capture history. Three main pre-capture terraces (referred to as Terraces A-C) can be traced following the proto drainage across the basin and through the
southern mountain ranges. Younger, post-capture terraces
(referred to as Terraces D-E), relate to modern drainage
directions.

Three contrasting zones in the evolution of the river
system can be defined by the capture. These are: (1)
Upstream of the capture site. Here, the upper reaches of
the Aguas show limited, early, episodic incision, into
which later, deep, capture-induced incision has progressed headwards; (2) Downstream of the capture site,
along the proto Aguas/Feos. Here, early episodic incision was followed by very little post-capture incision.
The modern Feos drains a much smaller catchment than
did the proto Aguas/Feos; (3) Downstream of the capture
site, along the lower Aguas. This zone experienced early
base-level induced incision, with limited post-capture
incision by the (now) main stream of the Aguas system.

3 METHODOLOGY

Within each zone of the study area, all terrace remnants
were mapped (Fig. 2), and classified as far as possible, on
the basis of visual continuity. Heights were surveyed
above the modern stream, whose elevation itself was
estimated from the new (1985) 1: 25,000 topographic
maps. The heights of the base of the terrace gravels,
where exposed, and of the terrace surface, were measured
by leveling up from the modern stream bed. Where
possible, the constituent terrace sediments were examined and described. Clast counts were made at a
number of sites, to augment data already presented (Harvey and Wells 1987), to characterise potential changes in
provenance characteristics. Each sample comprised in
excess of 100 clasts, minimum clast size 2 cm b axis.

Previous work had suggested that terraces A-C could
be distinguished from the younger terraces by their
redder Bt horizons and the presence of pedogenic carbonate horizons (Harvey and Wells 1987). Where the
morphological evidence was in doubt, soil development
was used to suggest a preliminary correlation for the
terrace remnants, which was further considered after
completion of the soil analyses. Soil profile descriptions
were carried out in the field, at sites where the morphological evidence was clear, in order that the soil chronosequence could be characterised. Samples were collected in
the field for laboratory analyses by procedures previously
shown to (Birkeland 1974), or that could be expected to
(Thompson and Oldfield 1986), highlight age-dependent
properties. A total of 18 complete soil profile descriptions
were carried out, and at a further 40 sites colour of the B
horizon was recorded (Table 1). At several sites multiple
samples were collected for laboratory analysis of down-
profile magnetic mineral, particle size and/or iron oxide
properties (Table 1). At a number of other sites B horizon
samples were collected for iron oxide and/or magnetic
mineral analyses (Table 1).

Table 1. Soil sample sites, for locations see Figure 2. At all sites
Bt horizon colour was determined; other data as indicated by
notes.

Rambla de los Chopos
Terrace A?: CH4,
Terrace B: CH3, CH6, CH7,
Terrace C: CH1(4), CH2(5), CH5, CH8, CH9,
Terrace D1: CH10

Rambla de Cinta Blanca
Terrace B: CB1, CB6(7),
Terrace C: CB2,
Terrace D1: CB3, CB4, CB5,

Rambla de Moras
Terrace A: RM2(7),
Terrace B: RM5,
Terrace C: RM1(6), RM3, RM4, RM6,

Rambla de Sorbas
Terrace A: RS4(7),RS6, RS7, RS8, RS11,
Terrace B: RS1, RS2(5), RS3(1), RS5(7), RS9, RS10,

Rio de Aguas (upstream reach)
Terrace B: RA5, RA6, RA12,
Terrace C: RA3(1), RA4(2),
Terrace D1: RA2(1), RA10(5), RA11(5),
Terrace D2: RA7(5), RA8(5), RA9(5), RA13(5),
Terrace D3: RA1(1),
Terrace E: RA14(5),

Rio de Aguas (downstream reach)
Terrace C?: LA1, LA3, LA4,
Terrace D1: LA2,

Ramble de los Feos
Terrace B: LF4,
Terrace D1: UF1, UF2, LF1(1), LF2,
Terrace E: LF3(3)

Notes: (1) Full sample site (profile, particle size, Fe extractions,
magnetics: all down profile); (2) As (1) but magnetics data only
on B horizon; (3) Full profile + Fe extractions down profile,
magnetics on B horizon; (4) Full profile description; (5) As (4)
but not illustrated here; (6) As (5) but also incl Fe extractions
and magnetics from B horizon; (7) B horizon Fe extractions and
magnetics.

In the laboratory (Miller 1991), homogenised sub-
samples of c. 20 g of soil, were taken from the bulk
samples, each representing a restricted vertical zone in
the soil profile. Conventional methods were used for
particle size analysis, dry sieving for the coarser fractions
(> 2 μm), and pipette methods for the finer fractions
(Krumbein and Pettijohn 1938).

The wet sequential extraction method was used to
determine the content of pedogenic iron oxides, from
samples taken down-profile from 7 sites and from Bt
horizons from an additional 5 sites. The pyrophosphate
iron oxide extraction procedure (McKeague 1967) was
used to determine the organically complexed (Fe_2O_3p)
iron, the oxalate iron (Fe_2O_3o) extraction procedure
(McKeague and Day 1966) to determine the paracrystal-
line iron oxides, and the dithionite citrate-bicarbonate
method (Mehra and Jackson 1960) to determine the total
secondary or free iron oxides (Fe_2O_3d).

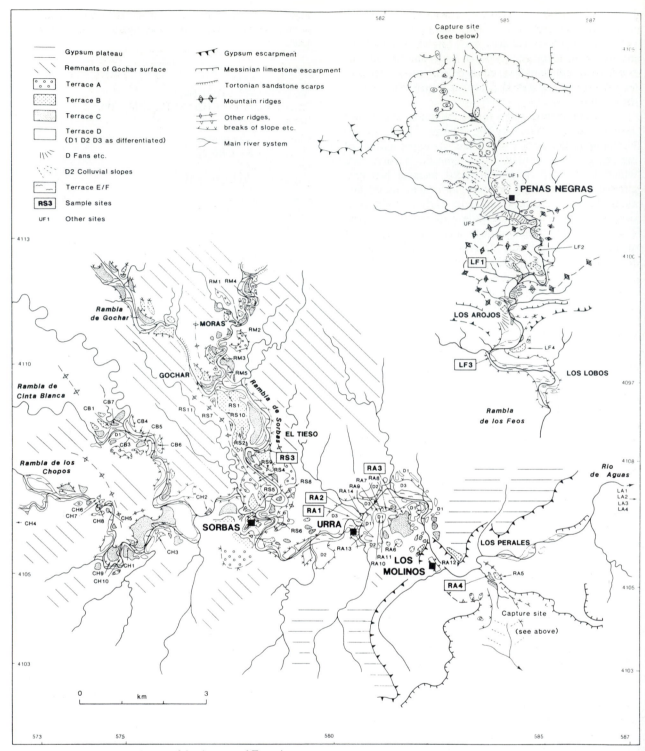

Figure 2. Map of the river terraces of the Aguas and Feos river systems.

Mineral magnetic measurements were carried out on samples down-profile from 5 sites and on an additional 7 individual Bt horizon samples. The measurements included magnetic susceptibility (χ), Anhysteretic Remanent Magnetisation (ARM), Saturated Isothermal Remanent Magnetisation (SIRM) at 1 Tesla (T), and the application of forward (+20 mT, +300 mT) and backfields (–20 mT, – 40 mT, –100 mT, –300 mT), normalised with SIRM, following the standard procedures outlined by Thompson and Oldfield (1986).

4 THE TERRACE SEQUENCE

The terraces were mapped from the Aguas headstreams through to the Feos Valley (Fig. 2), and down the lower Aguas. There, little terrace development is evident until near La Huelga (location: Fig. 1). The long profile relationships are shown on Figure 3.

The terrace sediments have been examined in all three zones of the system, and differences identified between the zones in terms of pre- and post-capture sediment characteristics. Terraces A–C show aggradational thicknesses of up to c. 20 m between Sorbas and Los Molinos, but up to c. 10 m elsewhere. The gravels rest unconformably on basement or Neogene rocks. They are usually well cemented both at the surface, by strongly developed calcretes derived from pedogenic carbonate horizons, and at the base at the gravel/bedrock contact. Terrace D can usually be subdivided into two. The older D1 terrace usually comprises up to c. 10 m of gravels, sometimes weakly cemented or capped by a thin, immature calcrete. The younger D3 terrace is more extensive but thinner (generally < 3 m), and comprises non-cemented gravels. The carbon-dated Holocene terrace E (Harvey and Wells 1987) is less extensive and comprises loose gravels and silts. Between Sorbas and Los Molinos the D terraces can be subdivided into three groups, with the middle D2 group comprising very thick (> 25 m) fine sediments deposited in association with the diapiric activity at Urra, that followed the post-capture erosional unloading of the Messinian gypsum (Mather et al. 1991). In the upper Feos Valley, sediments of D1 age include extensive colluvial silts north of Penas Negras (Fig. 2) and silt and gravel alluvial fans at the tributary junctions through the Sierra Alhamilla.

The main sediment sources feeding the Aguas headstreams are the metamorphic rocks of the Sierra de los Filabres. The differing importance of this sediment source upstream and downstream of the capture site is expressed in the pre- and post-capture sediments in the terrace gravels (Table 2). In the upstream reach there is little difference between the pre- and post-capture sediments, either in terms of size or in terms of lithology (Table 2). The distinctive green spotted biotite granitic gneiss, outcropping in the Filabres north and west of Uleila (Fig. 1), has been traced in decreasing proportions of the total clast assemblage, in the A–C terraces downstream from the upper Aguas into the Feos system (Harvey and Wells 1987), and through to the south of the Sierra de

Table 2. Clast lithologies from river terrace deposits pre- and post- capture, upstream and downstream of the capture site (percentage occurrence).

		U.Aguas Urra		Feos Los Arojos		L.Aguas La Huelga	
Terrace*		C	D3	B	D1	C	F
Green spotted gneiss	(F)	34	38	20	2	0**	7
Schists	(B)	28	20	10	11	35	33
Black phyllites	(A)	0	0	15	26	0	0
Quartz	(B)	27	23	7	6	7	9
Sandstone	(T, Mt)	0	5	4	9	15	6
Dark limestones	(T)	0	0	11	20	15	11
Pale limestones	(Mm)	9	13	32	23	15	18
Marl	(Mt,m)	0	0	0	0	4	11
Misc.	(X)	2	0	2	4	9	5

Potential source areas: (A) Sierras de Alhamilla/Cabrera; (B) Basement, Misc.; (F) Sierra de los Filabres; (M) Miocene rocks, m, Messinian, t, Tortonian; (T) Triassic rocks; (X) misc others e.g. gypsum (Mm), Quaternary calcrete, or unidentified rocks.
*Terraces B ,C are pre-capture; Terraces D, F (modern floodplain sediments) are post-capture.
**After an extensive search none could be found.

Alhamilla (Mather 1991). Along the Feos, post-capture sediments are much finer than the earlier sediments (Harvey and Wells 1987) and contain only rare reworked Filabride clasts (Table 1). In contrast in the lower Aguas, no distinct Filabride clasts can be found in the pre-capture terrace sediments, but they are present in the post-capture sediments (Table 1).

Upstream of the capture site the modern valleys are entrenched below either stripped surfaces cut across Messinian limestone or gypsum, or remnants of surfaces that represent the final Gochar-age basin filling sequence. The terraces, inset into these surfaces, record the Quaternary incisional history of the basin. In the far west of the basin, in the headwaters of the Rambla de los Glapos, there is virtually no incision. Elsewhere in the headstreams the total depth of Quaternary incision is 20-40 m, increasing downstream to c. 90 m at Sorbas, and c. 150 m at the capture site at Los Molinos.

The episodic nature of the incision is apparent in the terrace sequence A–C. All three terraces are present throughout the headstream areas mapped, and are separated altitudinally by up to c. 20 m. They become divergent downstream, perhaps indicating increasing rates of incision downstream. There is then a huge incision through the early stages of terrace D, especially downstream of Sorbas (Fig. 4). This represents the accelerated capture-induced incision working headwards from the capture site.

There are differences in the sequence between the various headstreams. In the Rambla de Sorbas system terrace B is more extensive than terrace C; the opposite is true in the Glapos system. There were three areas of uncertain correlation on the basis of the morphological evidence alone (Figs 2, 3). These areas were on the Rambla de Cinta Blanca, in the Moras area (terraces B and C), and through Urra, where the terrace heights have

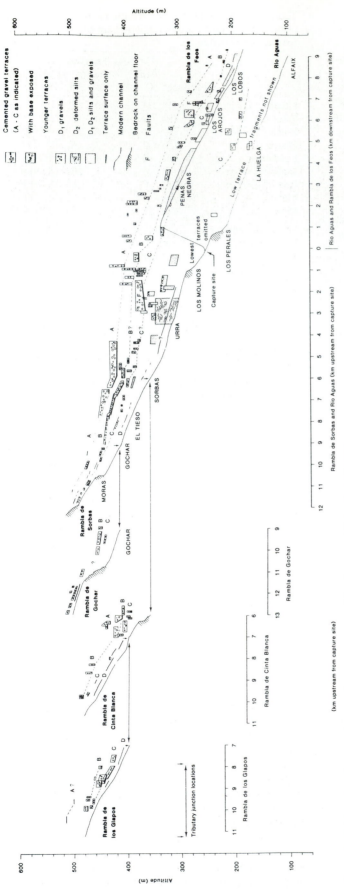

Figure 3. Height range diagram showing the river terraces of the Aguas and Feos river systems.

Figure 4. Quaternary river terraces at Sorbas; photograph taken from 1 km NE of Sorbas looking SE, showing relationships to underlying Messinian to Pliocene bedrock. Key: Bedrock: s Sorbas Member; z Zorreras Member; g Gochar Formation. Quaternary terraces A-E, F is modern valley floor.

been complicated by diapiric activity during D2 times (Mather et al. 1991). In these cases, the most likely ages were assigned later, taking into account the soil evidence.

Downstream of the capture site, along the Feos Valley, the A-C sequence is similar to that upstream of the capture. All three terraces can be traced across the southern margin of the Sorbas basin and through the mountains. There, terrace B at least, has been deformed by faulting (Harvey and Wells 1987). South of the mountains, terrace A is impossible to follow and may well be deformed and buried by younger deposits. Faulted fluviatile gravels overlying Gochar-age (?) fan delta deposits in the Polopos/Lower Feos area, may be the equivalent of this terrace (Mather 1991).

There has been almost no dissection since terrace C times in the Feos Valley. Terrace D deposits bury terrace C. The abundance of terrace D-age colluvial and tributary junction alluvial fan deposits attests to the underfit nature of the beheaded Feos in post-C times.

In the lower Aguas the picture is almost the reverse of that in the Feos Valley. From the capture site downstream to La Huelga, no A-C terraces are present. The whole area is deeply incised, as a result of the headwards erosion through the Abad marls, that brought about the capture in terrace C times. Red soils only occur on the uppermost hillslopes on the south side of the valley, above a zone of landslips induced by the rapid incision. The north side of the valley is occupied by rapidly eroding badlands below the gypsum caprock escarpment (Harvey 1987). Downs-

tream of La Huelga an extensive terrace with a red soil appears to be the local equivalent of terrace C. The clast content indicates a local and Cabrera source area, indicating a pre-capture origin (Table 1). Terrace fragments, apparently equivalent to D1, D3 and E terraces can be traced intermittently from Los Molinos, through the post-capture course of the lower Aguas, downstream to beyond La Huelga.

5 SOIL DEVELOPMENT

5.1 *Soil profile descriptions*

Undisturbed soil profiles on terraces A-C often exceed 1.5-2 m in thickness. Their Bt horizons normally show hues of Munsell 2.5YR colouration, though locally this may range between 10R and 5YR. These soil profiles generally show stage III carbonate accumulation (nomenclature after Machette's 1985 modifications of Gile et al. 1966). Soils on terrace D1 are much thinner (usually < 1.5 m), rarely show 5YR colour, and have stage I to II carbonate accumulation. Younger D terrace soils are even thinner (< 80 cm), with less colour (normally 10YR hues) and at best stage I carbonate morphology. Terrace E soils are very immature, with little or no colour differentiation of horizons and no carbonate accumulation.

Of the 18 soil profiles described in the field (Table 1), 9 are used here to characterise the soil chronosequence (Figs 5, 6). Full profile details are given for 4 profiles,

A.M. Harvey, S.Y. Miller & S.G. Wells

Table 3. Soil profile characteristics from four representative soil profiles.

Horizon	Depth (cm)	Colour (dry)	Texture	Structure	Consistence	Clasts	CaCO$_3$	Clay coatings	Porosity, pores	Roots
RS3 (Terrace B)										
Bt	0-19	2.5YR 4/6	SCL	sm blockey to platey	Firm friable	30-40%, 2-4 cm	None	On ped faces	Medium v comm v fine	Common f fib
Btw	19-32	2.5YR 3/6	CL	v sm blocky	Firm	30-40%, c. 5 cm	None	On peds in pore spaces	Medium common f-vf	Common f-m fib
Btw2	32-58	2.5YR 3/6	CL	Blocky	Moderate-strong	40% up to 10cm	None	Peds bonded and coated	Medium common m-f	Rare m fib common f fib
K (stage IV)	58-103	7.5YR 8/2	Calcrete		Indurated	40%	Solid nodular carbonate	–	–	Rare
K	103-150		Calcrete		Less indurated	80%	Solid nodular carbonate	–	–	–
C	150-250+		Z		Loose friable	60%	In matrix			
CH1 (Terrace C)										
Coll	0-47	5YR 4/4-5/4	SL	–	Firm	50% <5 cm	None	None	Medium common f fib	Common f fib m woody
Bt	47-60	5YR 4/6	SCL	sm blocky	Indurated	20-30% <2 cm	None	None	Medium common f fib	Common f fib
Btw	60-90	5YR 4/8	Gritty SC	sm angular	Firm brittle	20-30% v sm	None	Around clasts	Medium common f fib	Less common f fib
Bk (stage II-III)	90-133	–	ZL	–	Indurated	40-70%	Laminar	–	Low common-few	Rare
C	133-202	–	Gravels	–	Less indurated	>70%	Diffuse	–	–	–
RA2 (Terrace D1)										
Coll	0-43	10YR 6/4	SL	–	Firm	20%, 2-3 cm	Within matrix	None	High v common fiine	v few f fib
B	43-88	7.5YR 5/4	LS	m blocky	v firm	30%, 5-6 cm	Within pores and on clast bases	None	High common fine	Common f-vf fib
Btw	88-119	7.5YR 5/4-5/6	Gritty	Blocky	Indurated	10% <1 cm	Fine laminated and on ped faces	None	Medium common f fib	Rare
Bk (stage II)	119-198	7.5YR 5/4	LS	Blocky	Indurated	Rare	Laminar	None	Medium common f fib	Rare
K (stage II)	198-336 336+	10YR 7/3 Fluvial gravels	ZL	–	Indurated	Rare	Nodular decreases with depth	None	Medium abundant fine	None
RA1 (Terrace D3)										
A	0-16	10YR 5/3-5/4	ZL	Indet	Firm friable	10-20% <6 cm	None	None	High v common fine	Common f fib
B	16-42	10YR 5/4-6/4	ZL	–	Firm compact	<5%	None	None	Medium few fine	Few f fib
Bk (stage I)	42-80 80+	10YR 5/4-6/4 Fluvial gravels	SL	sm m blocky	Compact friable	<10% <20 cm	Within matrix	None	Medium common fine	Rare f fib

Abbreviations: Coll = colluvium; S = sand(y); C = clay(ey); Z = silt; L = loam; calc = calcrete, m = medium; sm = small; f = fine; v = very; vf = very fine; fib = fibrous.

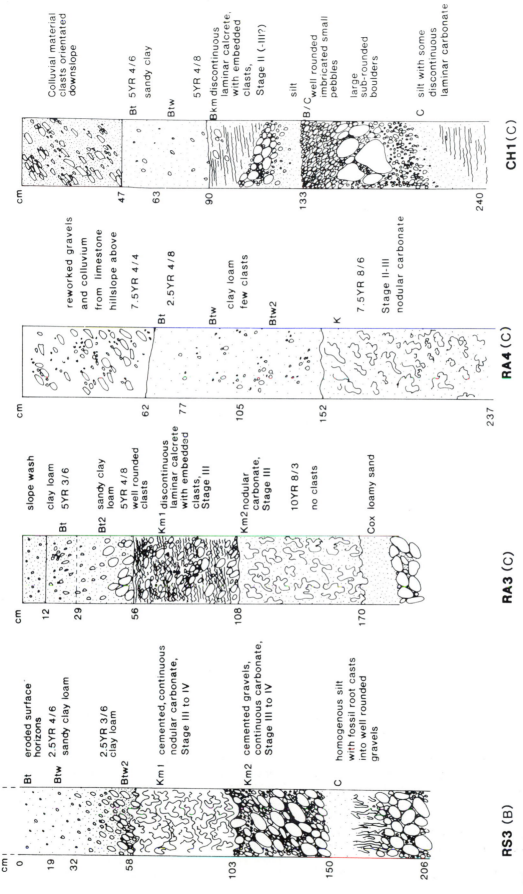

Figure 5. Examples of soil profiles on terraces A-C. Notations in brackets relate to terrace stage.

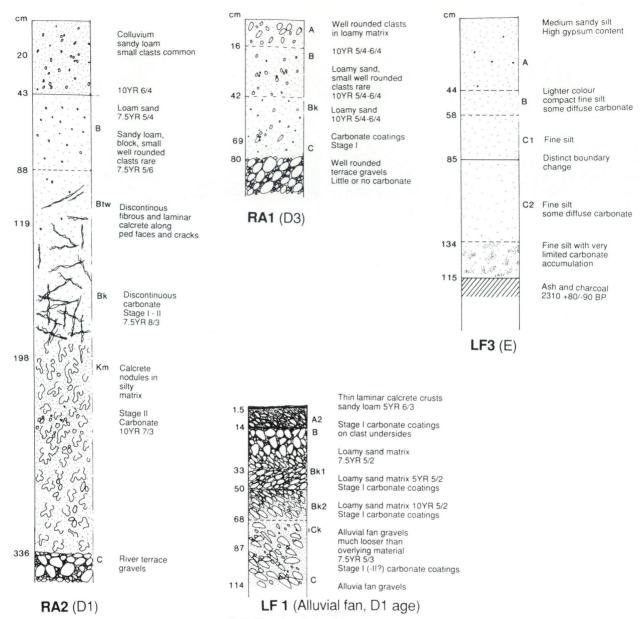

Figure 6. Examples of soil profiles on terraces D-E. Notations in brackets relate to terrace stage.

representative of the age-range (Table 3). No complete and accessible exposure was available for description of a terrace A soil profile, however at a number of sites it was possible to collect samples of the Bt horizon for colour determination or for later laboratory analysis. Profile RS3, north of Sorbas, was chosen as representative of terrace B soils (Fig. 5, Table 3). At this site c. 5 m of terrace B gravels rest unconformably on Gochar conglomerates. The surface of the profile has been truncated by erosion but > 60 cm of the Bt horizon remain, including that with the strongest colour (2.5YR 3/6). Soil texture fines downwards to a clay at c. 35 cm depth. None of the B horizon contains $CaCO_3$ but a strongly cemented petrocalcic horizon occurs below, displaying stage III to IV carbonate accumulation.

Several profiles illustrate terrace C soils. Profile RA3

(Fig. 5) is from a thick accumulation of stage C gravels overlying folded Upper Messinian, Sorbas member, laminated marls, between Urra and Los Molinos. It has a thick, Bt horizon (56 cm) attaining a colour of 5YR 4/8, over an even thicker (> 110 cm) stage III carbonate horizon. Though not showing the same degree of cementation as that in terrace B profiles, this calcrete horizon was sufficiently indurated to undergo brittle deformation during the diapiric phase at Urra (Mather et al. 1991). Profile RA4 (Fig. 5), from the floor of the capture col south of Los Molinos, provides an interesting contrast. The carbonate accumulation in the profile is less well developed than that at RA3, exhibiting c. 85 cm of nodular carbonate of stage II morphology. At site RA4, however the Bt horizon is thicker (> 90 cm) than that at RA3, and reaches 2.5YR 4/8 colouration. Most soil profiles on terrace C show less maturity than that exhibited by the Bt horizon of RA4 or by the calcrete at RA3. Profile CH1 (Fig. 5, Table 3) is more typical. The Bt horizon (< 50 cm) shows only 5YR 4/8 colouration and stage II (perhaps just reaching stage III) carbonate accumulation (c. 30 cm).

Soil profiles on terrace D are much less developed than those on the older terraces. Profile RA2 (Fig. 6, Table 3), from the D1 terrace at Urra illustrates the maximum soil development found on D terraces. A thick (> 150 cm) reddish Bt horizon (7.5YR 5/6) becomes increasingly calcareous towards its base and overlies a thick nodular stage II carbonate (> 130 cm). Unlike the stage II to III nodular carbonate horizons of terrace C soils these are discrete nodules rather than the fused nodular carbonate, typical of the older soils.

Profile LF1 (Fig. 6) is also taken from a soil developed on older D fluvial deposits. However, it may only be loosely comparable with the other sites, in that it is developed on the distal surface of an alluvial fan (of D1 age), rather than strictly on a river terrace. It does however, make a useful comparison with the true river terrace situations. The soil, developed in fluvial gravels, is thinner than RA2 (total thickness < 85 cm), and has no distinct carbonate horizon, with little more than stage I clast coatings and a thin laminar calcrete crust at the surface. This, together with the presence of the strongest colouration (locally 5YR 6/3) near the surface, suggests that the soil profile may have been subject to some erosional truncation.

Soil profiles on D2 and D3 terraces show less development than those on D1 surfaces. Profile RA1 (Fig. 6, Table 3) illustrates a typical D3 terrace soil profile. It is thin (< 80 cm), with little reddening (10YR 5-6/4), and a maximum of stage I carbonate accumulation. Also illustrated in Figure 6 is the terrace E profile from LF3, the site from which a ^{14}C date demonstrates a Holocene age for this terrace (Harvey and Wells 1987). This profile is developed in gypsum-rich alluvial silts, and shows only pale (10YR 7/3) colouration and little carbonate accumulation.

In summary, the soil profiles show distinct age-related characteristics. Soil profile depth tends to increase with age. Stage of carbonate accumulation ranges from almost nothing on terrace E to stage IV on terrace B. The strongest soil colour of the Bt horizon increases in redness with age. However, soils on each terrace show a range of characteristics, some of which overlap between the terraces. Parent material is unlikely to be a major factor in this variation, in that most of the main soil profiles described are located in the upper Aguas basin, with similar parent materials. Parent material variation could account for pre- and post capture differences between the Feos and the Lower Aguas valleys. The range of soil profile properties observed within the terrace stages of the upper Aguas basin suggests that local factors may influence soil profile development, and that each terrace surface may span a wide age-range.

Table 4. Mean Redness Indices (after Hurst 1977) for B horizons in soils on each terrace group.

Terrace	A	B	C	D1	D2-E*
n	7	16	16	12	7
Mean	14.0	12.7	9.1	4.2	1.0
S.D.	1.15	2.34	2.56	1.89	0.50

*Individual values for Terrace D2: 1.3, 0.53, 0.46; D3: 0.70; E: 0.70, 0.40.

5.2 *Soil colour*

Many previous studies of soil chronosequences in dry regions have identified soil reddening (rubification) with age (e.g. Birkeland 1974; Ajmone-Marson et al. 1988; McFadden and Hendricks 1985; Schwertmann 1988; Torrent et al. 1980; Torrent and Cabedo 1988), the 5-10YR colours reflecting the dominance of haematite (Schwertmann and Taylor 1977). The redness can be expressed numerically, using the Redness rating of Hurst (1977), similar to the Rubification ratio of Torrent et al. (1980), where:

Redness Rating = Hue × Chroma / Value from Munsell soil colours, using numerical values of 10 for 10R hues down to numerical values of 2.5 for 7.5YR hues. In this study, in order to increase the sensitivity at the lower end, a numerical value of 1 was given to 10YR hues rather than the 0 of the Hurst (1977) scheme.

Colours were recorded for the Bt horizons of 58 soils on terraces ranging in age from A to E (Table 1), and redness ratings calculated. The results, as expected, show an increase in mean redness with soil age (Table 4), albeit with some overlap between terrace stages. The few terrace sites whose correlations were uncertain on geomorphic grounds alone, were allocated to the most appropriate group on the basis of soil colour, by nearest group mean.

5.3 *Soil particle size*

Many studies of dry region soils have demonstrated an age-related increase in clay content in the Bt horizon (e.g. Birkeland 1974; McFadden and Hendricks 1985; McFadden and Weldon 1987). Our results (Table 5) demonstrate increasing clay content with soil age, and suggest more pronounced down profile accumulation in the older soils.

5.4 *Sequential iron oxide extractions*

Many authors have demonstrated changes in iron oxide properties as soils become older, with progressively more going into less active forms (Arduino et al. 1986), the Fe_2O_3d content increasing (McFadden and Hendricks 1985), and the activity ratio Fe_2O_3d/Fe_2O_3d decreasing with soil age (Schwertmann and Fischer 1973; Singer 1977; Ajmone-Marson et al. 1988; Alexander 1974; Torrent et al. 1980; McFadden and Weldon 1987; Torrent and Cabedo 1986; Diaz and Torrent 1989). The results of the sequential iron oxide extractions carried out here (Table 6) show the expected pattern of weak irregular increases

Table 5. Particle size characteristics, from selected soil profiles.

Terrace	Site	Horizon	Depth (cm)	% sand	% silt	% clay
B	RS3	Bt	0-19	55	25	20
		Btw	19-32	41	37	23
		Btw2	32-58	38	17	46
C	RA3	Bt	12-27	44	22	34
		Bt2	27-56	49	23	28
		Cox	>170	83	9	8
C	RA4	Btw	77-105	33	32	35
		Btw	105-125	25	42	33
		Btw2	125-152	33	35	31
D1	RA2	A	0-20	55	31	13
		B	20-43	83	8	9
		Btw	43-88	66	25	9
D3	RA3	A	0-16	47	40	13
		B	16-42	54	36	11
		Bk	42-80	79	19	1
D1 Fan	LF1	A2	2-14	63	30	7
		Bk1	14-33	69	24	7
		Bk2	33-50	78	18	4

in Fe_2O_3p and Fe_2O_3o with age, a clear increase in Fe_2O_3d and resulting decrease in activity ratios. The one slight anomaly, site LF1, the site on alluvial fan gravels rather than strictly on river terrace gravels (see above), has higher than expected Fe_2O_3d values. This may be because of limited transport of the sediment by fluvial processes prior to deposition, and therefore persistence of crystalline iron oxides from older soils.

5.5 Mineral magnetic measurements

Mineral magnetic analyses can indicate the presence, or concentrations, of different families of iron oxides, and therefore might be expected to show some age-dependent variation. In the results of the analyses carried out here however, there is little obvious pattern to the down-profile data (Fig. 7). Soil age, as expressed by Bt horizon development does not appear to be strongly reflected by either the primary data or by any of the usual ratios between the primary variables (Thompson and Oldfield 1986), nor by the reverse field ratios. What does appear to be expressed by these data is the distribution of carbonate down-profile, apparently influencing magnetic susceptibility, frequency dependent susceptibility (inversely), ARM and SIRM, and hence the ratios between them. The relatively 'hard' backfield measurements reflect dominance by an antiferromagnetic component, which in this environment is likely to be haematite.

When the Bt horizon data for the 12 sites are compared (Table 7) some apparently age-related patterns emerge. Both susceptibility and SIRM appear to increase initially from terrace E to a peak, at the D1 stage for SIRM, and at the C stage for susceptibility, then to show a steady decrease through to terrace A stage (Fig. 8). ARM shows the same trend, but more weakly, and χ_{fd} the reverse trend, but very weakly. The reason for these trends is

Table 6. Sequential iron extractions for selected soil profiles, horizons as indicated on Table 4 for multi-sample sites; for single-sample sites, samples are from B horizons.

Terrace	Site	Fe_2O_3p	Fe_2O_3o	Fe_2O_3d	Activity ratio Fe_2O_3o/Fe_2O_3d
A	RS4	0.005	0.33	2.58	0.12
	RM2	0.005	0.31	1.32	0.23
B	RS3	0.011	0.33	1.36	0.24
		0.021	0.44	1.66	0.26
		0.023	0.46	1.72	0.27
	RS5	0.006	0.29	2.23	0.13
	CB6	0.005	0.51	3.02	0.17
C	RA3	0.009	0.34	1.59	0.21
		0.014	0.38	2.23	0.17
		0.012	0.28	0.76	0.30
	RA4	0.019	0.45	1.66	0.27
		0.024	0.48	1.41	0.34
		0.019	0.38	1.26	0.30
	RM1	0.005	0.24	1.00	0.22
D1	RA2	0.005	0.20	0.60	0.34
		0.012	0.32	0.90	0.35
		0.008	0.29	1.01	0.29
D3	RA1	0.004	0.27	0.55	0.49
		0.005	0.29	0.51	0.57
		0.002	0.08	0.27	0.31
E	LF3	0.004	0.15	0.43	0.66
		0.003	0.07	0.15	0.46
		0.007	0.27	0.41	0.34
D1 Fan	LF1	0.002	0.10	0.57	0.17
		0.005	0.26	1.57	0.17
		0.005	0.15	1.37	0.11

Table 7. Mineral magnetic measurements on B horizons for selected soil profiles.*

Terrace	Site	χ	χ_{fd}	ARM	SIRM
A	RS4	110.1	9.17	150.4	9179
	RM2	104.7	5.22	74.3	10192
B	RS3	82.7	4.92	83.5	7270
	RS5	118.6	7.73	144.9	9437
	CB6	173.2	6.67	124.0	8132
C	RA3	135.4	2.12	128.2	13259
	RA4	220.3	7.59	368.0	15692
	RM1	161.7	4.88	177.3	16378
D1	RA2	134.6	1.22	145.3	15052
D3	RA1	75.9	9.83	121.1	7601
E	LF3	23.2	4.50	34.4	2620
D1 Fan	LF1	107.0	7.75	118.9	18021

* Magnetic measurements (see Fig. 7 and, Thompson and Oldfield 1986, for units and definitions).

uncertain, and any suggestions made here can be only speculative. The trends appear to reflect pedological processes that progress rapidly during the earlier stages of soil formation, then equilibrate or change in style. Progressive accumulation in, then slow leaching from the B horizon, or chemical change within the B horizon, could be such processes, though whether these trends may be reflections of clay, and/or iron oxide and/or carbonate behaviour is unknown. The magnetics results as a whole

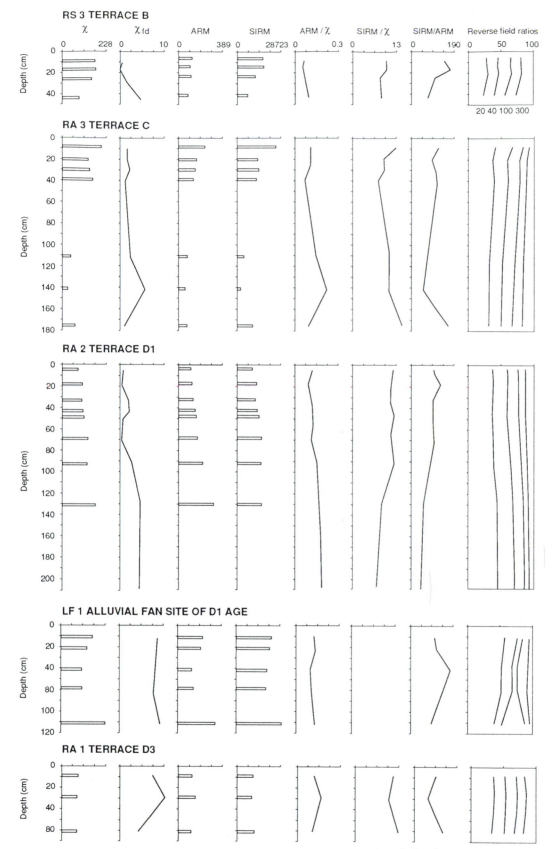

Figure 7. Down profile mineral magnetics analysis. χ: magnetic susceptibility ($10^{-8}\text{m}^3\text{kg}^{-1}$); χ_{fd}: frequency dependent susceptibility, % ($\chi\text{LF}-\chi\text{HF} \times 100$); ARM: Anhysteretic remanent magnetisation ($10^{-6}\text{Am}^2\text{kg}^{-1}$); SIRM: Saturated isothermal remanent magnetisation ($10^{-6}\text{Am}^2\text{kg}^{-1}$) at 1 Tesla (T); Reverse field ratios: Demagnetisation parameters measured at backfields of –20mT, –40mT, –100mT, –300mT, normalised with SIRM. Methodology as outlined by Thompson and Oldfield (1986).

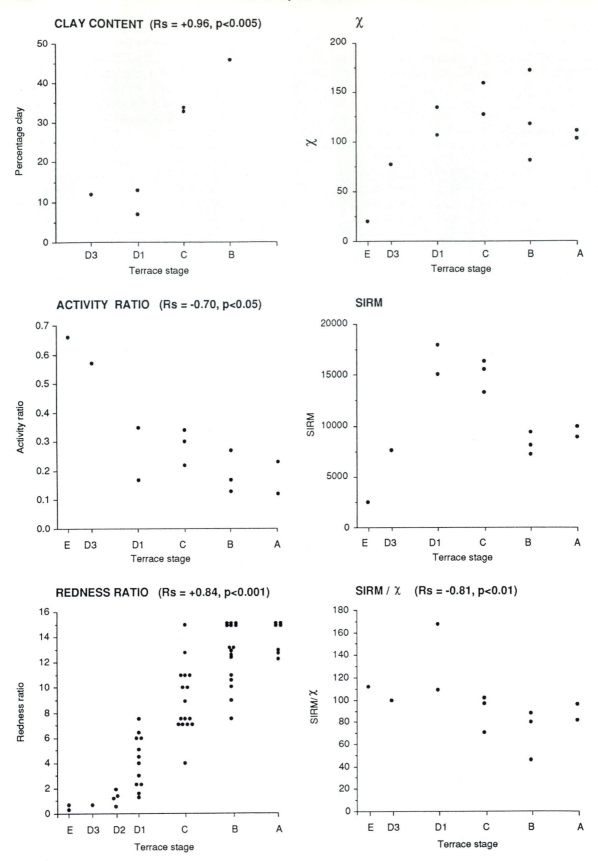

Figure 8. Summary soil properties vs terrace stage A-E.
Rs: Spearman's Rank Correlation Coefficient; p: significance, derived from Student's t tests. For definitions of parameters see text.

are less clear than might have been expected, especially in the light of the recent Tunisian results of White and Walden (1994) which suggest a consistent increase in the concentration of magnetic minerals through time.

5.6 *Soil properties: Summary and interpretation*

Several age-dependant soil properties have been identified. These include soil profile development characteristics, such as horizon thickness and calcium carbonate accumulation through stages I to IV (Gile et al. 1966; Machette 1985). In this area the equivalents of Machette's (1985) stages V and VI show differences related to cementation, brecciation and recementation partly irrespective of absolute carbonate volumes, and therefore appear to reflect exposure history rather than accumulation history. Other age-dependant soil properties include colour, expressed by Redness rating (after Hurst 1977), clay accumulation and iron oxide properties, expressed by the activity ratio, and to a certain extent, magnetic mineral properties. Summary plots of the quantitive measures are given in Figure 8, which also includes a plot of the SIRM/χ ratio, incorporating the observed trends of the two most important individual magnetic properties. In Figure 8 the soil properties are plotted against **relative** soil age, because of the uncertainty of absolute soil age. Apart from the complex behaviour of the two magnetic properties described above, all the remaining indices show clear trends with increasing relative soil age. Again, because of ordinal rather than absolute age data, rank correlation rather than product moment correlation has been used to express these trends. The Spearman rank correlations quoted in Figure 8 are all significant to at least the 5% level, and suggest relationships between the time sequence represented by the terrace sequence E-A and the selected soil properties. However, whether these relationships are linear or nonlinear cannot be determined on the basis of ordinal age data alone.

The sequence E-A, is a sequence without a precise timescale. It is possible to suggest at least the maximum likely limits of the timescale involved. The whole sequence postdates deposition, deformation and uplift of the Gochar Formation (Mather 1991), thus giving a maximum age of Early to mid-Pleistocene (c. 1.6-0.7 Ma). Most of the previous work on the calibration of dry-region chronosequences relates to the American West, and although there is a broad similarity of modern climate between southeast Spain and parts of the American West, it is probable that Pleistocene climates, and therefore soil forming regimes differed between the two regions. Through much of the Pleistocene, southern Spain appears to have been drier than today (Amor and Florschutz 1964), whereas much of the American West appears to have been wetter (Bull 1991). Therefore at best, reference to comparable American work can only provide the most approximate suggestion for calibration of the chronosequence reported here. McFadden and Hendricks (1985), working in California, suggest c. 700,000 years for the development of Redness ratings of 15. Birkeland (1984) and Machette (1985) suggest c. 100,000 years for the

development of stage III and > 150,000 years for stage IV carbonate. All these criteria suggest that soil formation on terrace C gravels may have started at least 100,000 years ago. Harvey and Wells (1987) calculate a Harden (1982) Index for soil RA4, (terrace C) and suggest an age, whatever climatic assumptions are made, and despite later burial, of at least several tens of thousands of years for this terrace soil. If the age of terrace C is approximately of the order suggested above, terrace B probably dates from the mid-Pleistocene, and terrace A is older by an unspecified amount, but itself must postdate the Early Pleistocene end-Gochar period. Such a timescale, however approximate, accords with dates suggested by Dumas (1969) and Dumas et al. (1978) for calcrete crusted depositional surfaces in southeast Spain, based on calcrete thickness, induration and complexity criteria (Harvey 1990).

If the suggestion for the approximate age of terrace C is of the right order, terrace D would appear to be younger than c. 100,000 years, terrace D1 perhaps correlating with the widespread 'Early Wurm' increase in sediment production recorded throughout the Mediterranean region (Butzer 1964). Colluvial deposits with carbonate accumulation and a weak calcrete crust similar to those of terrace D1, can be assigned such an age in several other localities in southeast Spain, in relation to the Quaternary marine sequence (Zazo et al. 1981; Harvey 1990).

Terrace D3 appears to be much younger. McFadden and Hendricks (1985), suggest ages of less than 13,000 years for Redness ratings less than 1. On the basis of the field relationships terrace D3 clearly pre-dates the radiocarbon-dated (2310 +80/–90 BP, DIC-3331) Holocene terrace E (Harvey and Wells 1987). It may be latest Pleistocene in age, perhaps corresponding to Vita Finzi's (1969, 1972) older valley fill, recognised widely throughout the Mediterranean region. If so, terrace D3 would appear to relate to the last Pleistocene cold, dry phase in the Mediterranean region (Sabelberg 1977; Rhodenburg and Sabelberg 1980), dating perhaps from c. 20,000 to c. 10,000 years BP.

6 DISCUSSION

The observed soil properties can be used to differentiate sets of terraces over Quaternary timescales, and to provide the framework for a relative chronology of the fluvial sequence. On the basis of the field criteria alone, the best age indicators seem to be a combination of soil colour and carbonate accumulation characteristics. Where one criterion gives an overlapping age-range estimate the other can be used to suggest differentiation between terraces.

The soil characteristics are not however, sensitive enough to identify any time-transgressive behaviour of the individual terraces. This implies that during the progressive incision of the drainage over the Quaternary as a whole, switches between dissection and aggradation took place relatively rapidly, more rapidly than can be detected from variations in the soil characteristics. These dissec-

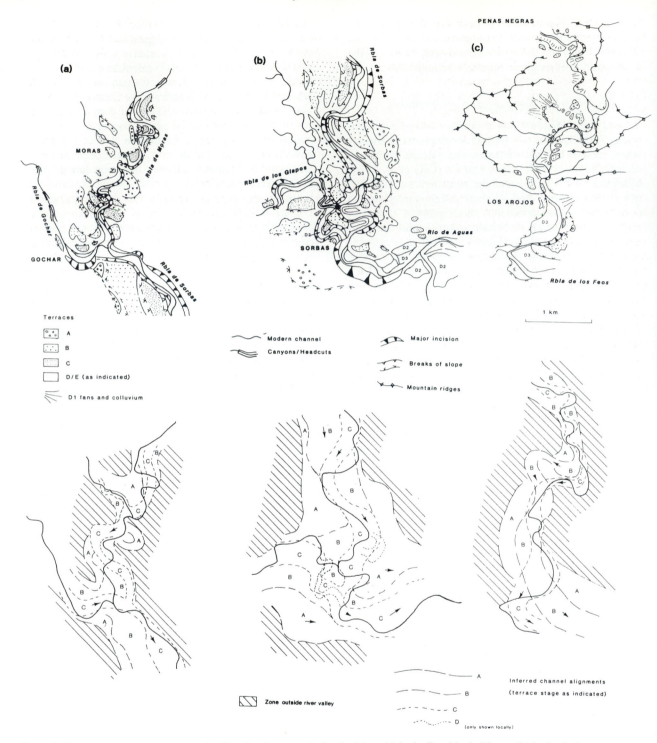

Figure 9. Example terrace sequences and valley development during incision: (a) in the Rambla de Moras, (b) in the Sorbas area, and (c) in the Feos Valley; above: river terraces, below: reconstructed valley alignments.

tional and aggradational changes occurred throughout the basin, almost irrespective of the local tectonic context. This suggests response to basin-wide controls of sediment supply, probably climatic controls in the case of the older terraces, and perhaps human-induced controls in the case of terrace E, rather than response to major tectonically- or capture-induced changes in base-level. The major exception of course, is the Aguas/Feos capture, that took place at Los Molinos between terraces C and D1 (perhaps c. 100,000 years BP). The overall tectonic framework, and the morphological setting of and response to this capture, appear to have controlled the context within which the shorter-term aggradation/ dissection sequence of terrace formation has operated.

Terrace mapping and correlation on the basis of the soil evidence can be used to clarify the Quaternary fluvial sequence and changing morphology within the several zones of the drainage basin. Within the upstream zone, capture-induced incision has worked headwards up the Rambla de Sorbas system, to major knickpoints at Moras and above Gochar. Above Moras (Fig. 9a) the total depth of Quaternary incision is < 50 m, most of it occurring between terraces A and C. Terraces D and E are of limited extent and not deeply cut into terrace C. Throughout the period of incision, lateral erosional activity has been important, producing meandering valleys.

Further downstream at Sorbas (Fig. 9b), the total depth of Quaternary incision exceeds 90 m, about half of which post-dates terrace C and is related to the headwards migration of capture-induced incision. As at Moras, continued lateral activity has been important during incision. The deep canyon on the Rambla de los Chopos, upstream of its confluence with the Rambla de Sorbas, is related to the cutoff of the abandoned meander encircling Sorbas village, that took place in D3 times. Downstream from Sorbas to the capture site, the complex terrace sequence is related to the diapirism of the Urra area, described elswhere (Mather et al. 1991).

At the capture site a total of > 150 m of incision has taken place, c. 90 m of which postdates terrace C. The absence of the older terraces for much of the way downstream to La Huelga, and the instability of the slopes on both valley sides (Harvey 1987) indicate very rapid incision during terrace C times, associated with the headwards migration of dissection towards the capture zone. In gross terms this situation reflects the geomorphic response to the differential uplift of the Sorbas basin in relation to the Vera basin. A zone of dissection has progressed upstream, from perhaps the Alfaix area (Fig. 1) in the early Quaternary, through the capture zone in the late Quaternary and is now in the upper part of the basin. In the lower Aguas zone, near La Huelga, dissection since

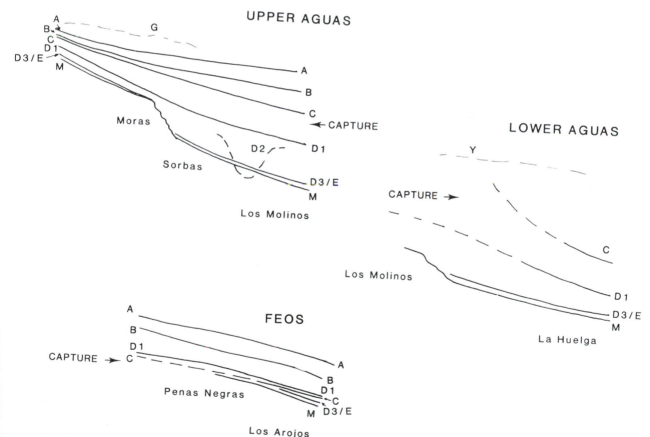

Figure 10. Simplified model of Quaternary incision and river terrace development in the three zones of the Aguas/Feos river systems, southeast Spain. A B C D1 D2 D3 E: river terrace surfaces, G: Pre-dissection Gochar surface, Y: Gypsum plateau surface, M: modern channel.

terrace C-equivalent is c. 50 m. Throughout the lower Aguas, intermittent older D and more continuous younger D and E terraces suggest much lower rates of incision during the Late Pleistocene, but increasing valley widening, especially in downstream reaches, through lateral migration.

In the Feos Valley the intermittent incision from terrace A to terrace C time follows a similar pattern to that identified upstream of the capture site. There has been almost no incision since the formation of terrace C. Indeed in many places terrace C sediments are buried by younger sediments. In the broad open valley upstream of Penas Negras (Fig. 2) there is a thick colluvial fill of early D age. Through the mountain section, terrace C occurs on the valley floor, locally buried by axial D1 and D3 terrace gravels, D1-age tributary junction alluvial fans and unspecified D-age colluvial deposits (Fig. 9c). The beheaded system appears to have lacked the necessary discharge and stream power for continued incision or much lateral migration.

The contrasting geomorphic sequences in the three zones, defined by the capture at Los Molinos, illustrate the differential roles of climate, tectonics and capture in the regional geomorphic development. The sequences can be summarised by a simple conceptual model (Fig. 10). There are contrasts between two styles of dissection, reflecting differing relationships between the three factors above. The first style exhibits progressive incision, punctuated by episodic aggradation and dissection. This style is characteristic of the upper Aguas and of the Feos Valley prior to the capture event. It appears to reflect overall incision during the Quaternary in response to early Quaternary uplift of the basin. The terrace sequence appears to be primarily controlled by climatically-induced (?) episodic variations in sediment supply. Since the capture beheaded the drainage, the Feos has been almost inactive. In the headwaters, the terraces are weakly divergent downstream, becoming parallel to each other further down the system (Fig. 10).

The second style of dissection exhibits major base-level related incision. This style is characteristic of the middle and lower Aguas, with downstream terrace divergence in the area above the capture site and convergence from there downstream (Fig. 10). This style reflects the response to major base level change, controlled by the regional pattern of differential tectonic uplift, locally modified and intensified by major river capture.

7 ACKNOWLEDGEMENTS

The authors are grateful to J Ritter (University of Pennsylvania) for field assistance, especially in the early stages of the work. SYM is grateful to the University of Liverpool for a Ph.D studentship and funding for her field work. We are grateful to the Visiting Scholars Fund (SGW) and the Research and Development Fund (AMH) of the University of Liverpool for grants towards the costs of the field work. We are also grateful to the Drawing Office and Photographic sections of the Department of Geography of the University of Liverpool for producing the diagrams.

REFERENCES

Ajmone-Marson, F., E. Barberis & E. Arduino 1988. A soil chronosequence in North Western Italy: morphological, physical and chemical characteristics. *Geoderma* 42:51-64.

Alexander, E.B. 1974. Extractable iron in relation to soil age in terraces along the Truckee River, Nevada. *Soil Science Society of America, Proceedings* 38:121-124.

Amor, J.M. & F. Florschutz 1964. Results of the preliminary palynological investigation of samples from a 50 m boring in southern Spain. *Boletin Real Sociedad Espanola de Historia Natural (Geologica)* 62:251-255.

Arduino, E., E. Barberis, F. Ajmone-Marson, E. Zanini & M. Francini 1986. Iron oxides and clay minerals within profiles, as indicators of soil age in northern Italy. *Geoderma* 37:45-55.

Birkeland, P.W. 1984. *Soils and Geomorphology*. Oxford, O.U.P.

Birkeland, P.W. 1985. Quaternary soils of the Western United States. In J. Boardman (ed), *Soils and Quaternary Landscape Evolution*. Chichester: Wiley.

Bull, W.B. 1990. Stream-terrace genesis: implications for soil development. *Geomorphology* 3:351-367.

Bull, W.B. 1991. *Geomorphic Responses to Climatic Change*. Oxford: O.U.P.

Butzer, K.W. 1964. Climatic-geomorphic interpretation of Pleistocene sediments in the Eur-African sub tropics. In F.C. Howell & F. Bouliere (eds), *African Ecology and Evolution*. London: Methuen, 1-25.

Diaz, M.C. & J. Torrent 1989. Mineralogy of iron oxides of two soil chronosequences of central Spain. *Catena* 16:291-299.

Dumas, M.B. 1969. Glacis et croutes calcaires dans le levant espanol. *Association Geographes Francais, Bull.* 375:34-47.

Dumas, M.B., P. Gueremy, R. Llenaff & J. Raffy 1978. Geomorphologie et neotectonique dans la region d'Almeria (Espagne du sud-est). Relief et Neotectonique des pays Mediterraneens, *Paris, Centre National de Recherche Scientifique*, Publication de la recherche cooperative sur programme 461:123-170.

Gerson, R. 1982. Talus relicts in deserts: A key to major climatic fluctuations. *Israel Journal of Earth-Sciences* 31:123-132.

Gile, L.H., F.F. Peterson & R.B. Grossman 1966. Morphological and genetic sequences of carbonate accumulation in desert soils. *Soil Science* 101:347-360.

Harden, J.W. 1982. A quantitative index of soil development from field descriptions: examples from a chronosequence in central California. *Geoderma* 28:1-28.

Harden, J.W. & E.M. Taylor 1983. A quantitative comparison of soil development in four climatic regimes. *Quaternary Research* 20:342-359.

Harden, D.R., N.E. Biggar & M.L. Gillam 1985. Quaternary deposits and soils in and around Spanish Valley, Utah. In D.L. Weide (ed), *Soils and Quaternary Geology of the Southwestern United States*. Geological Society of America, Special Paper 203, 43-64.

Harrison, J.B.J., L.D. McFadden & R.J. Weldon 1990. Spatial soil variability in the Cajon Pass chronosequence: implications for the use of soils as a geochronological tool. *Geomorphology* 3:399-416.

Harvey, A.M. 1987. Patterns of Quaternary aggradational and dissectional landform development in the Almeria region, southeast Spain: a dry-region, tectonically active landscape. *Die Erde* 118:193-215.

Harvey, A.M. 1990. Factors influencing Quaternary alluvial fan development in southeast Spain. In A.H. Rachocki & M. Church, *Alluvial Fans: a Field Approach*. Chichester: Wiley, 247-269.

Harvey, A.M. & S.G. Wells 1987. Response of Quaternary fluvial systems to differential epeirogenic uplift: Aguas and Feos river systems, southeast Spain. *Geology* 15:689-693.

Hurst, V.J. 1977. Visual estimates of iron in saprolite. *Geological Society of America, Bulletin* 88:174-176.

James, P.A. & D.K. Chester this volume. Soils of Quaternary river sediments in the Algarve.

Knuepfer, P.L.K. & L.D. McFadden 1990. *Soils and Landscape Evolution*. Amsterdam: Elsevier.

Krumbein, W.C. & F.J. Pettijohn 1938. *Manual of Sedimentary Petrology*. New York: Appleton.

Machette, M.N. 1985. Calcic soils of the southwestern United States. In D.L. Weide (ed), *Soils and Quaternary Geology of the Southwestern United States*. Geological Society of America, Special Paper 203:1-21.

Mather, A.E. 1991. Late Cenozoic drainage evolution of the Sorbas Basin, southeast Spain. University of Liverpool Ph.D. Thesis.

Mather, A.E. 1993. Basin inversion: some consequences for drainage evolution and alluvial architecture. *Sedimentology* 40:1069-1089.

Mather, A.E. & A.M. Harvey, this volume. Controls on drainage evolution in the Sorbas basin, southeast Spain.

Mather, A.E., A.M. Harvey & P.J. Brenchley 1991. Halokinetic deformation of Quaternary river terraces in the Sorbas Basin, south-east Spain. *Zeitschrift fur Geomorphologie*, Suppl.-Bd. 82:87-97.

McFadden, L.D. & D.M. Hendricks 1985. Changes in the content and composition of pedogenic iron oxyhydroxides in a chronosequence of soils in southern California. *Quaternary Research* 33:189-204.

McFadden, L.D. & R.J. Weldon 1987. Rates and processes of soil development on Quaternary terraces in Cajon Pass, California. *Geological Society of America, Bulletin* 98:280-293.

McKeague, J.A. 1967. An evaluation of 0.1 m pyrophosphate and pyrophosphate dithionite in comparison with oxalate as extractants of the accumulation products in podzols and some other soils. *Canadian Journal of Soil Science* 47:95-99.

McKeague, J.A. & J.H. Day 1966. Dithionite and oxalate extractable Fe and Al as acids in differentiating various classes of soils. *Canadian Journal of Soil Science* 46:13-21.

Mehra, O.P. & M.L. Jackson 1960. Iron oxide removal from soils and clays by a dithionite-citrate system buffered with sodium bicarbonate. *Clays and Clay Mineralogy* 7:317-327.

Miller, S.Y. 1991. Soil chronosequences and fluvial landform development: Studies in S.E. Spain and N.W. England. University of Liverpool, Ph.D. Thesis.

Pope, K.O. & T.H. Van Andel 1984. Late Quaternary alluviation and soil formation in southern Argolid: its history, causes and archaeological implications. *Journal of Archaeological Science* 11:281-306.

Rhodenburg, H. & U. Sabelberg 1980. Northwestern Sahara margins: Terrestrial stratigraphy of the Upper Quaternary and some palaeoclimatic implications. In E.M. Van Zinderen Bakker Sr. & J.A. Coetzee (eds), *Palaeoecology of Africa and the Surrounding Islands* 12:267-276.

Ruegg, G.J.H. 1964. Geologische onderzoekingen in bet bekken von Sorbas, SE Spanje. Internal report, Geological Institute. University of Amsterdam.

Sabelberg, U. 1977. The stratigraphic record of late Quaternary accumulation series in southwest Morrocco and its consequences concerning the pluvial hypothesis. *Catena* 4:209-214.

Schwertmann, U. 1988. Some properties of soil and synthetic iron oxides. In J.W. Stucki, B.A. Goodman & U. Schwertmann (eds), *Iron in Soils and Clay Minerals*. Dordrecht: Reidel Publ. Co.

Schwertmann, U. & W.R. Fischer 1973. Natural 'amorphous' ferric hydroxide. *Geoderma* 10:237-247.

Schwertmann, U. & R.M. Taylor 1977. Iron Oxides. In J.B. Dixon & S.B. Weed (eds), *Minerals in Soil Environments*. Madison, Wisconsin, Soil Science Society of America, 145-180.

Singer, A. 1977. Extractable sesquioxides in six Mediterranean soils developed on basalt and scoria. *Journal of Soil Science* 28:125-135.

Thompson, R. & F. Oldfield 1986. *Environmental Magnetism*. London: Allen and Unwin.

Torrent, J. & A. Cabedo 1986. Sources of iron oxides in reddish brown soil profiles from calcarenites in southern Spain. *Geoderma* 37:57-66.

Torrent, J., U. Schwertmann, H. Fechter & F. Alferez 1983. Quantitative relationships between soil colour and haematite content. *Soil Science* 136:354-358.

Torrent, J., U. Schwertmann & D.E. Schulze 1980. Iron oxide mineralogy of some soils of two river terrace sequences in Spain. *Geoderma* 23:191-208.

Vita Finzi, C. 1969. *The Mediterranean Valleys*. Cambridge: C.U.P.

Vita Finzi, C. 1972. Supply of fluvial sediment to the Mediterranean during the last 20 000 years. In D.J. Stanley (ed), *The Mediterranean Sea, A Natural Sedimentation Laboratory*. Stroudsberg, Pa., 43-46.

Weide, D.L. (ed) 1985. *Soils and Quaternary Geology of the Southwestern United States*. Geological Society of America, Special Paper 203.

Weijermars, R., Th.B. Roep, B. van den Eekout, G. Postma & K. Kleverlan 1985. Uplift history of a Betic fold nappe inferred from Neogene-Quaternary sedimentation and tectonics (in the Sierra Alhamilla and Almeria, Sorbas and Tabernas basins of the Betic Cordilleras, south-east Spain). *Geologie en Mijnbouw* 64:397-411.

White, K. & J. Walden 1994. Mineral magnetic analysis of iron oxides in arid zone soils, Tunisian Southern Atlas. In A.C. Millington & K. Pye (eds), *Environmental Change in Drylands: Biogeographical and Geomorphological Perspectives*. Chichester, John Wiley.

Zazo, C., J.I. Goy, M. Hoyos, B. Dumas, J. Porta, J. Martinell, J. Baena & E. Aguirre 1981. Ensayo de sintesis sobre el Tirrheniense Peninsular Espanol. *Estudios Geologicos* 37:257-262.

CHAPTER 24

Mediterranean Quaternary River Environments – some future research needs

JOHN LEWIN

Institute of Earth Studies, UCW Aberystwyth, Aberystwyth, Dyfed, UK

MARK G. MACKLIN

School of Geography, University of Leeds, Leeds, UK

JAMIE C. WOODWARD

Department of Environmental and Geographical Sciences, Manchester Metropolitan University, Manchester, UK

In exploring the development of Quaternary river environments in the Mediterranean, the papers in this volume have naturally exposed many shortcomings in their site-specific studies. But beyond these, and attempting to generalize for the Mediterranean as a whole, it is possible to identify a number of recurring and important research problems and themes for the future.

The research agenda for the 1990s has included a strong concern for the impact of environmental change on human welfare. It has involved in particular the development of global change models which may be used to predict or retrodict patterns resulting from global warming or ice-age cooling. These may be tested by evidence from ocean, lake or alluvial sediments. The Mediterranean basin is one area where such evidence may be sought, and as we have seen, it is also a region where the impacts of environmental change can be profound. What is required essentially are high-resolution records which are well-dated, and in which long term and yet subtle changes in sedimentation can be unequivocally related to particular environmental changes – some abrupt, but others more gradual or cyclic. As this volume has shown, such apparently straightforward requirements are usually very far from being fully satisfied.

A first need is for more fully and reliably dated alluvial sequences. There still are very few river catchments or sites of which any great satisfaction may be expressed concerning the dating of their alluvial bodies. More radiometric dates are needed, noting also the drawbacks and qualifications that are required when interpreting ^{14}C dates. Allowing also for the paucity of organic materials in many Mediterranean alluvial sediments, more widespread use of luminescence, magnetic dating and other techniques would also prove extremely helpful.

A second need is for more comprehensive coverage of river sedimentary environments. On the whole, most chapters in this volume have dealt with rivers of moderate dimensions. We have, however, very few studies of steepland landscapes of the Alpine Mediterranean, where linkages between slope stability and adjacent steep-gradient and often boulder-bed streams are manifest. Here sensitivity to recent climatic changes (for example in response to the so-called Little Ice Age of post-Medieval times) may well be recorded. Moreover, many of the bedrock gorges in the Mediterranean region contain thick sequences of slackwater sediments which offer exciting possibilities for the reconstruction of palaeo-flood characteristics during critical periods of environmental change. At the other extreme are floodplain wetlands and estuarine environments which are similarly under-represented. There is considerable potential at these sites for a much more complete Holocene sedimentary record, but this has only rarely and partially been achieved to provide a high-resolution analysis of environmental change. Furthermore, such depositional zones may allow direct integration of records of climate and vegetation change with catchment response in terms of alterations in the rate of supply and source of suspended sediment.

A third requirement for future studies is to set alluvial chronologies for particular sites into an appropriate understanding of basin sediment dynamics. Recent research in geomorphology has shown just how complexly river basins may respond to external changes. Comparatively simple events – whether of climatic, volcanic or human origin – may work their way through river systems in a very complicated fashion. It should be appreciated that analysis of an apparent 'signal' from a single site may not prove to be the same as that from other sites upstream or downstream, whilst it is also true that the preservation of well-dated alluvial sequences is also localized, and even possibly misleading in so far as relationships with global or region-scale 'causes' are concerned. Recent developments in the use of sediment budgets offer an attractive means of routing the movement and storage of sediment within the drainage basin delivery system – and one possible way of linking studies of contemporary processes and sediment yield with longer-term fluvial histories.

A further need concerns the better understanding of the timing, spatial patterning, style and amount of tectonic activity. Evidence from the Mediterranean shows quite clearly the significance of recent earth movements, with some hundreds of metres of vertical movement during the Quaternary Period. We are beginning to appreciate how early-stage river systems can become set in a variety of landscape patterns and how drainage systems and their associated depositional zones develop. This is a most valuable antidote to earlier notional and often simplistic ideas about landscape evolution, but one which requires close collaboration between students of tectonics and geomorphologists.

An equal degree of collaboration is required between archaeologists and specialists in studying the 'natural' environment. Artefacts provide evidence for the dating and interpretation of alluvial sequences, and yet they themselves may require environmental reconstructions for their own interpretation in an alluvial setting. Recent studies have indeed shown how rich an archive of archaeological materials alluvial deposits may themselves provide. However, there is a danger of circular arguments here, drawing conclusions as to the human nature and origin of environmental change from alluvial units rich in artefacts, for example. This volume has shown the potential and actual value of collaborative work in interpreting fluvial histories and environments, and for extracting the maximum and most reliable evidence. The minute-scale analysis which may be second-nature to the archaeologist is exemplary to geomorphologists, who in turn may provide regional meaning to the setting of archaeological sites. The work of palynologists and other biostratigraphic specialists is invaluable to both.

The agenda of problems and prospects so far discussed has, perhaps, a familiar ring to it – more dates, more sites, more collaboration. This does underline the fact that river environments, like others, are complex entities and that answers to relatively simple questions usually do require exhaustive and highly technical research programmes.

But this should not obscure the need to ask big questions in the first place. Are there key events whose riverine impact may potentially be identified? Was an alluvial unit fundamentally generated by climatic or human causes? What does the future hold for river environments given certain parameters for environmental change? The need to use the minutiae of site investigators for broader purposes remains of the greatest importance. Human welfare demands that big questions be answered through the painstaking execution and interpretation of detailed research.

For the present, we believe this volume has significantly pushed forward the frontiers of research in numerous ways. The wealth of evidence it contains shows the variable significance of tectonic, climatic and anthropogenic factors from site to site. The rich environmental archive of the Mediterranean continues to offer an attractive challenge to those who in the broadest sense are concerned with river palaeoenvironments, and it is a pleasure to have been able to present work by so many scientists from different backgrounds. We hope more will be stimulated to apply new techniques and to ask new questions of the Mediterranean, personified by Shelley as 'lulled by the coil of his crystalline streams'. Where better to work?

List of contributors

James T. Abbott, Department of Geography, The University of Texas at Austin, Austin, Texas 78712-1098, USA

Tjeerd H. van Andel, Department of Earth Sciences, University of Cambridge, Downing Street, Cambridge, CB2 3EQ, United Kingdom

J.M. Anketell, Department of Geology, University of Manchester, Oxford Road, Manchester, M13 9PL, United Kingdom

Jean-Louis Ballais, Institute de Geographie, Université d'aix-Marseille, 29 Avenue R. Schuman, 13621 Aix-en-Provence, France

Graeme W. Barker, School of Archaeological Studies, University of Leicester, University Road, Leicester, LE1 7RH, United Kingdom

David K. Chester, Department of Geography, The Roxby Building, University of Liverpool, Liverpool, L69 3BX, United Kingdom

Richard E. Collier, Department of Earth Sciences, University of Leeds, Leeds, LS2 9JT, United Kingdom

K. Gaki-Papanastassiou, Department of Geography and Climatology, University of Athens, Panepistimiopolis, 17237 20 Grafou, Athens, Greece

K. Gallis, Ephoria of Antiquities, Archaeological Museum, Larissa, Greece

S.M. Ghellali, Faculty of Earth Science, Al Fateh University, Tripoli, Libya

David D. Gilbertson, Institute of Earth Studies, Penglais, UCW Aberystwyth, Aberystwyth, Dyfed, SY23 3DB, United Kingdom

Jose Luis Goy, Departamento de Geologia, Facultad de Ciencias, Universidad de Salamanca, 37008 Salamanca, Spain

Adrian M. Harvey, Department of Geography, The Roxby Building, University of Liverpool, L69 3BX, United Kingdom

C.O. Hunt, Department of Geographical and Environmental Sciences, University of Huddersfield, Queensgate, Huddersfield, HD1 3DH, United Kingdom

J.A. Jackson, Bullard Laboratories, Madingley Road, University of Cambridge, Cambridge, CB3 0EZ, United Kingdom

Peter James, Department of Geography, The Roxby Building, University of Liverpool, L69 3BX, United Kingdom

Catherine Kuzucuoglu, Laboratoire de Geographie Physique, URA 141 CNRS, 1 Place A. Briand, 92195, Meudon Cedex, Paris, France

M.R. Leeder, Department of Earth Sciences, University of Leeds, Leeds, LS2 9JT, United Kingdom

John Lewin, Institute of Earth Studies, Penglais, UCW Aberystwyth, Aberystwyth, Dyfed, SY23 3DB, United Kingdom

Mark G. Macklin, School of Geography, University of Leeds, Leeds, LS2 9JT, United Kingdom

Hampik Maroukian, Department of Geography and Climatology, University of Athens, Panepistimiopolis, 17237 20 Grafou, Athens, Greece

Anne Mather, Department of Geography, University of Plymouth, Drake Circus, Plymouth, Devon, PL4 8AA, United Kingdom

S.Y. Miller, Department of Geography, The Roxby Building, University of Liverpool, Liverpool, L69 3BX, United Kingdom

David G. Passmore, Department of Geography, University of Newcastle upon Tyne, Newcastle upon Tyne, NE1 7RU, United Kingdom

M. Provansal, Institut de Geographie, Université d'Aix-Marseille, 29 Avenue R. Schuman, 13621 Aix-en-Provence, France

Neil Roberts, Department of Geography, University of Technology, Loughborough, Leicestershire, LE11 3TU, United Kingdom

Pablo G. Silva, Departamento de Geologia, Facultad de Ciencias, Universidad de Salamanca, 37008 Salamanca, Spain

Rodney Stevens, Geologiska Institutionen, Chalmers Tekniska Hogskola, Och Göteborgs Universitet, S-412 96 Göteborg, Sweden

G. Toufexis, Ephoria of Antiquities, Archaeological Museum, Larissa, Greece

Salvatore Valastro Jr., Radiocarbon Dating Laboratory, Balcones Research Centre, University of Texas at Austin, Texas, USA

Per O. Wedel, Geologiska Institutionen, Chalmers Tekniska Hogskola, Och Göteborgs Universitet, S-412 96 Göteborg, Sweden

S.G. Wells, Department of Earth Sciences, University of California, Riverside, California, USA

E. Wenzens, Geographisches Institut, Heinrich-Heine-Universitat Düsseldorf, D-4000 Düsseldorf, Germany

G. Wenzens, Geographisches Institut, Heinrich-Heine-Universitat Düsseldorf, D-4000 Düsseldorf, Germany

Jamie C. Woodward, Department of Environmental and Geographical Sciences, John Dalton Building, Manchester Metropolitan University, Chester Street, Manchester, M1 5GD, United Kingdom

Caridad Zazo, Departamento de Geologia, Museo Nacional de Ciencias Natural, CSIC, Jose Gutierrez Abascal 2, 28006 Madrid, Spain

Index